ENVIRONMENTAL IMPACT STATEMENTS

ENVIRONMENTAL IMPACT STATEMENTS

A Practical Guide for Agencies, Citizens, and Consultants

DIORI L. KRESKE

John Wiley & Sons, Inc.

New York ■ Chichester ■ Brisbane ■ Toronto ■ Singapore

Copyright © 1996 by John Wiley & Sons, Inc.

Library of Congress Cataloging-in-Publication Data
Kreske, Diori L., 1952-
    Environmental impact statements : a practical guide for agencies,
citizens, and consultants / Diori L. Kreske.
        p.   cm.
    Includes bibliographical references and index.
    ISBN 0-471-13741-3 (cloth : alk. paper)
    1. Environmental impact statements—United States.   I. Title.
TD194.55.K74   1996
333.7'14—dc20   95-52682

Printed in the United States of America

10 9 8 7 6 5 4 3 2 1

To My Mother

PREFACE

Several books are available on environmental impact statements (EISs) that discuss the technical process of analyzing environmental impacts or discuss the regulations and court decisions which provide guidance on what agencies must do to comply with the regulations. Both topics, the technical impact analysis and the legal guidelines, are important to understand and become proficient in preparing or managing EISs. A third and equally important consideration is the role of individuals, agencies, groups, and consultants in the EIS process. In essence, it is the people in the process—their widely diverse objectives, concerns, and political agendas—that make it so dynamic. If it weren't for the people involved, one EIS would be the same as the next and the EIS process would be predictable.

This book suggests an approach to managing an EIS or environmental review process and to preparing EISs that focuses on the roles of those involved in the process, and it places the EIS process in context with the overall planning process and other environmental laws. "EIS process" is used loosely to mean the entire environmental review process required by the National Environmental Policy Act (NEPA) or the comparable state environmental policy acts (SEPAs). At least 16 states have environmental policy acts, and another 11 states have limited environmental review requirements. Other documents besides EISs (such as environmental assessments) also are used in the process of environmentally reviewing proposed projects.

EISs are "action-forcing devices"[1] to ensure that agencies consider the environmental consequences of their actions. Many agencies struggle to make EISs more useful to decisionmakers and the public, and they also struggle to streamline the process so it isn't unnecessarily long and costly. My belief is that placing the EIS process in the context of defining participants' roles, along with understanding their responsibilities, will reduce conflicts and duplication of effort as well as result in a more useful process. The basic purpose of the process, after all, is to get the correct information to the correct people so that environmentally informed decisions can be made that, hopefully, result in less environmentally damaging projects than would occur without the information.

The goal of the environmental review process is to reduce environmental impacts, not produce reams of data or impressive scholarly works. A state agency representative once said to me, "You're just proposing a lot of mitigation measures to reduce impacts below significance so you can get away without doing an EIS." However, the goal of the process is not to produce more EISs, nor is it to avoid preparing EISs. If mitigation measures are part of a proposed action, or an agency commits to environmental mitigation measures so that mitigation is more than just a possibility, and therefore none of the environmental impacts would be significant, there is no need to prepare an EIS. The saving of time and money, concurrent with avoiding significant environmental impact, is a practical benefit to everyone (including taxpayers).

Federal, state, and local laws dictate the requirements of EISs and the environmental review process, depending on the agency with jurisdiction and type of proposed project. The information in this book, however, is not dependent on a particular set of environmental laws. Although laws and regulations are discussed in general, and the federal Council on Environmental Quality (CEQ) regulations in particular, the reader should also consult the specific EIS regulations of the agency with responsibility for the proposed project or action under review.

---

[1] 40 CFR §1502.1

This book covers a broad range of topics related to EISs in general. If you have specific or technical questions about writing EISs or the EIS process, please send your questions to the author at P.O. Box 99804, Seattle, Washington 98199. Include a stamped, self-addressed #10 envelope for replies. Legal advice cannot be provided.

Diori L. Kreske
Seattle, Washington

ACKNOWLEDGMENTS

Having spent over 20 years as a team member and interdisciplinary team leader in the preparation of environmental impact statements, I hadn't realized how accustomed I'd become to working with people on a daily basis. In contrast to the stimulation of solving problems and accomplishing goals within a diverse and supportive network of professionals, writing a book is a lonely endeavor. Therefore, I am indebted—more than I had expected to be—to those who took time out of their extremely busy schedules to discuss my ideas and provide valuable suggestions. In particular, I wish to thank Mr. Richard (Rick) Krochalis, Director of the City of Seattle's Department of Construction and Land Use; and Ms. Nancy Stehle, environmental consultant in Boulder, Colorado and former Director for Environment, Office of the Secretary of the Navy in Washington, D.C.

Many employees of local, state, and federal agencies responded to my requests for information, discussed their environmental and planning programs, and discussed how they are coping with regulatory reform in a climate of declining budget and staff resources. These conversations support my focus on the practical considerations of implementing environmental regulations. Theory and scientific methodology are meaningless if the environmental programs are not accomplishing their goals and if environmental regulations are discarded because they have become burdens on society or their value is no longer recognized.

Those who provided much information regarding their agencies' programs include Mr. John (Jack) Mills, Mr. Richard Crowe, and Mr. Alden Sievers of the U.S. Bureau of Land Management; Mr. Joseph Montgomery of the U.S. Environmental Protection Agency; Ms. Jamia Hansen-Murray and Mr. Doug Schrenk of the U.S. Forest Service; and Mr. Ray Clark of the President's Council on Environmental Quality. Mr. Terry Rivasplata of the California Office of Planning and Research, as well as Ms. Tess Bennett of the California Resources Agency, answered numerous questions regarding the California Environmental Quality Act. Mr. Hugh O'Neill of the Washington State Department of Ecology was generous with his time in describing permit assistance programs—programs provided by several states to (a) inform those who require permits of the environmental process and (b) streamline the environmental and planning processes in some cases.

Many of the concepts in this book are based on ideas originally presented by Mr. Joseph Wellington in a number of papers he wrote, while he was a Lieutenant Colonel in the U.S. Marine Corps, regarding the National Environmental Policy Act. I am very grateful to Mr. Wellington for granting me permission to use his ideas and graphics. I also wish to thank others who granted permission for the use of their materials, including the International Institute for Environment and Development in London and the Environmental Impact Assessment Centre at the University of Manchester, United Kingdom.

I am also grateful to those who helped produce this book. My thanks go to Ms. Wendy John, whose specialty is technical graphics and who, through patience and persistence, was able to translate complex ideas represented by my very rough sketches into figures and tables. Last, but not least, I wish to thank my publisher, John Wiley & Sons, who made this book possible and whose editors provided moral and technical support and suggestions for its improvement.

CONTENTS

ACRONYMS

ACHP	Advisory Council on Historic Preservation
AHPA	Archeological and Historic Preservation Act
AIRFA	American Indians Religious Freedom Act
ARPA	Archeological Resource Protection Act
BIA	Bureau of Indian Affairs
BLM	Bureau of Land Management
BOR	Bureau of Reclamation
CAA	Clean Air Act
CE	Categorical Exclusion or Exemption
CEQ	Council on Environmental Quality
CEQA	California Environmental Quality Act
CWA	Clean Water Act
DNS	Determination of Nonsignificance
DOA	Department of Agriculture
DOC	Department of Commerce
DOD	Department of Defense
DOE	Department of Ecology (Washington state)
DOE	Department of Energy (federal)
DOI	Department of the Interior
DOT	Department of Transportation
DS/SN	Determination of Significance and Scoping Notice

EA	Environmental Assessment
EIA	Environmental Impact Assessment
EIR	Environmental Impact Report
EIS	Environmental Impact Statement
EPA	Environmental Protection Agency
ESA	Endangered Species Act
FAA	Federal Aviation Administration
FCC	Federal Communications Commission
FERC	Federal Energy Regulatory Commission
FHA	Federal Housing Administration
FHWA	Federal Highway Administration
FONSI	Finding of No Significant Impact
FR	Federal Register
FRA	Federal Railroad Administration
FTA	Federal Transit Administration
GSA	General Services Administration
HUD	(Department of) Housing and Urban Development
IDCA	(U.S.) International Development Cooperation Agency
LO	Lack of Objections
MEPA	Michigan Environmental Protection Act
MMPA	Marine Mammals Protection Act
MMS	Minerals Management Service
NAGPRA	Native American Graves Protection and Repatriation Act
NAT	Notice of Action Taken
NBS	National Biological Survey
NEPA	National Environmental Policy Act
NHPA	National Historic Preservation Act
NMFS	National Marine Fisheries Service
NOAA	National Oceanic and Atmospheric Administration
NOI	Notice of Intent

NOP	Notice of Preparation
NPS	National Park Service
NRCS	Natural Resources Conservation Service (formerly SCS)[2]
ROD	Record of Decision
SCS	Soil Conservation Service
SEPA	State Environmental Policy Act (Washington)
SEPAs	State Environmental Policy Acts (in general)
SEQRA	State Environmental Quality Review Act (New York State)
SHPO	State Historic Preservation Officer
SOW	Scope of Work
SR	State Register
USACOE	U.S. Army Corps of Engineers
USAF	U.S. Air Force
USAID	U.S. Agency for International Development
USCG	U.S. Coast Guard
USFS	U.S. Forest Service
USFWS	U.S. Fish and Wildlife Service
USGS	U.S. Geological Survey
USMC	U.S. Marine Corps
USN	U.S. Navy

---

[2]Name was changed to Natural Resources Conservation Service in 1995.

ENVIRONMENTAL IMPACT STATEMENTS

INTRODUCTION

*In nature there are neither rewards nor punishments—
there are consequences.*

Robert Ingersoll

WHY UNDERSTAND ROLES?

Imagine being an agency employee, a potential developer, or a private citizen who is asked to participate in a public event. You are told the time and place, you are invited to "provide your input," and you are given a deadline to provide your input (the event is over). But you don't know why you have been invited, what the event is to accomplish, what everyone else is doing there, and most importantly you don't know what you are expected to do or how you are expected to do it. It is like being invited to play in a ball game and not knowing whether it is baseball or football. In this analogy, the "event" is the environmental impact statement process which most certainly has rules and regulations for how it is conducted, but only a few participants are informed about their roles and the rules that guide them. Those who don't understand the process, or their part in it, develop incorrect assumptions and expectations that lead to conflicts and delays.

This book discusses an approach to understanding the roles of participants in the environmental review process, to managing the process, to reviewing, and to preparing environmental impact statements (EISs). Since 1970, guidelines and regulations have provided the basic requirements for preparation of an EIS, but a great deal of flexibility, as well as ambiguity, is also in the regulations. The courts have provided interpretations of the regulations, clarifying some parts while leaving other parts ambiguous. In some cases the various courts have made conflicting decisions. Thus, over the course of the last 25 years, individuals who have managed and prepared EISs have developed their own style and method of EIS preparation that work best for them. The topic of this book is one of many possible approaches to understanding and working with EISs. Its basic tenet is that if you approach the EIS process with an understanding of your role and the roles of others in the process, you will become more effective in accomplishing your goals whether you are an author or a reviewer of EISs and whether you are a federal, state, or local agency representative, a private citizen, a developer, or an environmental activist.

ORGANIZATION OF THIS BOOK

General Organization

This book (unlike others on the subject) is not organized by major types of environmental documents such as by a chapter on environmental assessments and another chapter on environmental impact statements. Nor is it organized by chapters relating to the environmental elements such as air, water, plants, or animals. Instead of devoting a chapter to one topic at a time, the discussions in this book generally follow the EIS process. Whether it is federal or state, the overall emphasis is on process and the interaction and roles of people within the process.

Because this book is process-oriented, some redundancy is unavoidable—and not inappropriate because events in the process are interrelated. The process takes place in phases or steps primarily because laws and regulations require an evolutionary development of environmental review. The phases or steps in the process will vary to some degree depending on the type of proposed action that is being evaluated, the location of the action (its context), and the policies and procedures of the agencies with responsibility for the EIS process. All of the subjects discussed in the first two chapters are discussed in greater detail in the following chapters because the early chapters are introductory (and therefore do not provide all the details on a topic) and because some concepts and topics are applicable in more than one situation.

Because this book is about EISs in general (i.e., not limited to federal or state EISs), the examples of EIS requirements and procedures are from agencies at various levels of government and at various locations in the United States. EIS regulations provide agencies wide latitude in their EIS methodologies and processes. Most agencies with EIS responsibilities have detailed EIS procedures which are sampled in this book. The examples illustrate some similarities and dissimilarities between agency requirements, and they illustrate how some agencies conduct the EIS process.

Most concepts presented in this book start from a rudimentary level, and readers may be well aware of, for example, basic types of environmental documents [e.g., environmental assessments (EAs), findings of no significant impact (FONSIs), EISs] or basic ecological principles. Participants in an EIS process, however, have a wide variety of educational backgrounds, and some will be familiar with certain concepts but not others. Therefore, all concepts necessary to understand the EIS process are discussed either first as an introduction, and then later in greater detail, or as a brief discussion that is only sufficient to understand how the subject relates to an EIS.

Chapters 1 through 7 provide historical background regarding the need to evaluate environmental impacts, introduce readers to the EIS process, and discuss the process in relation to state and federal requirements and the courts. Traditionally, environmental impacts have been viewed primarily from a local perspective. However, agencies and individuals are realizing that environmental impacts are more effectively evaluated from the perspective of watersheds and ecosystems. Instead of focusing on a single resource (e.g., animals, plants, water), ecosystem analysis evaluates how the resources interact with each other. While ecosystem analysis is not the subject of this book, the concept is discussed because of its importance in evaluating environmental impacts. Global and transboundary environmental concerns, as well as environmental impact assessments of other countries, are briefly discussed to provide an even larger perspective.

Chapters 8 through 14 describe how to build an EIS team and an EIS's framework, how to manage the EIS process and its preparation, how to involve the public, and how to write and review EISs. Suggestions or tips are provided throughout these chapters to make an EIS participant more effective in his or her role, to make various tasks easier, and to avoid potential problems.

Chapter Discussions

Chapter 1 introduces basic concepts: the public's concerns for the quality of the environment that led to the passage of many environmental laws and the EIS process that is the topic of this book. The purpose of EISs and the roles of the major participants in the EIS process including the public, the agencies, and the courts are discussed in Chapters 1 and 2. Terminology is also described because the EIS process uses terms that are unique in their meaning or their usage. One must understand the terms as they are used in an EIS process in order to understand the subject of EISs. Roles, terminology, and the specific requirements of the EIS process as defined by regulations and the courts are actually repeated throughout the book and build upon earlier discussions. The reason for this approach is to increase the reader's familiarity with the subject by focusing on the EIS process, and roles of those in the process, as it progresses from start to finish.

An EIS process does not function in a vacuum. An EIS is part of, and affected by, other regulatory processes. One of the most common areas of confusion is with regard to the expectations of an EIS—what an EIS does or does not do. Chapter 3 describes the EIS process in relation to agency land use or natural resource planning processes, other environmental regulations, and the project planning process. Authors and public reviewers of EISs should understand the relationship between the EIS process and other processes to know the limitations of EISs as well as their purpose.

> **An EIS process does not function in a vacuum.**

Chapter 4 describes proposed actions or projects; the actions are the subject of analyses in an EIS. Types of actions that are evaluated in EISs are described in Chapter 4. How the actions are defined or described (discussed in Chapter 8) is an important consideration when deciding what type of EIS to prepare (discussed in Chapter 7), what actions to include in an EIS, and how to conduct the analysis (Chapter 12). A proposed action may be an agency proposal or a private proposal, a new action, or a continuing action. An agency action may be to decide whether to issue a permit for a privately proposed action. The relationship between a proposed action and an agency action needs to be clear when analyzing the impacts of actions.

Chapter 5 briefly highlights differences and similarities between state and federal EISs, and it touches on global environmental impacts and the international interest in environmental impact assessments. Chapter 6 discusses the substantial influence of the courts in the EIS process and how litigation and court decisions have affected the way agencies view EISs. Both chapters are brief discussions to introduce readers to other important aspects of EISs (that are not within the focus of this book). Each chapter alone is the subject of numerous publications.

With Chapter 7 we begin to focus on the preparation and management of EISs. Chapter 7 addresses the major types of EISs such as programmatic and project-specific EISs. Agencies have a wide range of responsibilities, and the actions required to carry out those responsibilities can be broad program-level actions or site-specific projects and everything in between. A range of types of EISs are available to match the particular circumstances of a proposed action. A programmatic EIS, for example, is designed to save time and paperwork by eliminating redundancy and discussing only those issues that are "ripe" for discussion.

Chapter 8 describes how to build the framework of an EIS and identifies important preliminary issues that an EIS team should discuss prior to starting an EIS process. This chapter emphasizes the need to thoroughly define the proposed action and its alternatives and to obtain internal agreement (within the agency and EIS team) on the EIS's scope, methodology, and any basic assumptions or requirements that will be part of the project or the analysis.

Chapter 9 deals with the nuts and bolts of managing an EIS from the standpoint of EIS project managers and team members. Developing and maintaining contract scope (as well as EIS scope), budget, schedule, and quality control are some of the tasks involved in managing EISs. Also, communication, leadership, and other skills are required to effectively manage EIS teams. The ideal interpersonal and management skills of EIS team members are somewhat subjective and will vary depending on the circumstances, but they are presented as considerations when building an EIS team (discussed in Chapter 10). Chapter 10 discusses the basics of EIS preparation starting from what to title an EIS, to outlining an EIS, identifying necessary information, building an interdisciplinary team, and conducting a kickoff meeting.

Chapter 11 addresses public involvement: when to involve the public, how to make the most of public input, and how to conduct public meetings or hearings. Public involvement is an integral part of an EIS process, and an EIS team should decide early in the process what its goals are for involving the public and how to accomplish those goals. Some agencies prepare public involvement plans as part of their EIS preparation plan. The public can provide valuable information in terms of identifying significant issues, possible alternatives to the proposed action, and sources of information—for it is information that an EIS process actively seeks. The human nature aspect of public involvement must also be recognized. Public input is based on values and perceptions, which are sometimes difficult to adequately address in EISs that are designed to be logical, methodical, and scientific planning documents. Values are neither right or wrong, black or white—they tend to be gray. They are not readily measurable and are hard to categorize. Nevertheless, the public's concerns must be addressed in EISs.

Chapter 12 discusses the purpose and the many facets of writing EISs. The content of Draft EISs and Final EISs are often completely different. A Draft EIS includes the environmental analyses of a proposed action and its alternatives, while a Final EIS may simply include responses to public comments on the Draft EIS. This chapter also addresses how to analyze impacts for significance based on, among other things, the condition of the existing environment, type of proposed action, and the intensity and context of the impacts (or the "so what?" of the analysis—*is* the impact "significant"?). Focusing the analysis on significant impacts or issues is a common theme throughout the book and is also described in this chapter. The identification, evaluation, and implementation of mitigation measures is another important aspect of impact analysis. Because most agencies are not required to adopt mitigation measures identified in an EIS, one should understand how mitigation measures can be made enforceable commitments.

EISs have some unique features: The format is one, the language of EISs is another. EISs reflect the planning process of which EISs are a part. Thus, "would" is used more frequently than "will," and an EIS author must sometimes be redundant between sections of an EIS or when comparatively describing the impacts of alternatives. While most EIS authors would not win awards for their prose, clarity and specificity in describing impacts is what counts. Scientific and professional standards, particularly with regard to the methodology of impact analysis, while not required to be the "best" methodology, must be appropriate and adequate for the circumstances. The fundamental question is, Are the conclusions in an EIS supported by the facts?

Chapter 12 also discusses the pros and cons of using an impact matrix to compare the impacts of alternatives. An impact matrix provides a visual aid for comparing impacts, but they are based on subjective values and cannot take all factors into account, such as the condition of the existing environment and impacts which are mixed (part adverse, part beneficial). One could argue that subjective values are inevitable even for analysis in narrative form. However, subjectiveness becomes more pronounced when values are condensed into one symbol on a matrix.

In writing EISs, cumulative impacts must be addressed. Chapter 12 describes the difference between cumulative actions and cumulative impacts, as well as methods for evaluating cumulative impacts. Cumulative impact analysis is part of an EIS, not a separate analysis or separate EIS. The open-ended definition of cumulative impacts provided by the federal Council on Environmental Quality (CEQ) regulations requires consideration of past, present, and future impacts and can lead to a never-ending analysis of impacts. Lack of clear direction on how to perform cumulative impact analysis results in inconsistent treatment from one EIS author to another. One must identify the scope and methodology for analyzing cumulative impacts in a similar fashion to the scope and methodology for analysis of environmental elements (environmental elements are the subjects of analysis such as air, water, animals, and cultural resources).

Chapter 13 provides suggestions for reducing time, effort, and paperwork. Many of the methods are identified by regulations which direct EIS authors and agencies to omit extraneous background data, focus on significant issues, and reduce the bulk of EISs by incorporating information by reference. Agencies with EIS responsibility (e.g., a state and federal agency) over a proposed action can combine their requirements and prepare joint EISs, saving the time and effort of multiple EISs, or adopt other agencies' EISs (or portions of EISs). The EIS team can also save time and effort by coordinating team members' efforts, including the production staff (e.g., editor, word processor, graphic artist), and maintaining an EIS's focus, scope, and schedule.

Chapter 14 discusses how reviewers can be effective in their review of EISs. Reviewers include internal reviewers (lead agency and EIS team) and public reviewers (individuals, groups, other agencies). The roles and goals of reviewers will vary depending on their interests, concerns, and purposes. Agencies and special interest groups will review an EIS with the goals of their organizations in mind. Although some reviewers may seek to discredit an EIS, and thereby stop or at least delay a proposed action, the focus of this chapter is not to evaluate whether or not a proposed action should be approved. The purpose of reviewing an EIS is to critique the document to ensure that the lead agency's decisionmaker understands the environmental consequences of choosing a proposed action or one of its alternatives. The environmental consequences or impacts must be based on rational and reasonable analyses of issues. Some of the factors to consider in this evaluation are listed in this chapter.

Because subjects related to EISs, but not within the focus of this book, receive only brief discussion, the Bibliography includes publications for additional information on subjects such as international impact assessment requirements and EIS case law. Lastly, information related to the subjects discussed in this book, and information that may be useful in preparing EISs, are available in the appendices. The appendices include a copy of the National Environmental Policy Act (NEPA) (Appendix A), the Council on Environmental Quality (CEQ) regulations (Appendix B), citations of federal agency regulations that implement NEPA (Appendix C), environmental data collection programs of federal agencies (Appendix D), and organization charts of selected federal agencies with NEPA responsibilities (Appendix E).

BASIC CONCEPTS AND PARTICIPANTS' ROLES

Environmental Assessments and Environmental Impact Statements

Before proceeding further, some basic concepts must be understood. Environmental impact statements (EISs), environmental assessments (EAs), and their state counterparts are prepared to comply with National Environmental Policy Act (NEPA) and state environmental policy acts (SEPAs) and their implementing regulations at the federal, state, and local level. EISs are prepared for proposed projects that have the potential to result in significant impacts to the environment. An EIS is the document that results from the EIS process. The process is as important as the EIS (as will be discussed later), and the emphasis of federal and state EIS regulations is on excellent decisionmaking by agencies, not excellent environmental paperwork.[3] This does not imply that environmental documents can be inadequate; it means that paperwork is secondary and that including environmental considerations in the decisionmaking process is paramount.

What is the decisionmaking process and who is the decisionmaker? It depends. Regulations discuss the "decisionmaker"; however, EIS regulations are written broadly because they apply to a wide range of agencies. Therefore, identifying the "decisionmaking process" and "decisionmaker" requires identification of the responsibilities and organizational structure of the particular agency involved in an EIS, as well as determining where decisions for the type of project under consideration are made in the organization. Generally, a decisionmaking process is the process by which an agency decisionmaker will take into consideration all of the factors (technical, economic, environmental) necessary to decide whether a proposed project should proceed. The decisionmaker is usually defined in an agency's delegation of responsibilities for certain actions. The difficulty comes when, for some actions, a single individual does not make the decision of whether to approve or proceed on a project. The project's approval may depend on a number of individuals (or committees), in which case one person will have to be designated as "the decisionmaker" for the EIS.

---

[3] 40 CFR §1500.1(c).

The terms "environmental impact assessment" and "environmental assessment" are often used generically to apply to the analysis of environmental impacts. However, these terms are also proper nouns such as (a) an environmental impact assessment (EIA) prepared in Canada and other countries and (b) an environmental assessment (EA) prepared under NEPA and some SEPAs in the United States. California calls its environmental impact assessment an Environmental Impact Report (EIR).

An EA is prepared for proposed actions when it is not known whether the proposal would result in significant impacts to the environment. States have similar documents (some are called "initial studies" or "environmental checklists") that are prepared to determine whether a proposed project has the potential to cause significant environmental impacts. Some states also call their documents "EAs."

An EA is not the same document as an "environmental assessment" prepared when evaluating the potential for a site to have been contaminated by past uses of the site. Such reports also have other names including Preliminary Site Assessments and Property Transfer Reports. These reports should not be called "environmental assessments" or "EAs" to avoid confusing them with a NEPA or SEPA EA.

Environmentalists and Environmentalism

"Environment," "environmentalism," and "environmentalists" have very different meanings for people depending on their values, background, and current circumstances. A person's perception of the environment, as well as his or her concerns for the environment, is subject to change as a result of an effect, or proposed change, on his or her personal environment (e.g., a neighborhood) or economic circumstance. Feelings can change from caring about environmental quality, or at least being neutral, to disliking the subject of the environment because of an environmental requirement that prevented a person from developing his or her property or caused the loss of livelihood. One person might feel proud if called an "environmentalist," while another would consider it an insult. The topic can be personal and very emotional.

An environmentalist has come to mean someone who has a political agenda to stop development. However, as long as the population continues to increase (approximately 2.8 million a year in the United

States during 1990 to 1993[4]), goods and services, housing, and transportation projects will continue to be necessary. An extreme posture to stop all development is unrealistic and untenable. On the other hand, citizens and land use agencies can decide that a proposed development is environmentally unsuitable for a specific location, or they can require changes to make a proposal more suitable.

Because of the emotional nature of the topic, some unbiased definitions are in order to establish common ground in discussing things related to the environment. *Webster's Tenth New Collegiate Dictionary* defines "environment" as:

a: the complex of physical, chemical, and biotic factors (as climate, soil, and living things) that act upon an organism or an ecological community and ultimately determine its form and survival

b: the aggregate of social and cultural conditions that influence the life of an individual or community.[5]

The Council on Environmental Quality states that the "human environment" shall be interpreted comprehensively to include the natural and physical environment and the relationship of people with that environment."[6]

Webster's also defines "environmentalism" and "environmentalist" as:

■ advocacy of the preservation or improvement of the natural environment; *esp*: the movement to control pollution

■ one concerned about environmental quality esp. of the human environment with respect to the control of pollution[7]

Using the definitions above, the vast majority of people in the country, or the world, could reasonably be called environmentalists who are concerned about the quality of the environment. Most people don't want to breathe polluted air or drink polluted water. Most people support

---

[4]Council on Environmental Quality, *Twenty-fourth Annual Report*, 1993.

[5]By permission from *Merriam-Webster's Collegiate Dictionary*, tenth edition, copyright 1994 by Merriam-Webster, Inc.

[6]40 CFR §1508.14.

[7]By permission from *Merriam-Webster's Collegiate Dictionary*, tenth edition, copyright 1994 by Merriam-Webster, Inc.

the movement to clean up polluted sites, or to set aside natural pre-
serves so that future generations can enjoy visiting a relatively natural setting. What separates environmentalists from anti-environmen-
talists, among other things, is the question of who pays the cost for
protecting the environment. The person who feels that he or she has to
pay in some manner equates an environmental action as a personal
and unreasonable cost, while those who only benefit from an environmental action see the cost impersonally (someone else is paying it) and
as worthwhile. The side of the fence you are on will depend on "whose
ox is being gored" (your's or a neighbor's).

The Cost of Protecting the Environment

Cleanup of pollution and preservation of environmental resources
have various costs associated with them. Costs can be in the the form
of an increase in operating expenses for businesses that must reduce
pollutants they generate. These costs may be passed on to the consumer by increasing the price of goods or services. On the other hand,
not all environmental costs result in negative economic impact. Many
businesses and industries are finding that by evaluating their use of
resources and production of wastes, along with retrofitting or updat-
ing their equipment, operational costs are reduced and their use of
fuel or water, for example, is reduced. Preservation of endangered species' habitats can preclude the removal of resources (such as timber or
minerals) that happen to be in their habitat, which may reduce the
need for labor. The cost is loss of jobs, which has become a political
issue of jobs versus the environment. These are simplifications of
course. Loss of jobs is not just related to environmental preservation,
but is influenced by many other factors, including a company's management practices. Nonetheless, the perception by many people is that
the government has chosen an animal's welfare, for example, over their
own.

An EIS process requires us to take a hard look at the consequences
of our actions on the environment. Because of the nationwide trend
toward depleting resources and allowing unchecked pollution, Congress
enacted environmental laws. Some think that environmentalists are
overreacting and that the condition of the environment is satisfactory.
However, negative impacts to the environment have accumulated over

the years, and it could take many years to correct environmental impacts, if they are correctable at all. Because regulations of the last 25 years have resulted in improvements to the environment, it is easy to forget some of the problems of the past. As a reminder, the following is a statement made in 1969 by Representative Michael A. Ferghan (D-Ohio):

> As the representative and citizen of a district which has the dubious distinction of claiming within its boundaries a river that periodically catches fire and which borders on a lake referred to as the "Dead Sea," I am particularly concerned with measures which would improve the condition of these and similarly afflicted areas.[8]

The EIS process, because it is forward-looking in its application, seeks to prevent environmentally degrading actions rather than allowing them to take place, thus leaving future generations faced with rectifying the problem. If we look ahead and correct our course of action, it will cost less in terms of money and quality of life than if we try to fix mistakes later. This is, some people say, a pay-as-you-go plan. The EIS process weighs the cost to the environment, in terms of impacts (not necessarily in dollars) of a proposed project, against a proposed project's benefits.

Lastly, there is no doubt that taking action on improving or protecting the environment results in costs to individuals, businesses, and taxpayers. Nevertheless, contrary to the opinion that what is good for plants and animals is at the expense of people, what is good for plants and animals is also good for people. When considering the total picture, a healthy environment is not only healthy for people, it is also healthy for the economy. People and businesses tend to gravitate to areas with a high quality of life, including environmental amenities. To maintain or enhance an area's environmental quality, while people who are drawn by those very qualities increase development pressure and cause conflicting demands on resources, is no simple task. The goal we are addressing, as stated in the National Environmental Policy Act is

---

[8]115 Congressional Record 26577.

> . . . to create and maintain conditions under which man and nature can exist in productive harmony, and fulfill the social, economic, and other requirements of present and future generations of Americans.[9]

Worth noting, in the above statement, is the intent to fulfill social and economic requirements of Americans (including future Americans); hence the law is not attempting to make environmental resources more important than people, but is seeking to achieve a balance between the two.

THE EIS PROCESS

To understand roles, one also must understand the EIS process. The process is discussed in more detail in the following chapters, however, for the purposes of this discussion, the major steps in the process include the following:

■ A proposal, also called a *proposed action* or *proposed project*

■ The lead agency's determination of whether an EIS is required for the project (a threshold determination); if so, then the following steps are necessary:

■ Public scoping (the public's input regarding subjects to be evaluated in an EIS)

■ The lead agency's decision on the scope of an EIS's analysis

■ Preparation of a Draft EIS

■ Public review and comment on the Draft EIS

■ Preparation of a Final EIS

■ The lead agency's decision on whether to issue a permit or approve the project (or whether to proceed if the proposal is the agency's)

Roles of Participants in the EIS Process

Those involved in the EIS process generally consist of the following:

---

[9]NEPA Section 101(a).

Participant	Role
Lead agency	The agency responsible for preparation of the EIS and for making a decision on the proposed action
The EIS team	An interdisciplinary team of specialists who may be employees of the lead agency or private consultants reporting to the lead agency, and support and agency staff
Project proponent	A private developer or landowner, or an agency proposing an action
Public reviewers of an EIS	Citizens and organizations interested in or affected by a proposed project, and other agencies with jurisdiction by law or expertise

The specific roles of participants in an EIS process will be discussed throughout the following chapters. An EIS process for a particular proposal will normally involve unique circumstances that require adjustments in the process to meet the needs of the situation. Therefore, flexibility of the team is essential, especially for long and complex projects, provided that any adjustments in roles and responsibilities are within the boundaries established by regulations. The following are commonly the basic roles of participants in an EIS process.

Lead Agency   The lead agency has overall responsibility for conducting an EIS process and for preparation of an EIS. The content of an EIS and its conclusions are those of the lead agency regardless of who prepares the EIS. If an EIS is prepared by someone other than the lead agency, the lead agency must provide independent review of the document and ensure that it meets the agency's standards. A lead agency makes decisions regarding the EIS's scope (the issues and alternatives analyzed in an EIS), the amount of public involvement, the time allowed for public review of the EIS, the data to be used, the accuracy of the data, and the appropriateness of the level of analysis.

A lead agency may be a federal, state, or local agency with authority to approve, permit,[10] or fund[11] a private proposal and with authority to act as a lead agency in the preparation of EISs. Some agencies have responsibilities to regulate private proposals as well as manage public land and natural resources. If a private developer or landowner requires approval, funding, or permits from an agency, that agency (or the agency with the most involvement when there is more than one agency approval required) becomes the lead agency with responsibility for the EIS on the private proposal. An agency may have its own proposed project that is required to carry out the agency's mission, and it may be the lead agency for the EIS.

The EIS Team EISs are prepared by individuals such as scientists, engineers, and planners. The required number and type of professionals depend on the range of issues (discussed under "Scope and Scoping" in Chapter 2) addressed by an EIS. The individuals preparing an EIS are often referred to as a team because it requires teamwork to prepare a document that is dynamic and usually complex. An EIS team consists of professionals with the credentials and experience to analyze the elements of the environment that are within the scope of an EIS. Preparers of EISs can be agency personnel, private consultants, university professors, or a combination of the foregoing. The lead agency and the project applicant or proponent also should be members of the team to provide the required guidance and project information, respectively.

EISs should be prepared by an interdisciplinary (ID) team. The difference between an interdisciplinary team and a multidisciplinary team is that interdisciplinary team members prepare an EIS in consultation with each other, whereas multidisciplinary team members analyze their respective topics and submit the analyses to an individual who then places them in one document. The primary drawback to a multidisciplinary approach is that it can result in inconsistencies and

---

[10]Limited permitting by a federal agency may not make a nonfederal action subject to NEPA. A large-scale nonfederal action is not necessarily federal simply because one portion of it requires a federal permit or approval.

[11]Funding must be substantial; also general revenue sharing may not be subject to NEPA (Fogleman, 1990).

contradictions within an EIS (see Chapter 10 for more information on interdisciplinary teams).

In analyzing potential environmental impacts, the team members' responsibilities are to provide objective analyses (not biased in support of or against a proposed project), thorough analyses commensurate with the level of significance of a potential environmental impact (not study an issue in detail if the issue is not significant), and analyses that are within the accepted professional standards of protocol or methodology. Facts presented in the analyses should be meaningful and relevant to the issues.

Project Applicant, Proponent, or Sponsor The terms *project proponent* and *project sponsor* are synonymous. If a permit is necessary for a proposed project, the "proponent" applies for a permit and is therefore called the "applicant." As stated previously, a project proponent can be an agency or a private entity that requires funding, approval, or a permit from an agency (such as a permit to fill wetlands from the U.S. Army Corps of Engineers or a construction permit from a city).[12] The primary role of a private project applicant or proponent in an EIS process is to provide information regarding the design, construction, and operation of a proposed action or project. The applicant identifies the purpose and need for the proposed project, lists the objectives of the project, provides available studies such as feasibility analyses, and supplies applicable engineering or architectural drawings. The lead agency, however, can request the applicant to provide specific types or details of project information that meets the agency's requirements and is understandable to the public.

An agency that proposes its own project has several roles as a proponent, EIS preparer, and lead agency with the responsibilities described above. The regulations specifically give agencies responsibility for EISs, even if it is for its own proposal, and do not consider the multiple roles of proponent and EIS preparer a conflict of interest. For federal EIS consultants, however, lead agencies must ensure that consultants have no interest in the outcome of a proposed action to avoid a potential conflict of interest.

---

[12]Not all SEPA (state) regulations extend the requirement to prepare EISs to local agency actions or to private actions.

Public Reviewers The public is usually involved in the scoping process (a process of providing input to a lead agency regarding the issues and alternatives to address in an EIS) and review of a Draft EIS. Lead agencies may choose to have the public more involved in the process than is specifically required by regulations. The public is sometimes invited to assist in identifying potential project alternatives (different ways to meet the project's purpose and objectives) and mitigation measures (methods for reducing potential environmental impacts) for significant environmental issues by having informal workshops or meetings during the course of EIS preparation.

Public review of an EIS is done by the following: private citizens, often those within close proximity of a proposed project (such as adjacent property owners); American Indian tribes (if a proposed project has potential to affect reservations); agencies having jurisdiction by law or expertise; and those who have requested notification (such as special interest groups or environmental organizations) on a given area or type of action.

Because an EIS process is not intended to take votes on whether reviewers support a proposal, the process does not seek opinions "for" or "against" a proposal. Rather, the public's role is to express any concerns by identifying significant issues, as well as to provide information and suggestions so that an agency decisionmaker can make an informed decision. From the perspective of an EIS process, the role of reviewers is to understand the proposed project, provide suggestions for alternative ways to meet the purpose and needs of the project, and critically review a Draft EIS to provide specific comments on a proposed project's potential environmental impacts, the methodology used in analyzing potential environmental impacts, the adequacy of project alternatives, and mitigation measures (see Chapter 14 for suggestions to critically review EISs).

THE OVERALL PROCESS AND ROLES

Figure 1.1 is a flow chart of the basic EIS process and roles of participants in the process. The federal process and examples of two state processes (California's and Washington's) are shown to illustrate the similarities among the basic processes. California and Washington are

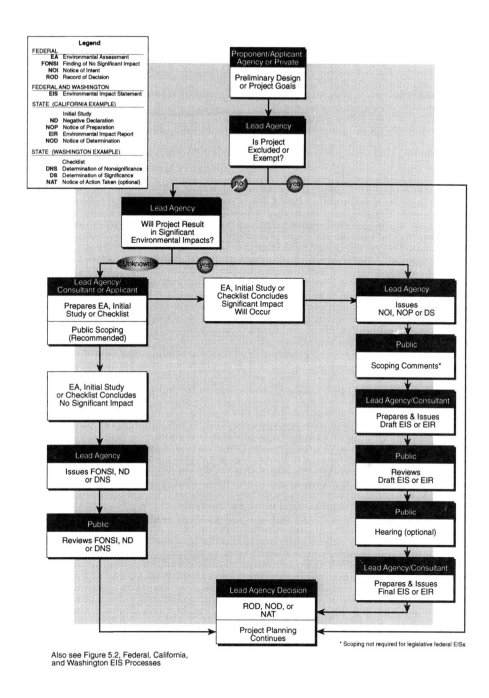

Legend

FEDERAL
- **EA** Environmental Assessment
- **FONSI** Finding of No Significant Impact
- **NOI** Notice of Intent
- **ROD** Record of Decision

FEDERAL AND WASHINGTON
- **EIS** Environmental Impact Statement

STATE (CALIFORNIA EXAMPLE)
- Initial Study
- **ND** Negative Declaration
- **NOP** Notice of Preparation
- **EIR** Environmental Impact Report
- **NOD** Notice of Determination

STATE (WASHINGTON EXAMPLE)
- Checklist
- **DNS** Determination of Nonsignificance
- **DS** Determination of Significance
- **NAT** Notice of Action Taken (optional)

Proponent/Applicant Agency or Private — Preliminary Design or Project Goals

Lead Agency — Is Project Excluded or Exempt? (no / yes)

Lead Agency — Will Project Result in Significant Environmental Impacts? (Unknown / yes)

Lead Agency/Consultant or Applicant — Prepares EA, Initial Study or Checklist / Public Scoping (Recommended)

EA, Initial Study or Checklist Concludes Significant Impact Will Occur

Lead Agency — Issues NOI, NOP or DS

Public — Scoping Comments*

EA, Initial Study or Checklist Concludes No Significant Impact

Lead Agency — Issues FONSI, ND or DNS

Public — Reviews FONSI, ND or DNS

Lead Agency/Consultant — Prepares & Issues Draft EIS or EIR

Public — Reviews Draft EIS or EIR

Public — Hearing (optional)

Lead Agency/Consultant — Prepares & Issues Final EIS or EIR

Lead Agency Decision — ROD, NOD, or NAT / Project Planning Continues

* Scoping not required for legislative federal EISs

Also see Figure 5.2, Federal, California, and Washington EIS Processes

■ **FIGURE 1.1.** Overall environmental review process and participants

19

provided as examples because of their detailed EIS regulations and substantial experience with EISs. The main differences between the basic federal and state processes are in the names of some of the documents and timeframes for actions that take place in the process. Depending on the agency jurisdiction for the project location, type of project, and type of approval being sought, the agencies involved in the process and details of procedures within the processes will vary.

The process begins when someone—an agency, a private individual, or an organization—proposes an action or a project. At this stage, the proposed action or project usually is conceptual or preliminary in design. The lead agency (federal or state[13]) determines whether the proposal is categorically excluded from the requirement to prepare environmental documentation. Categorical exclusions (federal) or categorical exemptions (California and Washington) are categories of actions that normally do not result in significant impacts and therefore do not require further environmental review. Some actions are also statutorily exempt (a federal or state law expressly exempts actions under the law from NEPA or a SEPA review). If the proposed project is excluded from further environmental review, the project continues in the planning process (see discussions of the planning process in Chapter 2 and Chapter 3).

If a proposal is not excluded from environmental review, the question a lead agency must answer is, Will the proposal result in significant environmental impacts? This is called the *threshold determination*. If the answer is "yes," the proposal will require preparation of an environmental impact statement. If the lead agency does not know whether the proposal would result in significant environmental impacts, an environmental assessment (federal), an initial study (California), or an environmental checklist (Washington) would be prepared to evaluate the proposal's potential environmental impacts and to determine whether any of the impacts would be significant. The EA, initial study, or environmental checklist may be prepared by the lead agency, a project applicant, or a consultant. If an EA, initial study, or checklist is prepared by an applicant or consultant, the lead agency reviews the document for adequacy. If the EA, initial study, or envi-

---

[13]Local agencies also prepare EISs in some states.

ronmental checklist concludes that the proposed project would not result in significant environmental impacts, the lead agency issues a finding of no significant impact (FONSI) (federal), negative declaration ("Negdec" or ND as shown in Figure 1.1) (California), or determination of nonsignificance (DNS) (Washington) which is publicly reviewed. If the EA, initial study, or environmental checklist concludes that the proposed project would result in significant environmental impacts, the lead agency issues a notice of intent (NOI) to prepare an EIS (federal), a notice of preparation (NOP) (California), or a determination of significance and scoping notice (DS/SN) (Washington), and the public scoping process for an EIS is begun.

The next major steps are (a) preparation of a Draft EIS by the lead agency or consultant and (b) distribution of the Draft EIS or EIR for public review. After public review of the Draft EIS or EIR, a Final EIS or EIR, which responds to comments on the draft document, is then prepared and publicly distributed.

FUNDAMENTALS

Anyone who believes in the wholesomeness of sausages and laws shouldn't watch them being made.

Anonymous

LAWS AND PROCEDURES

Concerns regarding the environment, pollution, congestion, and desires to preserve natural areas are not entirely new to Americans. Although during the Nation's early development it was not uncommon for land to be denuded of trees and streams or rivers to serve as a town's sewer system, as early as the 1800s America had environmental laws. Yellowstone National Park, established in 1872, was the first such land reserve in the country. The Rivers and Harbors Act of 1899, also known as the "Refuse Act," required a permit from the U.S. Army Corps of Engineers to dump material in the Nation's waterways. President Theodore Roosevelt and Gifford Pinchot, the first Director of the U.S. Forest Service, closed millions of acres to development. Gifford Pinchot established the concept of multiple use of public lands, while John Muir started the movement toward preservation of public lands and resources. The National Park Service was established in 1916, and the Wildlife Restoration Act, which was intended for management of wildlife habitat but dealt primarily with sport species, was passed in 1937. Still, a large amount of land seemed available for development further west, and the environmental laws were relatively few and rarely enforced. By the late 1960s, many felt that developers and mining, forestry, and cattle grazing (range) industries were allowed, or encouraged, to maximize production or profit at the expense of the quality of the environment. The pendulum had swung to an extreme. The environmental movement began to swing the pendulum the other way as the 1970s, which became known as the "environmental decade," resulted in the passage of numerous environmental laws. That period in time is described by Knight, in *The New Environmental Handbook*:

> In the wake of NEPA came an armada of laws. . . . In 1970, Congress passed the Clean Air Act and amended it twice later in the decade.
>
> The water pollution control act passed in 1972, and was later amended. Congress passed a law to protect drinking water, and other laws to regulate the strip mining of coal, coal leasing on federal lands, leasing practices on the outer continental shelf, deep water ports, liquefied natural gas terminals, and gas pipelines. Acts were passed that controlled toxic substances, solid waste recovery, and the use of rodenticides, insecticides and fungicides. Finally, two laws were passed that made the survival of life forms a national goal of highest priority—the Endangered Species Act and the Marine Mammal Protection Act.
>
> . . . An Environmental Protection Agency was created by President Nixon to house the environmental regulatory programs of the federal government under one roof. By the end of the decade EPA had become the largest regulatory agency and the administrator of the largest public works programs in the government, the water pollution control program.
>
> The Forest Service and Bureau of Land Management had the laws governing their very existences rewritten to reflect the changes in our perception of how natural resources should be managed. Protection for federal lands was expanded and enhanced. . . . New categories of protected lands were created, including the National Wild and Scenic Rivers, the National Wildlife Refuges, the National Recreation Areas, National Marine Sanctuaries, and the National Trails System. Vast new parks and wilderness areas were added to existing systems and whole ecosystems were put up for protection in Alaska. Management of the lands slowly improved too. Mining was halted in national parks. Predator con-

trol practices changed. Use of off road vehicles was limited, and concessionaires—the companies that provide housing, food and services in national parks—were put on a shorter leash. Major inventories of roadless areas on federal lands were undertaken. The Coastal Zone Management Act paid the states to plan for the use and protection of their coastal lands.[14]

The concern that has not changed to this day is how to have "progress," which comes with serious environmental costs, yet manage natural resources for the long-term and maintain the quality of life that is important to our existence. Environmental regulations are unpopular with some landowners, developers, and businesses because they curb or control development, and the use of natural resources. Goals of environmental protection or enhancement tend to be for the long-term and for the benefit of all citizens—sometimes at the expense of profit for the short-term and for a smaller group of citizens. Thus, federal EISs are required to evaluate "the relationship between short-term uses of man's environment and the maintenance and enhancement of long-term productivity."[15] In the past, no one knew that filling swamps, the breeding place of mosquitoes, had consequences such as decreasing water quality and increasing flooding, resulting in impacts on homes, farmland, and commerce, as well as wildlife. The environmental review process requires agencies to consider these types of environmental consequences when making decisions on their actions.

National Environmental Policy Act (NEPA)

Although the focus of this book is on an approach to EISs, and not on the specific legal requirements of EISs by each jurisdiction, an understanding of EISs would not be complete without a basic understanding of the intent of relevant legislation and the general relationship between federal, state, and local laws (more on this in Chapters 3 and 5). The first law dealing with EISs was the National Environmental Policy Act (NEPA) of 1969 (see Appendix A). NEPA did not require that states pass their own versions of this act. However, several states followed NEPA's lead and passed laws requiring EISs for certain actions. The state laws are similar to NEPA and have been called "little NEPAs,"

---

[14]Reprinted by permission, copyright 1980, Garrett DeBell.
[15]NEPA, Section 102(2)(c)(iv).

although the state laws have some differences in substance or procedure from the federal law. Because the federal and state requirements are similar and NEPA and the Council on Environmental Quality (CEQ) regulations have provided the basic framework for the EIS regulations of other jurisdictions, the federal regulations are discussed here. However, be aware that agencies at the federal, state, and local levels have internal procedures, regulations, or ordinances that may have specific requirements that go beyond requirements discussed in this book. Agencies will normally provide a copy of their EIS regulations on request, sometimes requiring payment for the cost of copying and postage. Refer to Appendix C, which lists citations of federal agency NEPA implementing regulations.

NEPA was signed into effect by President Nixon on January 1, 1970. This federal statute requires "a detailed (environmental impact) statement" for major federal actions significantly "affecting the quality of the human environment."[16]

The term "human environment" has no practical distinction from "environment" if one acknowledges that humans affect the environment and the environment affects humans (see definitions in Chapter 1). Humans and the environment are inseparable. For the most part, potential environmental impacts, whether to animate or inanimate objects (such as historical structures), are taken to mean impacts to humans, because the environment is the world in which everything exists and in which an effect on water, air, or animals eventually, or immediately, affects humans.

NEPA states that its purpose is

> To declare a national policy which will encourage productive and enjoyable harmony between man and his environment; to promote efforts which will prevent or eliminate damage to the environment and biosphere and stimulate the health and welfare of man; to enrich the understanding of the ecological systems and natural resources important to the Nation; and to establish a Council on Environmental Quality.[17]

NEPA further states that it is the responsibility of federal agencies "to use all practicable means, consistent with other essential considerations of national policy . . .," so that the Nation may:

---

[16]NEPA, Section 102(2)(c).
[17]NEPA, Section 2.

(1) fulfill the responsibilities of each generation as trustee of the environment for succeeding generations;

(2) assure for all Americans safe, healthful, productive, and esthetically and culturally pleasing surroundings;

(3) attain the widest range of beneficial uses of the environment without degradation, risk to health or safety, or other undesirable and unintended consequences;

(4) preserve important historic, cultural, and natural aspects of our national heritage and maintain, wherever possible, an environment which supports diversity, and variety of individual choice;

(5) achieve a balance between population and resource use which will permit high standards of living and a wide sharing of life's amenities; and

(6) enhance the quality of renewable resources and approach the maximum attainable recycling of depletable resources.[18]

Yet, despite the expansive statements in the act, unlike some environmental laws, NEPA and SEPAs have no "teeth"; that is, environmental considerations do not have to be elevated above other considerations; and there are no civil or punitive penalties, such as fines or imprisonment, for not complying with the law. The CEQ regulations established detailed requirements for agencies to consider environmental impacts in their decisionmaking process. The regulations also required that federal agencies develop their own regulations to ensure that the agencies' decisions are made in accordance with the policies and purposes of the act. The regulations gave agencies deadlines by which they were to have established procedures to supplement the CEQ regulations; however, many years after the deadline, some agencies still had not done so. Court actions by members of the public encouraged recalcitrant agencies to implement NEPA's requirements. And it is still the public, by and large, that serves the watchdog function to ensure that agencies comply with NEPA or a SEPA's requirements; and it is ultimately the courts that decide whether an agency has complied with the laws.

> **Environmental considerations do not have to be elevated above other considerations.**

---

[18]NEPA, Section 101(b).

ROLES OF CEQ, EPA, AND THE COURTS

Role of the Council on Environmental Quality

As stated previously, the Council on Environmental Quality (CEQ) was created by NEPA. The CEQ reports directly to the President, and it issues an annual report on the status of the environment. The CEQ is not a full regulatory body because it can neither veto nor control another agency's projects. If issues in an EIS have been taken to the courts, the courts give any opinions of the CEQ regarding those issues substantial deference or weight. Presidential Executive Orders (11514 and 11991) authorized the CEQ to issue regulations governing the EIS process, and the latter executive order made the regulations binding on federal agencies (executive orders are discussed in Chapter 3).

The CEQ regulations implement NEPA and provide the procedural and substantive requirements for compliance with NEPA. CEQ also acts as a referee between federal agencies when disagreements arise regarding, for example, which agency should be the lead agency for preparation of an EIS. Because federal agencies do not sue other federal agencies, as a matter of policy, an agency who cannot come to agreement with the lead agency on a serious environmental matter related to an EIS can refer the issue to the CEQ. Also, if the Environmental Protection Agency (EPA), upon review of an EIS, determines that a proposed federal action is environmentally unsatisfactory, the EPA may refer the matter to the CEQ. Generally, however, the CEQ is not involved in the day-to-day world of EIS preparation. CEQ's other areas of responsibility, such as development of national and international environmental policy and review of agency environmental quality programs, are topics dealt with at the policy level and will not be addressed here.

CEQ Regulations   The CEQ regulations, in Appendix B, are fairly straightforward. However, since the regulations were written for all federal agencies, which have a wide variety of responsibilities, they are necessarily broad and, in some cases, vague. Highlights from the CEQ regulations are noted throughout this book in relevant sections and are footnoted with their locations in the Code of Federal Regulations (CFR). The CEQ regulations are in 40 CFR Parts 1500 through 1508.

Role of the Environmental Protection Agency

The Environmental Protection Agency (EPA) has many responsibilities under several environmental laws. With regard to reviewing, rating, and filing federal EISs, however, the EPA's specific responsibility is derived from Section 309 of the Clean Air Act. Federal EISs are "filed" with the EPA (i.e., federal agencies send EISs to the EPA), who then publishes a notice of availability in the *Federal Register*. The date that the notice appears in the *Federal Register* is the date from which the number of days for public review are counted.

The EPA also reviews and provides written comments on federal EISs. The EPA rates Draft EISs on (1) the environmental effects of the proposed action and (2) the adequacy of the Draft EIS (see Figure 2.1). The highest ratings that may be received on an EIS are LO (Lack of Objections) on the impact of the action and Category 1 (Adequate) on the adequacy of the EIS. If an EIS's ratings are not so perfect, the ratings themselves do not provide sufficient information for the lead agency to know what deficiencies the EPA has identified in the EIS. The EPA's procedures specify criteria used by their reviewers to determine the appropriate ratings for an EIS, procedures for follow-up actions with the lead agency, and procedures for distribution of comments. In most cases the EPA provides detailed comments regarding its opinions on, for example, the range of alternatives, the methodology used in the analyses, and the adequacy of data.

The EPA may not only refer an unsatisfactory EIS or agency action to the CEQ, as discussed under CEQ's roles above, but may also take administrative action under environmental laws over which it has jurisdiction. The EPA may decide not to issue a permit, in a case where an EPA permit is required for a proposed project, or recommend that another agency not issue a permit under that agency's purview.

Congress is currently considering changes to the EPA's responsibilities and whether the CEQ will continue to exist.[19] The ramification of such changes, if they occur, in terms of how they will affect EISs or the EIS process is unknown at this time.

---

[19]Ray Clark, Council on Environmental Quality, personal communication, September 1995.

Environmental Impact of the Action

• LO — **Lack of Objections**

EPA has not identified environmental impacts requiring substantive changes to the proposal.

• EC — **Environmental Concerns**

EPA identified environmental impacts that should be avoided. EPA would like to work with the lead agency to reduce these impacts.

• EO — **Environmental Objections**

EPA identified significant environmental impacts that must be avoided. EPA intends to work with the lead agency to reduce these impacts.

• EU — **Environmentally Unsatisfactory**

EPA review identified adverse environmental impacts of sufficient magnitude that they are unsatisfactory from the standpoint of public health or welfare or environmental quality. If these impacts are not corrected in the Final EIS, the proposal could be referred to CEQ.

Adequacy of the Draft Environmental Impact Statement

• Category 1 — **Adequate**

Draft EIS adequately sets forth impacts of alternatives. No further analysis is necessary.

• Category 2 — **Insufficient Information**

Draft EIS is insufficient to assess impacts that should be avoided, or new reasonable alternatives were identified that could reduce environmental impacts and are within the spectrum of alternatives in the Draft EIS. The additional information or discussion should be included in the Final EIS.

• Category 3 — **Inadequate**

Draft EIS does not adequately assess significant environmental impacts, or new reasonable alternatives were identified that could reduce environmental impacts and that are outside the spectrum of alternatives in the Draft EIS. The Draft EIS should be revised and made available to the public as a Supplemental Draft EIS. The EIS could be referred to the CEQ.

■ **FIGURE 2.1.** EPA ratings of federal actions and EISs. *Source*: Modified from U.S. Environmental Protection Agency, *Policy and Procedures for the Review of Federal Actions Impacting the Environment*, 1984.

30

Role of the Courts

If someone feels that an agency has not complied with a requirement to prepare an EIS, or that an EIS is inadequate, and the agency has issued a Final EIS or made a decision to not prepare an EIS, lawsuits may be filed for resolution by the courts. Many agencies do not have an administrative appeal procedure. Therefore, following completion of a Final EIS and a decision on whether to proceed with the action, the only recourse for an aggrieved party is to file a lawsuit. As of 1992, courts have reviewed 2265 cases[20] involving NEPA alone. Because of NEPA's and SEPAs' ambiguity, the courts provide guidance by their interpretation of the laws. Chapter 6 discusses some of the conclusions of the courts with regard to procedural and substantive adequacy of EISs, and it also discusses standards under which agency actions are reviewed.

TERMINOLOGY

Words, phrases, or expressions when used in an EIS process can have meanings that are very different from the same words used in everyday speech. Many terms are "legalese," since the process is dictated by regulations and court decisions, while others are simply unique in their usage. Some of the terms are discussed in Chapter 1, in the section entitled "Basic Concepts and Participants' Roles," but others are described here and throughout the book. The following are the CEQ's definitions of types of documents in the federal EIS process (described in more detail in Chapter 5):

> *Environmental Assessment (EA).* A concise public document that analyzes the environmental impacts of a proposed federal action and provides sufficient evidence to determine the level of significance of the impacts.
>
> *Finding of No Significant Impact (FONSI).* A public document that briefly presents the reason why an action will not have a significant impact on the quality of the human environment and therefore will not require preparation of an environmental impact statement.

---

[20]Council on Environmental Quality, *Twenty-fourth Annual Report*, 1993.

Notice of Intent (NOI). A notice that an environmental impact statement will be prepared that describes the proposed action and possible alternatives; the agency's proposed scoping process including whether, when, and where a scoping meeting will be held; the name and address of a person within the agency who can answer questions about the proposed action and the environmental impact statement.

Environmental Impact Statement (EIS). The "detailed statement" required by Section 102(2)(C) of NEPA which an agency prepares when its proposed action significantly affects the quality of the human environment.

Record of Decision (ROD). A public document signed by the agency decisionmaker at the time of a decision. The ROD states the decision, alternatives considered, the environmentally preferable alternative or alternatives, factors considered in the agency's decision, mitigation measures that will be implemented, and a description of any applicable enforcement and monitoring programs.

Categorical Exclusion (CE). Categories of actions which normally do not individually or cumulatively have a significant effect on the human environment and for which, therefore, an EA or an EIS is not required.[21]

As noted above, there are other types of NEPA documents besides EISs (e.g., EAs, FONSIs, RODs) that are prepared depending on the type of proposed project, whether significant impacts would occur, and where one is in the environmental review process (see Figure 1.1). In addition to types of documents, terms that are used throughout the environmental review process and in preparation of EISs include the following:

Alternatives

Alternatives to a proposed action include the no-action alternative, other reasonable courses of action besides the proposed action, and mitigation measures not in the proposed action. The alternatives analysis compares the possible environmental impacts of one alterna-

---

[21]Council on Environmental Quality, *Twenty-fourth Annual Report*, 1993.

tive to another so that a decisionmaker can make a choice among alternatives while knowing the environmental consequences of his or her choice. A decisionmaker is not required to choose the least environmentally damaging alternative.[22]

> **A decisionmaker is not required to choose the least environmentally damaging alternative.**

No-Action Alternative   The no-action alternative must be included in all EISs.[23] This alternative represents the existing environmental conditions and provides an environmental baseline against which the potential impacts of a proposed action and other alternatives may be compared. The no-action alternative is also referred to as "maintaining the status quo," meaning that the proposed action would not proceed and existing conditions would not be changed. This is misleading, however, because "status quo" implies that no changes will take place and no impacts will occur. Actually, the status quo includes existing environmental impacts and ongoing actions, and does not preclude other or future actions from taking place that might change the status quo. The no-action alternative simply means that the proposal under consideration would not be implemented. The no-action alternative does not necessarily mean that no environmental impacts would take place because the existing conditions already may have environmental impacts, and the status quo could allow those impacts to continue or allow future environmental impacts to take place. For example, if a proposal to thin trees in a forest did not take place (i.e., the no-action alternative was selected), the potential for future forest fires, or for trees to become diseased, may be greater than if one of the "action" alternatives were selected.

> **The no-action alternative does not necessarily mean that no environmental impacts would take place.**

---

[22]40 CFR §1505.2(b); Robertson v. Methow Valley Citizens Council (1989).

[23]Exception to this rule is when Congress prohibits maintaining the status quo, or no action. CEQ, however, in its *Forty Most Asked Questions Concerning CEQ's National Environmental Policy Act Regulations*, states that the no-action alternative should be included even if an agency is under a court order or legislative command to act.

Other Reasonable Alternatives Other reasonable courses of action, or other reasonable alternatives, are alternatives that include modifications to the proposed action. For example, alternatives to the proposed project could include a reduction in the project's size, a modification of its design, or siting the project at another location. One court required an agency to consider an alternative that was completely different from the proposed action: energy conservation instead of offshore oil leasing. The purpose for modifying a proposed project is to create alternatives that have a range of environmental impacts. Some alternatives would have less environmental impacts than others or would affect different environmental resources. The key is the term "reasonable alternative," which generally is taken to mean (a) an alternative that meets a project's underlying purpose and objectives (perhaps not all, or at a reduced level) and (b) an alternative that can be accomplished (e.g., one that is technically feasible). An EIS is not required to have a certain number of alternatives or to address every conceivable alternative.[24] What is required is that a reasonable, but fairly comprehensive, range of alternatives be provided. The number in a "reasonable" range of alternatives can range from two (the proposed action and no-action alternatives) to a dozen or more depending on the complexity of the project and environmental issues. When a large number of alternatives are being considered in an EIS, a "short list" of alternatives can be developed. A short list would consist of, for example, five of the twelve alternatives that best represent the total range of possible alternatives, impacts, and mitigation measures. The process of evaluating a large number of possible alternatives against criteria developed by an EIS team to create a more manageable list of alternatives is discussed in Chapter 8.

> **An EIS is not required to address every conceivable alternative.**

Alternatives also can include mitigation measures that are not already part of a proposed action. Because mitigation measures include such actions as avoiding or reducing impacts, some of the alter-

---

[24]Vermont Yankee Nuclear Power Corp. v. Natural Resources Defense Council, Inc. (1978).

natives discussed above, such as changing the project location or reducing its size, would be mitigation measures as well.

Commentors and Commenting

Comments from public reviewers provide valuable information and feedback that can result in modifications to a proposed action to not only reduce environmental impacts but also produce a project that reflects public concerns and needs (see Chapter 14 for suggestions to make comments more effective). A lead agency requests comments from the public during scoping and during public review of a Draft EIS. Commentors include:

- Federal, state, and local agencies with jurisdiction by law or expertise
- American Indian tribes
- The project applicant, if any
- Those persons or organizations who may be interested in or affected by a proposal

Comments are considered by an EIS team when preparing a Final EIS. The EIS team responds to public comments in a Final EIS by

- Modifying alternatives including the proposed action
- Developing and evaluating alternatives not considered in the Draft EIS
- Supplementing, improving, or modifying the analyses
- Explaining why the comments do not warrant further response

Usually, the period of time allowed for public review of a Draft EIS is established by the lead agency. The regulations prescribe minimum Draft EIS public comment periods as 45 days for federal EISs and 30 days for some state EISs. The public can request an extension of the comment period, which is at least 15 days (for federal EISs and Washington's EISs; California has no extension provision). The extension is discretionary, and a lead agency can decide whether, or how long, to grant the requested extension. For large and complex EISs, lead agencies have been known to allow the public 60 or 90 days from the outset to comment on a Draft EIS.

For federal EISs, the EPA may reduce or extend prescribed comment periods if the lead agency or another federal agency provides compelling reasons in the national interest. If the lead agency does not concur with the extension of time, the EPA may not extend it for more than 30 days.[25]

An environmental assessment (EA) or state environmental checklist is not required to be publicly reviewed prior to becoming final. However, a federal finding of no significant impact (FONSI) or state threshold determination notice, prepared after an EA or checklist is completed, is required to be made available to the public. A FONSI includes a summary or a copy of the EA. Although public comments may be received on a FONSI and EA, there is no formal requirement for an agency to respond to the comments other than may be required by an agency's own policies and procedures. A decisionmaker takes the comments into consideration when making his or her decision. In some cases a decisionmaker will, based on public comments, change the original threshold determination.

An EIS also will be reviewed internally within the agency preparing the document, as well as among any cooperating agencies, before the document is ready for public review and comment. It is a good idea to have members of the interdisciplinary team (those who are preparing an EIS, discussed in Chapter 10) review and comment on each other's sections, particularly where environmental issues overlap; for example, impacts to water quality would have a bearing on impacts to fish.

Controversy

Public controversy regarding a proposed project can be a factor in an agency's threshold determination, the determination of whether significant impacts would occur from a proposal and require preparation of an EIS. The controversy, however, must be based on environmental impacts. Mere public opposition to a project does not require an EIS.[26]

Controversy can also exist among experts in a given field regarding the potential effects of an environmental impact. Another area of controversy or conflict may be between agencies who have different goals or management practices for a particular environmental resource,

---

[25]40 CFR §1506.10(d).
[26]40 CFR §1508.27(b)(4); Hanly v. Kleindienst (II) (1973).

or different preferences for ways to mitigate environmental impacts. These areas of conflict must be identified and discussed in an EIS as part of its duty to disclose controversy or responsible opposing views.[27] The regulations are silent on how to deal with irresponsible opposing views, however.

Decision

The decision under consideration in an EIS process is whether an agency should approve, permit, or fund a proposal. A decisionmaker takes factors into consideration in making his or her decision (such as engineering, feasibility, and cost) of which environmental impact is only one. An EIS is not required to contain all of the information that a decisionmaker must consider.

> **An EIS is not required to contain all of the information that a decisionmaker must consider.**

Ecosystem

An ecosystem is an interconnected community of living things, including humans, and the physical environment with which they interact. Ecosystem management is an approach to restoring and sustaining healthy ecosystems and their functions and values. It is based on a collaboratively developed version of desired future ecosystem conditions that integrates ecological, economic, and social factors affecting a management unit defined by ecological, not political, boundaries.[28]

Effects and Impacts

Effects and impacts are used synonymously as in "the environmental effects of a project" or "the environmental impacts of a project." Environmental effects or impacts include those on ecological or natural resources (e.g., water, air, wildlife, soil, plants, and their interaction) and on social or man-made resources (e.g., utilities, schools, transportation, aesthetic quality, and historic, cultural, and socioeconomic concerns).

---

[27]40 CFR §1502.9(a) and (b); §1502.12.
[28]Interagency Ecosystem Management Task Force, in Council on Environmental Quality, *Twenty-fourth Annual Report*, 1993.

When discussing categories of the environment, they are usually divided into two major areas: (1) the natural or physical environment and (2) man-made or built environment. None of the commonly used terms for environmental categories are completely accurate. Very few places are "natural" anymore because human impacts have altered, directly or indirectly, the environment in some way all over the world. The "physical environment" is a term used in the CEQ regulations; however, one could argue that the man-made environment is just as physical as the natural. The term "built environment" seems fairly innocuous, but in what category do impacts to air quality fall, for instance? Air isn't built by humans; however, the impacts on air are primarily from human activities in the built environment. In the end it becomes a matter of preference. Although categorization of environmental subjects may not be critical, it can affect the organization of an EA or EIS and the way the subjects relate to each other.

Direct or Indirect and Short- or Long-Term Effects

Effects or impacts can be direct or indirect, short-term or long-term. If a proposed project would cause impacts to occur at the same time and place that the project is taking place, it is a direct effect. Construction, for example, directly affects soil and vegetation within the "footprint" of activity. Indirect effects that would be caused by a proposed project are those that take place later in time or are farther removed in distance, but are still reasonably foreseeable. Growth-inducing effects are indirect effects. A project might cause a pattern of land use, or population growth rate, to change and result in related impacts to air, water, and other natural systems.

A short-term impact is one that takes place for a limited amount of time, whereas a long-term impact is one that is associated with the project for an extended period of time or for the foreseeable future. Because there is no definition of short-term or long-term impact (time being relative, and the perception of impacts subjective), one would make the evaluation on a case-by-case basis. Generally, construction impacts, such as noise from construction, are considered short-term, because the impacts end when construction is completed. Operational impacts, such as noise from traffic generated by the project while it is operational, would be considered long-term because the impacts would continue for the life of the project. If a project itself were temporary, such as a weekend training exercise, most of the impacts associated with the project might be short-term. The fact that most of a project's impacts would be temporary, however, does not necessarily mean that they would be insignificant (although they usually are; and some courts have held that temporary impacts are not significant and do not require EISs[29]).

Beneficial and Adverse Effects

Effects or impacts also can be beneficial or adverse.[30] There is no requirement that an environmental document only identify adverse environmental impacts, and lead agencies may consider beneficial effects when making a threshold determination. However, if a proposal would result in significant environmental impacts, even if beneficial, the CEQ regulations require that a federal lead agency prepare an EIS.

If a proposed project includes actions that would result in an improvement or enhancement of the environment, it is the duty of EIS authors to identify the beneficial as well as the adverse impacts. Because agencies are required to be unbiased when preparing environmental documents, agencies will sometimes minimize discussion of beneficial impacts for fear of appearing biased in favor of a project. The document should present all the facts, positive as well as negative, to the public and to the decisionmaker so that he or she can make an informed decision.

Cumulative Effects

Cumulative effect or impact is the environmental impact that results from the incremental impact of a project when added to other past, present, and reasonably foreseeable future actions regardless of who undertakes the other actions. This definition from the CEQ regulations sounds deceptively simple. Actually, the subject of cumulative impacts can be one of the most difficult to address in an EIS because of its open-ended nature. How wide do you look geographically, and how far do you look into the past or into the future? How do you find out what others (other private citizens as well as other agencies) have been doing in the area of analysis? How do you get the information from them, and how do you "add" data that may be in different units or based on widely varying methodologies?

---

[29]40 CFR §1508.27(b)(7) states, however, "Significance cannot be avoided by terming an action temporary. . . ."

[30]40 CFR §1508.8; §1508.27(b)(1).

The intent of the requirement to analyze cumulative impacts is to ensure that a proposed project's impacts are evaluated in light of impacts that are already in existence or that will exist in the future. A single project may not have a significant impact, but the impacts of one project added to other projects may result in significant impacts overall or in significant impacts on a particular environmental resource. Chapter 12 includes a discussion of the pitfalls and challenges of conducting cumulative impact analyses.

Labeling Impacts   Although the types of environmental impacts discussed above should be included in an EA or EIS, it usually is not necessary to label the impacts "adverse," "short-" or "long-term," or "direct" or "indirect." If all discussions of impacts were thus labeled as to type, the analysis would become cluttered and cumbersome. Also, you can avoid arguments with fellow team members as well as with the public on how an impact should be labeled if the impacts are not labeled. Label an impact only if it is necessary to give the reader a clearer understanding of the impact. If an EIS team thoroughly describes exactly what the impacts would be, quantifying impacts where possible, describing the timeframe of the impact, describing the effectiveness of mitigation measures, and providing other relevant data, the reader will know what types of impact they are.

Historic and Cultural Resources

Historic and cultural resources are structures, artifacts, or areas that represent a significant element of man's past activities. "Historic property" includes archaeological sites and culturally important places of significance in prehistory (during the thousands of years that American Indians occupied the continent before Europeans arrived) as well as properties used since the arrival of Europeans. Such properties are evaluated for their eligibility for listing on the National or State Register of Historic Places, or both. A historic property can be a district, site, building, structure, or object that possesses integrity of location, design, setting, materials, and workmanship as well as other relevant features identified by criteria in federal[31] and state laws. An EIS is

---

[31]National Historic Preservation Act of 1966; other related laws include Archeological and Historic Preservation Act of 1974; American Indians Religious Freedom Act of 1979; and Archeological Resources Protection Act of 1979.

required to evaluate a proposal's potential for affecting any historic resource in the study area and, if impacts would result, to discuss mitigation measures. The analysis is normally done by an archaeologist or historian, or both. Coordination with the State Historic Preservation Officer and (federal) Advisory Council on Historic Preservation, and any affected American Indian tribe, may be required to establish agreements on the appropriate mitigation measures. See Chapter 12 for a general approach to writing the cultural and archaeological resources section of an EIS.

Issues

Issues are the topics or subjects of EAs and EISs—such as when saying, "several significant issues were identified," or "further analysis is not warranted for insignificant issues." Issues usually are the environmental impacts under evaluation. The term can also be used in a broader sense such as a point of disagreement or contention as, for example, when someone makes an issue of an agency's policy or decisionmaking process regarding an environmental resource.

Mitigation Measures

Mitigation measures are methods used to mitigate, or reduce, adverse impacts to the environment. Some mitigation measures are requirements of existing federal, state, or local regulations. For example, local agencies may require sedimentation and erosion control plans before a project sponsor could obtain a construction permit. The proposed project itself may contain features that result in mitigation of potential environmental impacts. Lastly, EIS team members and public reviewers may identify additional mitigation measures that could be implemented by the lead agency or other agencies who condition their approvals with certain mitigation measures. Mitigation measures include avoiding, minimizing, rectifying, reducing, or compensating for an impact (see Chapter 12).

For each environmental impact discussed in an EIS, mitigation measures for the impact should be discussed. EISs are not required to mitigate every impact entirely, to a level of no impact (a difficult, if not impossible, prospect). However, the level of effectiveness of proposed mitigation measures should be evaluated so that a decisionmaker and the public will understand the effect of the project on the environment even with mitigation measures. Also, an EIS is not required to contain a detailed mitigation plan, and a lead agency does not have to adopt the mitigation measures contained in an EIS, unless the agency's procedures require it.[32] Other environmental laws may require a mitigation plan for a particular resource (such as wetlands) prior to an agency's issuance of permits (U.S. Army Corps of Engineers) or approvals, but it is not required to be in an EIS unless the lead agency's regulations require it.

> **EISs are not required to mitigate every impact entirely, to a level of no impact.**

Despite the fact that agencies are not forced by the courts to implement an EIS's mitigation measures, agencies should not be cavalier about them. A good faith effort must be made to identify specific actions for reducing environmental impacts if an EIS is to avoid or survive litigation. Also, the intent of the CEQ regulations to make agencies seriously consider mitigation is clear, for example, in requiring that an agency's record of decision (ROD) "State whether all practicable means to avoid or minimize environmental harm from the alternative selected have been adopted, and if not, why they were not. A monitoring, and enforcement program shall be adopted and summarized where applicable for any mitigation."[33]

Even though mitigation measures in an EIS do not obligate a project proponent to implement a given mitigation measure, unless it is one that is already required by law or part of a proposed action, mitigation measures are made a requirement by including them in a ROD or a finding of no significant impact (FONSI) or by making them conditions of a permit or other approval. Thus, to make mitigation measures conditions of approval, they must be clear and enforceable rather than stated as vague goals. The development of mitigation measures is a logical process in which there is a direct link between the potential impact resulting from a proposed project and the method for mitigating the impact. A methodology for identifying and evaluating the appropriateness of mitigation measures is discussed in Chapter 12.

---

[32]Robertson v. Methow Valley Citizens Council (1989).
[33]40 CFR §1505.2(c).

Planning Process

When considering EISs, two planning processes are at work. One is a regulatory type of planning, such as land use and natural resource management planning, conducted for a given area under an agency's jurisdiction. These plans govern land use (e.g., through zoning ordinances) and control development or other activities. Much of the analysis in an EIS is an evaluation of whether a proposed project is consistent with the objectives of applicable land use or natural resource plans and regulations for the area in which a project is proposed to take place.

Another type of planning is a project's planning process which begins when someone (a project applicant or proponent) has an idea for a project, develops a preliminary design or other concepts of the project, and approaches an agency to discuss the agency's requirements for obtaining permits or other approvals that would allow the project to go forward. An agency that is proposing its own project would go through a similar process, except that the project would be subject to the agency's design review and funding approval process.

EISs are to be prepared early in the planning process,[34] so that modifications can be made to the project that would reduce environmental impacts before the project plans have progressed to a point where modifications would become difficult to accommodate. An EIS is rarely the last step in a project's planning and approval process. Most actions or projects that require EISs are fairly large or complex and require multiple approvals and refinement of plans (e.g., design and engineering) that may have to go through several steps before project planning is final. Therefore, after an EIS is completed, the proposed project continues to be evaluated through the planning process, within the agency proposing a project, or by agencies having jurisdiction over a project. These agencies often have detailed and specific requirements of the project proponent, beyond the level that was required for an EIS. By the time the project has reached this stage in the planning process, the project has undergone considerable modification and evolved into a more "final design" that reflects public input and regulatory modifications.

---

[34]40 CFR §1500.5; §1501.2; and §1502.5.

> An EIS is rarely the last step in a project's
> planning and approval process.

Proposal, Proposed Action, or Project

The terms "proposal," "proposed action," and "proposed project" are used synonymously. Environmental evaluation cannot begin until there is a proposal by either a private party or an agency. The proposal must be sufficiently defined to allow meaningful analysis, but it should be at an early enough stage in a planning process that modifications can be made to the project to reduce environmental impacts. Proposed actions include actions approved by permit or other regulatory decision, as well as agency and agency-assisted activities.

Proposals or proposed actions subject to evaluation by EISs can be specific projects such as (a) construction or management activities in a defined geographic area, (b) broad programs that may result in specific projects after the program is implemented, (c) adoption of official policy, such as rules and regulations, and (d) adoption of formal plans which guide or prescribe uses of resources upon which future agency actions will be based.[35]

Those who are part of a team to prepare an EIS also refer to an EIS as a "project," or the project of preparing an EIS. Thus, the project manager (PM) of an agency or consulting firm is the individual assigned the project of guiding and directing preparation of an EIS.

Scope and Scoping

The scope of an EIS consists of the range of actions, alternatives, and impacts to be considered. Basically, the scope of an EIS is determined by the lead agency, preparers of the EIS, and the public, including other agencies. To determine the scope of an EIS requires identification of (a) significant issues, those environmental impacts of most concern to receive detailed analysis, (b) reasonable alternatives to the proposed action, (c) the geographic area of analysis (the study area), (d) other actions in the study area that should be included in the EIS, (e) a timeframe governing the analysis of impacts if appropriate, such

---

[35]40 CFR §1508.18(b).

as a planning period of 5 or 10 years into the future, and (f) any other relevant factors that affect the scope of analysis.

Scoping is the process of determining the scope of an EIS. Part of that process includes obtaining the public's opinion on what important environmental issues and what project alternatives should be addressed in the EIS. Scoping begins when an agency is considering an action to permit or to approve a project proposed by a private party or when an agency is considering whether to go forward with, or internally approve, its own proposed project. The public scoping process formally begins when an agency publishes a notice—such as a notice of intent by a federal agency, or a determination of significance by a state or local agency—in the *Federal Register* (for federal actions) and in newspapers or other publications. Also, lead agencies mail notices to public and private interested parties and have the option of posting notices on the site where the project is proposed to take place. The notice describes the proposed action and requests that the public provide comments to a designated agency contact by a certain date.

Significant Impacts

The requirement to prepare EISs hinges on whether a proposed project would result in significant impacts to the environment, a critical decision called the "threshold determination." However, significance is very subjective: What is significant to one person may not be significant to another. Another problem is how to know if an impact will be significant unless an EIS evaluates it. There are no hard and fast rules for determining whether an impact would be significant, however, the CEQ provides some guidance on whether actions would be significant. These considerations are context and intensity. For context, significance varies with the setting of the proposed action. For instance, the affected region, affected interests, the locality, and short- and long-term effects would be relevant. Intensity refers to the severity of the impact determined by conditions such as the type of impact, whether it affects public health or safety, unique geographic characteristics of the area, and whether other environmental laws would be violated. See Chapter 12 for additional discussion of significant actions and impacts.

At the other end of the scale, agencies have lists of actions that are excluded or exempt from environmental documentation (there would be no need to prepare an EA or EIS) if the actions clearly would not result in significant impacts. Between the two extremes of a categorical exclusion or EIS is a large gray area in which an action may not clearly fall into one category or the other. In those instances, an EA or a state's equivalent of an EA (such as an initial study or environmental checklist) to analyze the potential impacts would aid the decisionmaker in determining whether significant impacts would occur from a proposed action.

Study Area

The study area is the geographic scope of analysis in an EA or EIS. The size of the study area will depend on the type of proposed project, its size, and the physical extent of the impacts that would result from the project if it were implemented. The study area could be established on some agreed-upon natural boundaries, such as a valley or a watershed. Some agencies are emphasizing an ecosystem approach to planning and to managing natural resources which may affect the definition of study area boundaries. Because ecosystems do not follow jurisdictional or political boundaries, partnerships between federal, state, and local agencies, private citizens, and organizations are necessary in some areas to implement this approach. An ecosystem approach to defining a study area is more common for proposed actions primarily involving natural resources as opposed to actions within a predominantly built or urban environment. The reason is that built environments have few ecosystems or natural features left.

Absent natural boundaries, a reasonable distance from a project site such as an area within a one-mile radius from the project could be established as the study area. Or if a proposal is an agency proposal, the area within the agency's jurisdiction might be an appropriate study area (e.g., within a city, county, or national forest). The boundaries of the study area must include all or most of the potential project impacts. If most impacts would occur within a one-mile radius, but traffic impacts would occur as far as 20 miles away, a separate study area for traffic might be reasonable. Similarly, if a proposal would result in impacts outside an agency's area of jurisdiction, the study area should include those areas that would be affected by the proposal. For projects that include very tall structures, aircraft, or artillery, the study area may have to include airspace for a number of miles, or square miles, above the ground to analyze potential effects on aircraft operations or flight patterns.

RESPONSIBILITY: WHOSE EIS IS IT?

Regardless of whether an EIS is a federal or state document, and regardless of who prepares the document, an EIS is the responsibility of the lead agency. A federal agency is usually the lead agency for a federal EIS, and a state or local agency is the lead agency for a state document. A joint EIS can be prepared that meets both state and federal EIS requirements to reduce duplication of effort and save time. Both agencies would share the responsibility to ensure that the EIS and the EIS process meet the requirements of each; however, a federal agency is usually the lead agency. A lead agency may have the document prepared by consultants or prepared by in-house staff. In either case the EIS "belongs" to the lead agency and the method of analysis, format, opinions, content, conclusions, and everything else in the EIS are the agency's. An agency must not delegate the entire responsibility for an EIS, or for carrying out the EIS process, to a contractor through agreements (including leasing agreements or turnkey contracts). If an agency adopts another agency's EIS, the adopting agency must independently review the document, and provide any necessary public review, to ensure that the document meets its requirements prior to adopting the EIS.

Congress amended NEPA in 1975 to allow state agencies or officials to prepare federal EISs if the state agencies have statewide responsibility for managing federally funded transportation projects or program of grants to the states.[36] For example, a state highway department may prepare federal EISs for federally funded highway proposals as long as the Federal Highway Administration (FHWA) provides review and oversight to ensure compliance with NEPA. Congress also authorized delegation of NEPA responsibilities to local governments in the Housing and Community Development Act of 1974. In this case, however, local governments may determine the level of NEPA documentation and the adequacy of the document (the federal Department of Housing and Urban Development is not required to provide the same level of control and review of EISs as the FHWA).

---

[36]NEPA Section 101, subparagraph D.

PROCESS

The only thing that saves us from the bureaucracy is inefficiency.
An efficient bureaucracy is the greatest threat to liberty.

Eugene McCarthy

INTRODUCTION

As discussed in Chapter 1, an environmental review process can be described in very basic terms, but environmental review is only part of a picture that includes other processes and requirements of which environmental impact statements (EISs) are a part and which, in turn, affect the environmental review process (or what has been loosely termed the "EIS process"). This interaction of processes makes preparation and management of EISs challenging because an EIS is not about a single environmental concern or one set of regulatory requirements. The relationship of environmental laws and the planning process with the EIS process are discussed in this chapter.

Preparation of EISs combines the principles of science, law, and planning. The EIS process is part of governmental planning processes and is affected by other environmental laws such as the Endangered Species Act, the National Historic Preservation Act, and pollution control laws. Zoning ordinances and natural resource management plans control proposed actions and impose mitigation measures. States such as California, New York, and Washington require environmental review of state and local agency planning actions. Federal agency actions are subject to the National Environmental Policy Act (NEPA) in all states wherever a federal action may be proposed. Therefore, federal EISs must sometimes evaluate a federally proposed action's consistency with local plans whether or not states require review of state or local actions under a state environmental policy act (SEPA).[37]

PLANNING AND ENVIRONMENTAL REGULATIONS

Laws and regulations that control private and public activities can be broadly categorized as those that control land use (or activities in the air or on or in water), the use of natural resources, and pollution. An EIS must consider how a proposed action would or would not comply with all of these other regulations. Compliance with regulations does not take place via an EIS. The only law an EIS complies with is NEPA or a comparable state law (a SEPA). An EIS discloses the other laws and regulations that are applicable to a proposal, and it states whether and to what degree a proposed action would comply with those other laws.

Planning laws and regulations control existing and planned, or future, activities and attempt to further various planning agency goals. Regulations dealing with natural resources are considered both environmental and land use regulations, and pollution control regulations generally fall into the environmental category. The problem with categorizing regulations is that some of them don't fall neatly into only one category. But for the sake of illustration, Tables 3.1 through 3.3 categorize examples of federal environmental laws that are frequently evaluated in an EIS or environmental review process.

Table 3.1 lists some of the environmental statutes commonly addressed by EISs. The statutes that are applicable to most types of proposed actions include the National Historic Preservation Act and the

---

[37]Generally, local land use plans do not control activities on federally owned property. However, plans and policies of other agencies should be addressed in a federal EIS if issues or resources are of common concern, and a proposed action has the potential to affect those issues or resources.

■ **TABLE 3.1** Federal Resource Protection and Land Use Control
Statutes

Statute	Concern	Acronym
National Environmental Policy Act of 1969	Environmentally informed decisions, public involvement	NEPA
National Historic Preservation Act of 1966	Preservation of prehistoric and historic sites/structures	NHPA
Archaeological Resources Protection Act of 1979	Prehistoric artifacts including skeletal remains	ARPA
Marine Protection, Research, and Sanctuaries Act of 1972	Marine sanctuaries	MPRSA
Marine Mammal Protection Act of 1972	Marine mammal protection	MMPA
Magnuson Fishery Conservation and Management Act Amendments of 1990	Management of marine fisheries and mammals (driftnets ban)	Magnuson Act
Endangered Species Act of 1973	Protection of endangered/ threatened plant and animal species	ESA
Fish and Wildlife Coordination Act of 1958	Fish and wildlife conservation, interagency coordination	FWCA
Outer Continental Shelf Lands Act of 1953	Offshore oil development	OCSLA
Coastal Zone Management Act of 1972	Control of projects in coastal zone	CZMA
Farmlands Protection Policy Act of 1981	Conversion of farmlands	FPPA
Airport and Airway Development Act of 1970	Airport development impacts on communities	AADA
Department of Transportation Act of 1966 (section 4(f) and Federal-Aid Highway Act of 1968 (section 18(a))	Use of land for highways (especially parks and other reserves)	FAHA
Federal Land Policy Management Act of 1976	Multiple use of Department of the Interior public lands	FLPMA
Multiple-Use Sustained-Yield Act of 1960	Multiple use management on national forests	MUSYA
Forest and Rangeland Renewable Resources Planning Act of 1974	Comprehensive planning for USFS forest and rangelands	RPA
National Forest Management Act of 1976	Directs forest planning	NFMA
Wilderness Act of 1964	Management of wilderness areas	Wilderness Act

Source: Modified from Joseph A. Wellington, A Primer on Environmental Law for the Naval Services, *Naval Law Review*, 1989.

■ **TABLE 3.2** Federal Pollution Control Statutes

Statute	Concern	Acronym
Federal Water Pollution Control Act Amendments of 1972 (Clear Water Act)	Water pollution including filling wetland	FWPCA (CWA)
Marine Protection, Research, and Sanctuaries Act of 1972 (Ocean Dumping Act)	Ocean dumping of wastes/ dredge material	MPRSA
Clean Air Act Amendments of 1970	Air pollution	CAA
Noise Control Act of 1972	Noise pollution	NCA
Federal Insecticide, Fungicide and Rodenticide Act of 1947	Pesticide pollution	FIFRA
Federal Environmental Pesticide Control Act of 1972	Pesticide pollution	FEPCA
Resource Conservation and Recovery Act of 1976	Hazardous and nonhazardous waste management	RCRA
Toxic Substances Control Act of 1976	Chemical substances control	TSCA
Rivers and Harbors Act of 1899	Deposition of refuse (Section 13) in navigable waters	RHA

Source: Modified from Joseph A. Wellington, A Primer on Environmental Law for the Naval Services, *Naval Law Review*, 1989.

Endangered Species Act. Other laws are applicable depending on the type of proposal and the environmental resources it has the potential to affect. Some statutes, such as the Federal Land Policy Management Act, affect how an agency (the Bureau of Land Management) manages its land and natural resources.

An EIS must address federal pollution control statutes (some of which are listed in Table 3.2) if a proposed action has the potential to

■ **TABLE 3.3** Federal Environmental Restoration Statutes

Statute	Concern	Acronym
Comprehensive Environmental Response, Compensation and Liability Act of 1980	Contain and clean up releases of hazardous substances	CERCLA
Superfund Amendment and Reauthorization Act of 1986	Cleanup of contamination from past hazardous waste disposal	SARA

Source: Modified from Joseph A. Wellington, *Naval Readiness, Operational Training and Environmental Protection: Achieving an Appropriate Balance Between Competing National Interests* (unpublished), June 1988.

cause air or water pollution, or to generate hazardous or toxic waste. The main concern is public health and safety. Generally, if a proposal's pollutant emissions are within acceptable regulatory standards, the environmental effects of the proposal are not considered significant (although cumulative impacts could be significant). In any case, the impacts do not have to be significant to be addressed in an EIS.

Environmental restoration statutes, listed in Table 3.3, provide for "response" and "remedial" actions that are taken when there has been a release of a hazardous substance or to cleanup existing hazardous substances. Some agencies take the position that actions to restore the environment should not require EISs because the restoration statutes' processes are the "functional equivalent" of the EIS process (functional equivalence is discussed below under "The EIS Process and Other Environmental Regulations").

The President of the United States issues executive orders, some of which relate to environmental protection. Examples of executive orders relating to the environment (including land use and natural resources) are listed in Table 3.4. Presidential executive orders set forth national policies and provide direction and mandates to agencies to achieve those policies or goals. The most recent environmental executive order that affects EIS preparation is EO 12898, Federal Actions to Address Environmental Justice in Minority Populations and Low-Income Populations. Federal EISs must include an evaluation of a proposed action's potential impacts on minority and low-income populations.

Environmental statutes and executive orders are applicable to all federal agencies, however, certain agencies have a primary, or oversight, role to ensure compliance by other agencies and private parties. These are shown in Table 3.5.

The EIS Process and Other Environmental Regulations

Unlike most environmental laws which primarily focus on one environmental area of concern, NEPA and SEPAs require the evaluation of all environmental matters that might be affected by a proposed project. A lead agency must consider a proposed project's potential to violate federal, state, or local environmental laws when determining whether the project would have significant environmental impacts and therefore require the preparation of an EIS.

■ **TABLE 3.4** Presidential Executive Orders Pertaining to Environmental Protection

Number	Date	Title	President	Federal Register Cite Vol.	Page
11514	5 May 70	Protection and Enhancement of Environmental Quality	Nixon	35	4247
11593	13 May 71	Protection and Enhancement of the Cultural Environment	Nixon	36	8921
11644	8 Feb 72	Use of Off-Road Vehicles on Public Lands	Nixon	37	2877
11967	24 May 77	Exotic Organisms	Carter	42	26949
11988	24 May 77	Flood Plain Management	Carter	42	26951
11989	24 May 77	Off-Road Vehicles on Public Lands (amends 11644 above)	Carter	42	26959
11990	24 May 77	Protection of Wetlands	Carter	42	26961
11991	24 May 77	Protection and Enhancement of Environmental Quality (amends 11514 above)	Carter	42	26967
12088	13 Oct 78	Federal Compliance with Pollution Control Standards	Carter	43	47707
12114	4 Jan 79	Environmental Effects Abroad of Major Federal Actions	Carter	44	1957
12148	20 Jul 79	Federal Emergency Management (amends 11988 above)	Carter	44	43239
12316	14 Aug 81	Responses to Environmental Damage	Reagan	46	42237
12580	23 Jan 87	Superfund Implementation (amends 12088 above)	Reagan	52	2923
12898	11 Feb 94	Federal Actions to Address Environmental Justice in Minority Populations and Low-Income Populations	Clinton	59	7629

Source: Modified from Joseph A. Wellington, A Primer on Environmental Law for the Naval Services, *Naval Law Review*, 1989.

An EIS process is forward-looking and seeks to prevent adverse environmental impacts from taking place rather than fixing or cleaning up existing environmental problems caused by past actions. This doesn't mean that a proposed project could not, as part of mitigation, propose to improve or rehabilitate an environmentally degraded area, but that is not the main purpose of NEPA or a SEPA.

■ TABLE 3.5 Federal Environmental Protection Oversight Agencies

Type	Statute	Oversight Agencies
Resource protection/ land use control	NEPA	CEQ (EIS process), and EPA (EIS review and classification)
	NHPA	ACHP, NPS
	ARPA	Federal Land Manager[a]
	MPRSA	NOAA (permits)
	MMPA	USFWS, NMFS[b]
	ESA	ESC (exemptions)[c]
		USFWS, NMFS (biological opinions)
	CZMA	NOAA
Pollution control	FWPCA	EPA (NPDES permit)[d]
	(CWA)	COE (dredge/fill permit)
	RHA	COE (obstructions and disposal permits)
	MPRSA	EPA (non-dredge waste dumping permit)
		COE (dredge material dumping permit)
		USCG (surveillance of dumping)
	CAA	EPA (permit)
	NCA	EPA
	FIFRA	EPA
	RCRA	EPA (permits, identification numbers)
Environmental restoration	CERCLA	EPA
	SARA	EPA

Key: ACHP, Advisory Council on Historic Preservation; CEQ, Council on Environmental Quality; COE, U.S. Army Corps of Engineers; ESC, Endangered Species Committee; EPA, Environmental Protection Agency; NMFS, National Marine Fisheries Service; NOAA, National Oceanic and Atmospheric Administration; NPS, National Park Service; USCG, U.S. Coast Guard; USFWS, U.S. Fish and Wildlife Service.

For statutes, see Tables 3.1 through 3.3.

[a]The term "Federal Land Manager" means the Secretary of the Department, or the head of any other agency or instrumentality of the United States having primary management authority over public lands.

[b]The U.S. Fish and Wildlife Service and the National Marine Fisheries Service have jurisdiction over different species of marine mammals. For example, the USFWS has jurisdiction over polar bears, sea otters, walruses, and manatees, while the NMFS has jurisdiction over whales, porpoises, seals, and sea lions.

[c]The Endangered Species Committee can grant exemptions if necessary (e.g., for reasons of national security.) See 16 U.S.C.S. §1536(j).

[d]An "NPDES" (National Pollutant Discharge Elimination System) permit allows its holders to discharge a certain type/amount of pollutant into the waters of the United States.

Source: Modified from Joseph A. Wellington, A Primer on Environmental Law for the Naval Services, *Naval Law Review*, 1989.

Evaluation of other environmental regulations and discussion in an EIS does not constitute compliance with those other environmental regulations. Conversely, compliance with all applicable environmental regulations does not constitute compliance with NEPA or a SEPA. At first the compliance efforts may seem a duplication of effort, but the environmental regulations were established to meet a variety of purposes, and the specific processes required by each law can only be complied with by actually going through the process. In limited cases, "functional equivalence" has been granted by the courts when an agency's environmental evaluation and public participation procedures are equivalent to NEPA's. In some cases, Congress has provided statutory exemptions for an agency's duty to comply with NEPA. The Environmental Protection Agency (EPA) is exempt from NEPA for actions taken under the Clean Air Act and some actions under the Clean Water Act. The courts have applied functional equivalence exemptions for the EPA's actions under the Insecticide, Fungicide and Rodenticide Act. A court also held that NEPA does not apply to the EPA's procedure for permitting hazardous waste landfills under the Resource Conservation and Recovery Act. Federal agency functional equivalence to NEPA has generally been limited to programs administered by the EPA. A state court also applied the functional equivalence doctrine and held that a state agency was an environmental regulatory agency and did not have to prepare an EIS for an action under its statutory mandate to protect the state's water quality.[38]

The following discussion regarding requirements of environmental laws identifies the differences in their purposes and roles of participants. In understanding the roles of those involved in the environmental process in general (including NEPA), it is helpful to understand the roles and operation of the environmental regulatory process described by Joseph A. Wellington,[39] an attorney who has written several papers on governmental planning and environmental regulatory pro-

---

[38]Holtz & Krause, Inc. v. (Wisconsin) State Department of Natural Resources (1978).

[39]Mr. Wellington is currently with the firm of Carmody & Torrance in Waterbury, Connecticut.

cesses. The roles of those in the regulatory and environmental processes can be described in terms of "players," "levers," and "hammers." At the broadest level, players who have roles in protecting the environment include all three branches of the federal government. In the legislative branch, Congress passes environmental laws that establish societal norms. The executive branch implements the laws by developing regulations, providing regulatory oversight, and prosecuting those who violate the law. Finally, the judiciary resolves controversies arising from the rulemaking, oversight, and enforcement actions. On a daily basis, players who have a role in the environmental process include federal, state, and local agencies; environmental groups; citizen groups; legal foundations; and property owners potentially affected by a proposed project.

Environmental protection levers are those statutory tools that key players may use to influence the regulatory community (e.g., agencies). The levers and environmental statutes are shown in Table 3.6. Levers range from the requirement that federal facilities obtain environmental permits to the ability of an individual citizen to initiate a lawsuit in opposition to certain federal actions. Players have levers to influence agency actions through (a) permit systems, licensing, recordkeeping, and reports, (b) inspections and audits, (c) procedures for public comments on proposed projects, and (d) citizen suits and judicial review under the Administrative Procedure Act.

Environmental protection hammers are those tools that key players may use to force the regulated community to comply with the law. Hammers may be directed against federal facilities (e.g., military bases) or against federal officers, employees, and agents. Those environmental protection hammers applicable to federal facilities or actions are shown in Table 3.7; and hammers applicable to federal officers, employees, and agents are shown in Table 3.8. Hammers range from injunctive relief to fines and imprisonment. For NEPA, plaintiffs seek remedial relief, usually through an injunction (a court requirement to do or refrain from doing a specific act).

The EPA Administrator may approve the pollution control programs of states with environmental programs that are at least as stringent as federal requirements. The statutes that authorize the states to assume the lead in overseeing an EPA-approved program include

- Federal Water Pollution Control Act (FWPCA), State National Pollutant Discharge Elimination System Permit Program

- Clean Air Act (CAA), State Implementation Plans for National Primary and Secondary Ambient Air Quality Standards

- Resource Conservation and Recovery Act (RCRA), Authorized State Hazardous Waste Programs

■ **TABLE 3.6** Environmental Protection Levers Applicable to Federal Facilities

	Statutes														
	Resource Protection Land Use Control							Pollution Control						Environmental Restoration	
Federal Facility Action Required	NEPA	NHPA	ARPA	MPRSA (Marine Sanctuaries)	MMPA	ESA	CZMA	FWPCA	MPRSA (Ocean Dumping)	CAA	NCA	FIFRA	RCRA	CERCLA	SARA
Obtain permit			●		●			●	●	●			●		
Maintain records	●							●		●			●		
File reports	●							●		●			●		●
Make findings							●								
Submit to inspections								●		●			●		●
Coordinate with states	●	●				●	●	●							●
Coordinate with other federal agencies, councils, committees	●	●	●	●		●		●							●
Provide for public comment	●						●								●
Obtain exemption						●		●		●	●	●	●		
Statutory citizen suit						●		●	●	●	●		●	●	●
Judicial review under Administrative Procedure Act (APA)	●	●				●	●								

Source: Modified from Joseph A. Wellington, A Primer on Environmental Law for the Naval Services, *Naval Law Review*, 1989.

■ TABLE 3.7 Environmental Protection Hammers Applicable to Federal Facilities

Hammers	Resource Protection Land Use Control							Pollution Control						Environmental Restoration	
	NEPA	NHPA	ARPA	MPRSA (Marine Sanctuaries)	MMPA	ESA	CZMA	FWPCA	MPRSA (Ocean Dumping)	CAA	NCA	FIFRA	RCRA	CERCLA	SARA
State civil penalty													a		
Liability for remedial/response costs due to release of hazardous substance from federal facility													●	●	●
Liability for remedial/response costs due to release of hazardous substance into waters of the United States								●							
Economic sanctions for violation of injunction (contempt)	●		●		●	●		●	●	●			●		
Injunction	●		●		●	●		●	●	●			●		

[a]*Meyer v. Gracey et al.* (1986), held that a state could not impose a civil penalty against the United States under RCRA; *McClellan Ecological Seepage Situation (MESS) v. Weinberger* held that a citizen group could not collect civil penalties against the United States under RCRA because the RCRA waiver of sovereign immunity extends only to injunctive relief.

Source: Modified from Joseph A. Wellington, A Primer on Environmental Law for the Naval Services, *Naval Law Review*, 1989.

Relationship of the EIS Process and Planning Processes

Planning as discussed here is a governmental land use control process. It is a forward-looking process, as is the EIS process. The planning process, whether conducted by local or federal agencies, evalu-

■ **TABLE 3.8** Environmental Protection Hammers Applicable to Federal Officers, Employees, and Agents

	Statutes														
	Resource Protection Land Use Control							Pollution Control						Environmental Restoration	
Hammers[a]	NEPA	NHPA	ARPA	MPRSA[b]	MMPA	ESA	CZMA	FWPCA	MPRSA[c]	CAA	NCA	FIFRA	RCRA	CERCLA	SARA
Civil penalty				●	●			d		e			f		
Fine[g]			●		●	●		●		●	●			h	
Imprisonment[g]			●		●	●		●		●	●			h	
Injunction	●							●		●			●		
Economic sanction for violation of injunction (contempt)	●							●		●			●		
Compliance orders										●			i	i	i

[a]Environmental hammers (fines and imprisonment) may be brought to bear on individual federal employees, including military personnel, as individuals. See *United States v. Dee* (1988).

[b]Marine Sanctuary provisions of MPRSA.

[c]Ocean Dumping Act provisions of MPRSA.

[d]The FWPCA provides that "No officer, agent or employee of the United States shall be personally liable for any civil penalty arising from performance of his official duties for which he is not otherwise liable. . . ." 42 U.S.C. §1323(a).

[e]Section 118 of the Clean Air Act, 42 U.S.C. §7418, 644 F. Supp. 221 (E.D.N.C. 1986), contains language identical to that quoted in note d of this table.

[f]In *Meyer v. U.S. Coast Guard*, the Commandant of the Coast Guard was sued in both individual and official capacity. He was dismissed from the suit on other grounds prior to resolution of the question of whether he could have been held personally liable under RCRA for civil penalties growing out of delayed filing of an application for permit.

[g]Only sovereigns, such as the United States or a state, may initiate criminal prosecutions. Private interests and citizen groups are normally limited to injunctive relief.

[h]Section 103(b) of CERCLA, 42 U.S.C. §9603(b), requires any "person in charge" of a vessel or an offshore or onshore facility to give notification "as soon as he has knowledge" of a reportable release of a hazardous substance under penalty of fine and imprisonment.

[i]Compliance orders of administrative agencies are enforced in judicial proceedings, but the federal government does not sue itself and EPA may not unilaterally issue compliance orders to federal facilities under RCRA, CERCLA, and SARA.

Source: Modified from Joseph A. Wellington, A Primer on Environmental Law for the Naval Services, *Naval Law Review*, 1989.

ates an area for its potential suitability for various activities or types of development. Most people associate planning with cities, but any agency with responsibilities to manage land or natural resources must also have plans and policies for managing activities that use or affect those resources.

Planners historically were primarily concerned with the orderly development of towns and cities. Some planners came to realize that the best use of land is not always determined by calculating the least expensive method of development for the maximum use. In a book called *Design with Nature*, Ian McHarg, a landscape architect and planner, introduced an approach to planning and to designing projects that took into consideration natural features, constraints to development, and social values or amenities, which could be shown as transparent overlays on a map. A proposed project could then be sited or designed to reduce or avoid environmental impacts by avoiding areas having the most environmental and social concerns, thereby avoiding areas considered most unsuitable for development. A development project designed with natural constraints in mind met the requirements or purpose of a proposed project, while minimizing impacts to the environment and to society. The effect of this approach is very similar to the intent of NEPA or a SEPA.

Mr. McHarg's approach to planning also involved a more interdisciplinary approach than was usual for the time. He crossed an imaginary boundary by considering ecology in his planning and design profession, which is noteworthy when discussing the EIS and bureaucratic processes because of the tendency for specialists to work only within the confines of their educational background and occupational specialty. Subjects have become complex and specialized over time, and it would be impossible for one person to be knowledgeable regarding all specialties. For this reason the EIS process requires an interdisciplinary team of specialists to work together and obtain more enlightened approaches or solutions to problems. Ecology and planning in particular are interrelated in an EIS process.

Whether a proposed project includes features or constraints of local planning (such as ordinances) is relevant to determining whether impacts are significant and, therefore, whether an EIS is required (the threshold determination). If a proposed project complies with local ordinances, it is more likely that the wishes of the community regarding the way land is used are reflected in the design of the project. However, a project that has the potential to violate a zoning ordinance is not necessarily going to result in significant environmental impacts. The potential environmental impact would depend on the type of project and type of ordinance, as well as on the facts and circumstances of the situation.

Planning takes place at all levels (federal, state, and local) of government. Thus, the evaluation of a proposed project's consistency with plans and policies of agencies with jurisdiction over the project could involve more than one plan or type of plan. Some of the plans may have policies inconsistent with plans and policies of another agency. Although an EIS could simply disclose the inconsistencies between agency plans and policies, it is far more useful if there is dialogue between the agencies and the inconsistencies are resolved. Because an EIS process provides the impetus for dialogue, an EIS team leader or project manager is often the person who contacts the agencies, arranges and conducts meetings, and acts as a facilitator to resolve differences or negotiate acceptable compromises.

If an EIS is prepared for a proposed project, the EIS accompanies the proposal through the decisionmaking or planning process. The EIS is considered, along with economic and technical considerations, by the decisionmaker in making his or her decision on whether to fund, permit, or otherwise approve a proposal.

Local Planning Organizations

Decisions in an EIS process at the local level are often made by the director of a planning department or city council. Knowing how an agency is organized and who makes the decisions can be useful when participating in an EIS process. Suggestions regarding agency programs and procedures can be more specific when one understands an agency's decisionmaking process. Agency organizations vary greatly depending on the size of the city or county, its functions, legislative mandates, and many other variables. Figure 3.1 is an example of a city's organization chart. Within this city is a department of construction and land use where permits are issued, among other functions. Its organization chart is shown in Figure 3.2. As with many planning agencies that are streamlining their procedures, these organization charts may soon be out of date as agencies find more efficient ways to organize their functions. Figure 3.3 is an example of a county planning agency that has combined the planning process and the environmental process in an attempt to provide more responsive service to the community. To obtain information regarding an agency's decisionmaking process and organization, call the agency's public information office (if they have

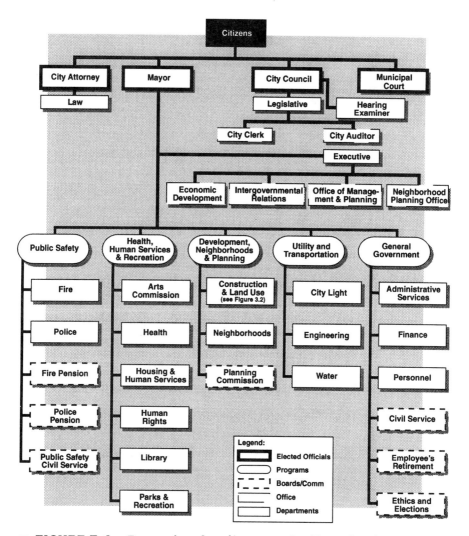

■ **FIGURE 3.1.** Example of a city organization chart. *Source*: City of Seattle 1995–1996. Adopted Budget.

one) or a planner who can explain the process or send information about the agency's functions and organization, or both.

Zoning

Zoning is a commonly used legal device for implementing a community's land use plan, and it primarily controls development on privately owned land. A basic power of the state and its political subdivisions is to enact legislation protecting the public health, safety, and

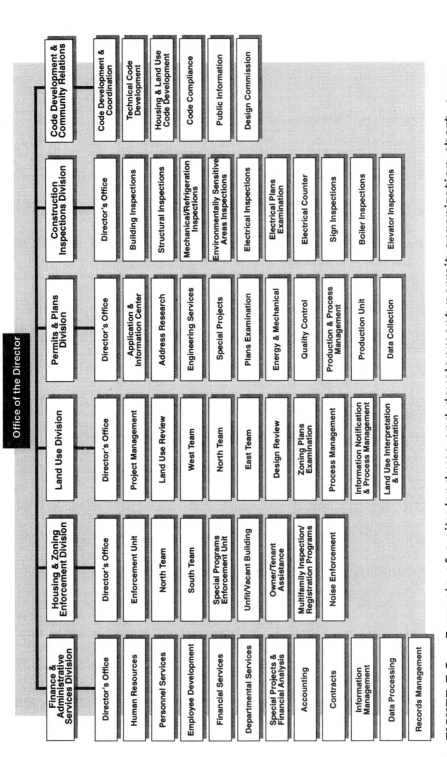

■ FIGURE 3.2. Example of a city land use and development permit organization chart. *source:* Seattle Department of Construction and Land Use, 1994.

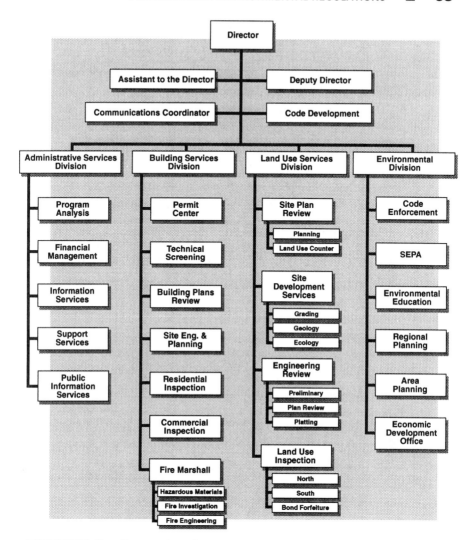

■ **FIGURE 3.3.** Example of a county planning agency organization chart. *Source:* Modified from King County, Washington, Department
of Development and Environmental Services, May 2, 1995.

general welfare of its citizens. Zoning regulates such things as the
height and bulk of buildings, the area of a lot which may be occupied,
the size of required open spaces, the density of population, and the
use of buildings and land for trade, industry, residences, or other
purposes.

AGENCIES WITH EIS RESPONSIBILITIES

Environmental and planning processes vary by the type of agency and its mission, as well as with regard to where in an agency's departmental hierarchy an EIS is being prepared. Agencies that most frequently deal with EIS processes are those that have planning and environmental responsibilities, such as

■ Land and natural resource managing agencies
■ Property managing agencies
■ Agencies involved with construction
■ Agencies that issue permits and licenses
■ Agencies that grant and lend funds

The following are examples of federal agencies that frequently conduct environmental reviews in carrying out their responsibilities. Many other federal, state, and local agencies have similar responsibilities.

Land and Natural Resource Managing Agencies

Land managing agencies are those with responsibilities to manage activities that use land or natural resources in a manner consistent with laws, regulations, and the agencies' policies. Examples of such agencies at the federal level include the Bureau of Land Management, U.S. Forest Service, National Park Service, and military departments (Army, Air Force, Navy, Marine Corps). Also, American Indian tribes manage land and resources for which they are stewards on their reservations. Millions of acres of land and its resources are managed by federal agencies (see Table 3.9). Public land managed by federal agencies total over half a billion acres,[40] or 28.6 percent of the land in the United States. The amount of public land has gradually decreased (almost 20 percent of public domain land has become private since 1900).[41] Some public lands are available for public uses such as recreation, mining, cattle grazing, or harvesting timber; some are available for limited public use; and some are withdrawn from the public domain for the exclusive use of the agency. About one-tenth of all federal lands have protected designations such as wilderness areas, but even

---

[40]Council on Environmental Quality, *Twenty-second Annual Report*, 1991.
[41]Council on Environmental Quality, *Twenty-fourth Annual Report*, 1993.

■ TABLE 3.9 Land Managed by Federal Agencies[a]

Department	Agency	Land (Million Acres)
Interior	Bureau of Land Management	270
Agriculture	Forest Service	191
Interior	Fish and Wildlife Service	92
Interior	National Park Service	80
Defense	Army, Air Force, Navy, Marine Corps	25
		Total 658

[a]The amount of land in federal ownership has been decreasing. For example, the Bureau of Land Management (BLM) managed 450 million acres of public land in 1976 (Council on Environmental Quality, *Seventh Annual Report*, 1976), 40 percent more land than in 1991.

Source: Council on Environmental Quality, *Twenty-second Annual Report*, 1991.

protected lands can provide multiple uses such as recreation and wildlife habitat.[42]

States also manage forests, parks, and wildlife reserves that have a range of allowable public uses. States and the private sector have two-thirds of the land in the United States.[43] Use of private land is primarily controlled by local agencies. Environmental review may be required for proposed agency or private actions on state-managed land, as well as on private land if agency approvals, permits, or funds are required and if the SEPA review requirements of a state extend to local agencies.

Property Managing Agencies

Property managing agencies are those involved in the purchase, sale, or lease of buildings or real estate necessary for an agency's operation. Examples of such agencies at the federal level include the military departments, U.S. Postal Service, and General Services Administration.

Agencies Involved with Construction

Agencies involved with construction include the military departments for all types of construction, including the U.S. Army Corps of

---

[42]Council on Environmental Quality, *Twenty-second Annual Report*, 1991.
[43]*idem.*

Engineers for construction of various types of civil works projects (historically dams), Bureau of Reclamation for irrigation projects, and Department of Transportation for highway construction.

Agencies That Issue Permits and Licenses

Federal agencies that issue permits and licenses subject to NEPA review include the land managing agencies, U.S. Army Corps of Engineers, U.S. Coast Guard, and Federal Energy Regulatory Commission.

Agencies That Grant and Lend Funds

Agencies that grant and lend funds include the Department of Transportation, Department of Housing and Urban Development, Economic Development Agency, and Federal Housing Administration.

Although agencies with similar responsibilities comply with the same environmental laws, the internal policies and procedures for a particular environmental concern may differ from agency to agency. Policies and procedures regarding how an agency will manage a resource, develop plans for that resource, prepare planning documents, and comply with environmental regulations are developed at the upper level of the bureaucratic hierarchy (e.g., the department level) and at each major level or agency.

Figure 3.4 shows some of the departments within the federal government, and the agencies within those departments (the various levels of the hierarchy), that have planning and environmental duties. A law (such as NEPA) passed by Congress, or an executive order issued by the President (such as EO 11990 on wetlands), directs the heads of departments (such as the Secretary of the Department of the Interior) to implement the requirements of the law or executive order. The departmental level (the Secretary) then issues direction to the agencies within that department to implement the mandate, and this process continues until directions are received at all levels within a department. Organization charts for selected federal agencies that have environmental duties are in Appendix E. States have similar types of agencies, and many carry out duties delegated from the federal level.

■ **FIGURE 3.4.** Selected federal agencies with environmental roles. *Source*: Modified from Joseph A. Wellington, A Primer on Environmental Law for the Naval Services, *Naval Law Review*, 1989.

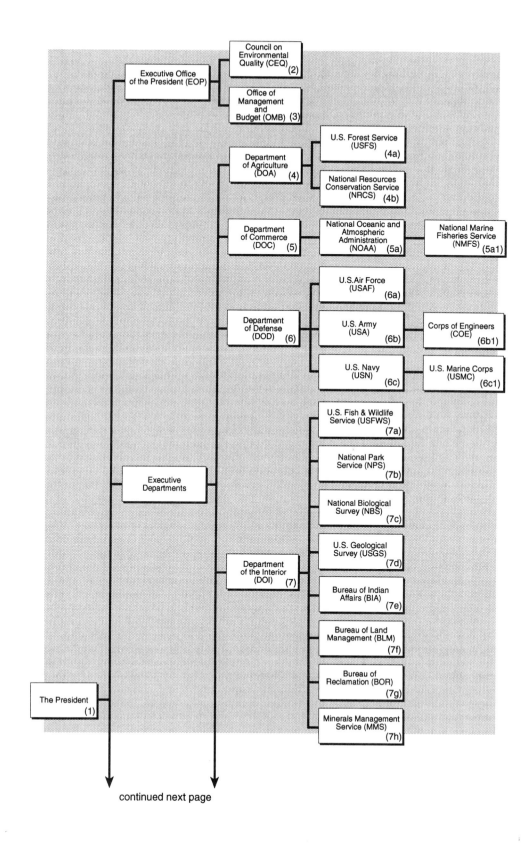

continued next page

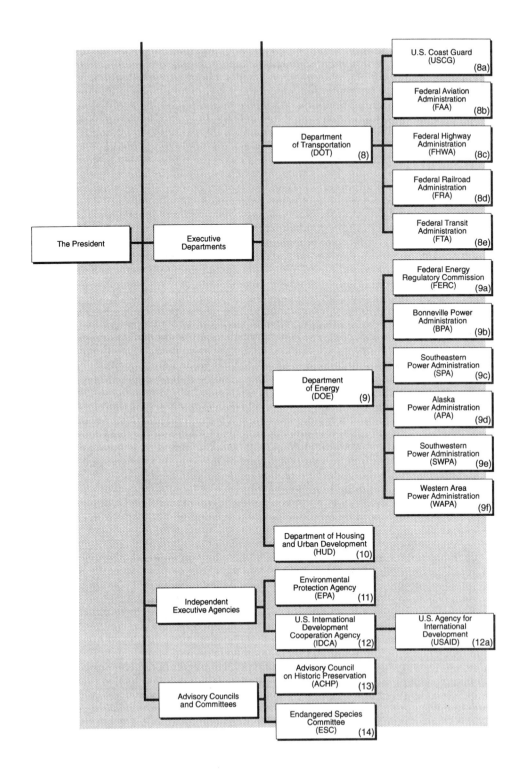

■ FIGURE 3.4. (Continued)

An environmental impact statement is not one of the delegated duties, however, because the states that enacted state environmental policy acts (SEPAs) did so at their own initiative.

To understand roles, it is useful to know the purpose and responsibilities of agencies involved in an EIS process. The following brief descriptions of the agencies shown in Figure 3.4 are excerpts from *The United States Government Manual 1994/95*. Because of their large number, many agencies, including most independent federal agencies (those commissions created by Congress such as the Nuclear Regulatory Commission and Federal Trade Commission), are not listed here. The numbers for agencies described below correspond to the numbers in the figure.

Federal Agency Responsibilities

(1) **The President** The President is the administrative head of the executive branch of the government, which includes numerous agencies, both temporary and permanent, as well as the 14 executive departments. The President is constitutionally charged with ensuring the execution of the Nation's laws, and he issues executive orders (see Table 3.4) including those regarding the environment.

(2) **Council on Environmental Quality (CEQ)** The CEQ consists of three members appointed by the President with the advice and consent of the Senate, and one of the members is designated by the President as Chairperson. The Council oversees implementation of NEPA, develops and recommends to the President national policies that further environmental quality, performs a continuing analysis of changes or trends in the national environment, appraises programs of the federal government to determine their contributions to sound environmental policy, and assists the President in the preparation of an annual environmental quality report to Congress. The CEQ's roles with regard to the EIS process are discussed in Chapter 2.

(3) **Office of Management and Budget (OMB)** The OMB evaluates, formulates, and coordinates management procedures and program objectives within and among federal departments and agencies. It also controls the administration of the federal budget while routinely providing the President with recommendations regarding budget proposals and relevant legislative enactments. Senator Jack-

son (the primary author of NEPA) originally intended the OMB to have a more active role in overseeing NEPA implementation by federal agencies, but this has not been the case. The CEQ has the bulk of responsibility for agency oversight.

(4) Department of Agriculture (DOA) Programs of the DOA include agricultural research, management of natural resources, animal and plant health, and food safety inspection. The Department provides loans and financing assistance for small community and rural development through the Farmers Home Administration, the Rural Development Administration, and the Rural Electrification Administration (a credit agency of the DOA). Food and Consumer Services and International Affairs and Commodity are also large programs within the Department. The DOA has a Science and Education program that includes the Agricultural Research Service, Cooperative State Research Service, Extension Service, and National Agricultural Library. The U.S. Forest Service and Natural Resources Conservation Service (formerly the Soil Conservation Service) are agencies frequently involved in the EIS process.

(4a) U.S. Forest Service (USFS) The Forest Service manages public land under the sustainable, multiple-use management concept to meet the diverse needs of the public. The agency manages 155 national forests, 20 national grasslands, and 8 land utilization projects on over 191 million acres in 44 states, the Virgin Islands, and Puerto Rico. Public uses of USFS land include recreational activities, timber harvesting, mining, and grazing. The land is also managed for wildlife habitat, water supplies, and wilderness. Approximately 34.6 million acres are set aside as wilderness, and 175,000 acres are designated as primitive areas where timber will not be harvested.

(4b) Natural Resources Conservation Service (NRCS) The NRCS has responsibility for developing and carrying out a national soil and water conservation program in cooperation with landowners, operators, and other land users and developers, with community planning and regional resource groups, and with other federal, state, and local agencies. The agency also assists in agricultural pollution control, en-

vironmental improvement, and rural community development. The soil and water conservation program is carried out through technical assistance to local conservation districts and to local sponsors of watershed protection projects and resource conservation and development projects, as well as through consultative assistance to other individuals and groups. About 3000 conservation districts cover more than 2 billion acres in the 50 states, Guam, Puerto Rico, and the Virgin Islands.

(5) **Department of Commerce (DOC)**   The DOC serves and promotes the nation's international trade, economic growth, and technological advancement. The Department works to improve the understanding and benefits of the Earth's physical environment and oceanic resources, provides social and economic statistics and analyses for business and government planners, provides research, administers programs to prevent unfair trade and competition, and provides research and support for the increased use of scientific, engineering, and technological development. The DOC includes the National Oceanic and Atmospheric Administration, which in turn includes the National Marine Fisheries Service.

(5a) National Oceanic and Atmospheric Administration (NOAA) The NOAA's mission is to (a) explore, map, and chart the global ocean and its living resources and to manage, use, and conserve those resources, (b) describe, monitor, and predict conditions in the atmosphere, ocean, sun, and space environment, (c) issue warnings against impending destructive national events (hurricanes, floods, tsunamis, tornadoes), (d) assess the consequences of inadvertent environmental modification over several scales of time, and (e) manage and disseminate long-term environmental information. The NOAA is also responsible for the administration of the Coastal Zone Management Act (CZMA). The agency offers federal matching grants to coastal states to develop and implement coastal plans that balance coastal development and environmental protection. As of 1991, 29 of the 35 eligible states and territories have federally approved CZMA plans.[44]

---

[44]Council on Environmental Quality, *Twenty-second Annual Report*, 1991.

(5a1) National Marine Fisheries Service (NMFS) The NMFS conducts an integrated program of management, research, and services related to the protection and rational use of living marine resources and their habitats and protects marine mammals and endangered species.

(6) Department of Defense (DOD)
The DOD is responsible for providing the military forces needed to deter war and protect the security of the Nation. The major elements of these forces are the Army, Navy, Marine Corps, and Air Force. Under the President, who is also Commander-in-Chief, the Secretary of Defense exercises authority, direction, and control over the Department. Each military department (the Department of the Navy includes naval aviation and the United States Marine Corps) is separately organized under its own Secretary and functions under the authority, direction, and control of the Secretary of Defense. Orders to the military departments are issued through the Secretaries of these departments, or their designees, by the Secretary of Defense or under authority specifically delegated in writing by the Secretary or provided by law.

All of the following military departments manage land and natural resources (on 25 million acres of federal lands and facilities) and conduct training operations and civil works projects which must comply with environmental regulations. Thus, actions of the military departments must comply with NEPA as well as the requirements of other environmental regulations.

(6a) United States Air Force (USAF) or Department of the Air Force (DOAF) The Department of the Air Force is responsible for defending the peace and security of the country through control and exploitation of airspace.

(6b) United States Army (USA) or Department of the Army (DOA) The Army's mission is to organize, train, and equip active duty and reserve forces for the preservation of peace, security, and the defense of the Nation. The Army's mission focuses on land operations. The Army also administers programs to protect the environment, improve waterway navigation, flood and beach erosion control, and water resource development. It provides military assistance to federal, state, and local agencies including natural disaster relief assistance.

(6b1) U.S. Army Corps of Engineers (COE) The Commanding General of the COE serves as the Army's Real Property Manager, performing the full cycle of real property activities (requirements, programming, acquisition, operation, maintenance, and disposal); manages and executes engineering, construction, and real estate programs for the Army and United States Air Force; and performs research and development in support of these programs. He manages and executes Civil Works Programs including research and development, planning, design, construction, operation and maintenance, and real estate activities related to rivers, harbors, and waterways, and he is in charge of administration of laws for protection and preservation of navigable waters and related resources such as wetlands. He also assists in recovery from natural disasters.

(6c) United States Navy (USN) or Department of the Navy (DON)
The primary mission of the Department of the Navy is to protect the United States by the effective prosecution of war at sea including, with its Marine Corps component, the seizure or defense of advanced naval bases; to support, as required, the forces of all military departments of the United States; and to maintain freedom of the seas. The Department of the Navy includes the U.S. Coast Guard when it is operating as a Service in the Navy.

(6c1) United States Marine Corps (USMC) The Marine Corps is organized, trained, and equipped (combined arms and supporting air components) for service with the (naval) fleet in the seizure and defense of naval bases, and it also conducts land operations for the prosecution of naval campaigns.

(7) Department of the Interior (DOI) As the Nation's principal conservation agency, the DOI has responsibility for most public lands and natural resources. This includes the following: fostering sound use of land and water resources; assessing and protecting fish, wildlife, and biological diversity; preserving the environmental and cultural values of the national parks and historical places; and providing for outdoor recreation. The Department assesses mineral resources and works to ensure that their development is in the best interests of the public by encouraging stewardship and citizen participation in their

care. The DOI also has a major responsibility for American Indian reservation communities and for people who live in island territories under United States administration.

The jurisdiction of the DOI includes:

- Administration of over 500 million acres of federal land, as well as trust responsibilities for approximately 50 million acres of Indian lands
- Conservation and development of mineral and water resources, including minerals on the outer continental shelf
- Management of revenue from federal and certain Indian mineral leases
- Assessment, conservation, development, and utilization of fish and wildlife resources
- Coordination of federal and state recreation programs
- Preservation and administration of the Nation's scenic and historic areas
- Operation of Job Corps Conservation Centers and Youth Conservation Corps Camps, as well as coordination of other youth training programs
- Management of irrigation systems on arid lands in the West
- Management of hydroelectric power systems
- Assistance to the states in protecting society and the environment from the adverse effects of coal mining

(7a) United States Fish and Wildlife Service (USFWS) The USFWS is comprised of a headquarters office in Washington, D.C., seven regional offices in the lower 48 States and Alaska, and a variety of field units and installations. These include nearly 500 national wildlife refuges and 166 waterfowl production areas totaling almost 91 million acres, 78 national fish hatcheries, and a nationwide network of wildlife law enforcement agents. The agency is responsible for migratory birds, endangered species, certain marine mammals, and inland sport fisheries. Its mission is to conserve, protect, and enhance fish and wildlife and their habitats for the continuing benefit of the American people.

In the area of resource management, the USFWS is involved in the protection and improvement of land and water environments (habitat preservation), which directly benefits the living natural resources and adds quality to human life. Activities include

■ Surveillance of pesticides, heavy metals, and other contaminants
■ Studies of fish and wildlife populations
■ Ecological studies
■ Environmental impact assessment, including hydroelectric dams, nuclear power sites, stream channelization, and dredge-and-fill permits
■ Environmental impact statement review

Specific wildlife and fishery resources programs include

■ *Migratory Birds*: Wildlife refuge management for production, migration, and wintering; law enforcement and conduct of game bird population, production, and harvest surveys
■ *Mammals and Nonmigratory Birds*: Refuge management of resident species, law enforcement, protection of certain marine mammals, and technical assistance
■ *Coastal Anadromous Fish*: Hatchery production and stocking
■ *Great Lakes Fisheries*: Hatchery production of lake trout and fishery management in cooperation with Canada and the States
■ *Other Inland Fisheries*: Hatchery production and stocking of Indian lands, and technical assistance

The Service provides national and international leadership in the area of identifying, protecting, and restoring endangered species of fish, wildlife, and plants. This program includes

■ Development of the federal Endangered and Threatened Species List, conduct of status survey, preparation of recovery plans, and coordination of efforts nationally and internationally
■ Operation of national wildlife refuges
■ Foreign importation enforcement
■ Consultation with foreign countries

(7b) National Park Service (NPS) The Park Service administers an extensive system of national parks, monuments, historic sites, and recreation areas. The objectives of the agency are to administer the properties under its jurisdiction for the enjoyment and education of

citizens, to protect the natural environment of the areas, and to assist states, local governments, and citizen groups in the development of park areas, the protection of the natural environment, and the preservation of historic properties. There are more than 365 units in the National Park System, including national parks and monuments of noteworthy natural and scientific value; scenic parkways, riverways, seashores, lakeshores, recreation areas, and reservoirs; and historic sites associated with important movements, events, and personalities of the American past.

(7c) National Biological Survey (NBS) The Biological Survey, established in 1993, is a relatively new agency within the DOI. The mission of the NBS is to gather, analyze, and disseminate the biological information necessary for sound stewardship of the Nation's natural resources. To accomplish this mission, the NBS undertakes research, inventory, monitoring, information sharing, and technology transfer to foster an understanding of the biological systems and their benefits to society. The agency establishes partnerships with other federal, state, and local agencies, with museums and universities, and with private organizations in order to bring coherence to largely uncoordinated efforts.

(7d) U.S. Geological Survey (USGS) The Geological Survey's primary responsibilities are as follows: investigating and assessing the Nation's land, water, energy, and mineral resources; conducting research on global change; investigating natural hazards such as earthquakes, volcanoes, landslides, floods, and droughts; and conducting the National Mapping Program. To attain these objectives, the USGS prepares maps and digital and cartographic data, collects and interprets data on energy and mineral resources, conducts nationwide assessments of the quality, quantity, and use of the Nation's water resources, performs fundamental and applied research in the sciences and techniques involved, and publishes and disseminates the results of its investigations in thousands of new maps and reports each year.

(7e) Bureau of Indian Affairs (BIA) The principal objectives of the BIA are to encourage and assist Indian and Alaska Native people to manage their own affairs under the trust relationship to the federal

government and to facilitate, with maximum involvement of Indian and Alaska Native people, full development of their human and natural resource potential. The Bureau works with them in the development and implementation of programs for their economic advancement and for full utilization of their natural resources consistent with the principles of resource conservation.

(7f) Bureau of Land Management (BLM) The BLM is responsible for the total management of more than 270 million acres of public lands. These lands are located primarily in the West and Alaska; however, small scattered parcels are located in other states. The BLM's mission of land management was derived from almost 3000 public land laws which were often overlapping and conflicting—some dating back more than 170 years.[45] The Federal Land Policy and Management Act of 1976 repealed and replaced many obsolete or overlapping statutes. It provides a basic mission statement for the Bureau and establishes policy guidelines and criteria for the management of public lands and resources administered by the agency.

Resources managed by the BLM include timber, solid minerals, oil and gas, geothermal energy, wildlife habitat, endangered plant and animal species, rangeland vegetation, recreation and cultural values, wild and scenic rivers, designated conservation and wilderness areas, and open space. The agency's programs provide for the protection (including fire suppression), orderly development, and use of the public lands and resources under principles of multiple use and sustained yield. Land use plans are developed with public involvement to provide orderly use and development while maintaining and enhancing the quality of the environment. The BLM also manages watersheds to protect soil and enhance water quality, develops recreational opportunities on public lands, administers programs to protect and manage wild horses and burros, and, under certain conditions, makes land available for sale to individuals, organizations, local governments, and other federal agencies when such transfer is in the public interest. Lands may be leased to state and local agencies and to nonprofit organizations for certain purposes.

---

[45]Council on Environmental Quality, *Seventh Annual Report*, 1976.

The BLM oversees and manages the development of energy and mineral leases and ensures compliance with applicable regulations governing the extraction of these resources. In addition to minerals management responsibilities on public lands, the BLM is also responsible for subsurface resource management of an additional 300 million acres where minerals rights are owned by the federal government. The agency has responsibility to issue rights-of-way, in certain instances, for crossing federal lands under other agencies' jurisdiction. It also has general enforcement authority.

(7g) Bureau of Reclamation (BOR)   The mission of the BOR is to manage, develop, and protect, for the public welfare, water and related resources in an environmentally and economically sound manner. The Reclamation Act of 1902 authorized the Secretary of the Interior to administer a reclamation program that would provide the arid and semiarid land of the 17 contiguous western states a secure year-round water supply for irrigation. The reclamation program helped settle and develop the West through the development of a water storage and delivery infrastructure, which provides safe and dependable water supplies and hydroelectric power for agricultural, municipal, and industrial users, protects and improves water quality, provides recreational and fish and wildlife benefits, enhances river regulations, and helps control damaging floods. With this infrastructure largely in place, the reclamation program now focuses on resource management and protection.

(7h) Minerals Management Service (MMS)   The MMS assesses the nature, extent, recoverability, and value of leasable minerals on the Outer Continental Shelf. It ensures the orderly and timely inventory and development, as well as the efficient recovery, of mineral resources, encourages utilization of the best available and safest technology, provides for fair, full, and accurate returns to the Federal Treasury for produced commodities, and safeguards against fraud, waste, and abuse.

(8) Department of Transportation (DOT)   The DOT establishes the Nation's overall transportation policy. Under its umbrella are 10 administrations whose jurisdictions include the following: highway planning, development, and construction; urban mass transit; railroads; aviation; and the safety of waterways, ports, highways, and oil and gas pipelines. Decisions made by the DOT in conjunction with appropriate state and local officials affect other programs such as land planning, energy conservation, scarce resource utilization, and technological change. The central management concept of the agency is that operating programs are carried out by the operating administrations, which are organized generally by mode (e.g., air, rail, etc.). The Secretary and Deputy Secretary are responsible for the overall planning, direction, and control of all departmental activities.

(8a) U.S. Coast Guard (USCG) The Coast Guard is a branch of the Armed Forces of the United States at all times and is a service within the Department of Transportation, except when operating as part of the Navy in time of war or when the President directs. Some activities of the USCG include:

■ *Search and Rescue.* The Coast Guard maintains a system of rescue vessels, aircraft, and communications facilities to carry out its function of saving life and property in and over the high seas and the navigable waters of the United States. This function includes flood relief and removing hazards to navigation.

■ *Maritime Law Enforcement.* The Coast Guard is the primary maritime law enforcement agency for the United States. It enforces or assists in the enforcement of applicable federal laws and treaties and other international agreements to which the United States is party, on, over, and under the high seas and waters subject to the jurisdiction of the United States. The USCG works with other federal agencies in the enforcement of such laws as they pertain to the protection of living and nonliving resources and in the suppression of smuggling and illicit drug trafficking.

■ *Marine Environmental Response.* The Coast Guard is responsible for enforcing the Federal Water Pollution Control Act and various other laws relating to the protection of the marine environment. Program objectives are to ensure that public health and welfare and the environment are protected when spills occur.

■ *Bridge Administration.* The USCG administers the statutes regulating the construction, maintenance, and operation of bridges and causeways across the navigable waters of the United States to provide for safe navigation.

(8b) Federal Aviation Administration (FAA) The FAA is charged with

■ Regulating air commerce in ways that best promote its development and safety and fulfill the requirements of national defense

■ Controlling the use of navigable airspace of the United States and regulating both civil and military operations in such airspace in the interest of safety and efficiency

■ Installing and operating air navigation facilities

■ Developing and operating a common system of air traffic control and navigation for both civil and military aircraft

■ Developing and implementing programs and regulations to control aircraft noise, sonic boom, and other environmental effects of civil aviation

(8c) Federal Highway Administration (FHWA) The FHWA is concerned with the total operation and environment of highways systems, including highway and motor carrier safety. In administering its highway transportation programs, the agency considers the impacts of highway development and travel; transportation needs; engineering and safety aspects; social, economic, and environmental effects; and project costs. The FHWA uses a systematic, interdisciplinary approach in providing for safe and efficient highway transportation.

(8d) Federal Railroad Administration (FRA) The purpose of the FRA is to promulgate and enforce rail safety regulations, administer railroad financial assistance programs, conduct research and development in support of improved railroad safety and national rail transportation policy, provide for the rehabilitation of Northeast Corridor rail passenger service, and consolidate government support of rail transportation activities.

(8e) Federal Transit Administration (FTA) The missions of the FTA are

■ To assist in the development of improved mass transportation facilities, equipment, technique, and methods, with the cooperation of public and private mass transportation companies

■ To encourage the planning and establishment of areawide urban mass transportation systems needed for economical and desirable urban development, with the cooperation of public and private mass transportation companies

■ To provide assistance to state and local governments and their instrumentalities in financing such systems, to be operated by public or private mass transportation companies as determined by local needs

■ To provide financial assistance to state and local governments to help implement national goals relating to mobility for elderly persons, persons with disabilities, and economically disadvantaged persons

(9) Department of Energy (DOE) The DOE provides the framework for a comprehensive and balanced national energy plan through the coordination and administration of the energy functions of the federal government. The Department is responsible for the following: long-term, high-risk research and development of energy technology; the marketing of federal power; energy conservation; the nuclear weapons program; energy regulatory programs; and a central energy data collection and analysis program. The Department of Energy Act of 1977 consolidated major federal energy functions into one Cabinet-level Department, transferring to DOE all the responsibilities of the Energy Research and Development Administration, the Federal Energy Administration, the Federal Power Commission, and the Alaska, Bonneville, Southeastern, and Southwestern Power Administrations, as well as the power-marketing functions of the Department of the Interior's Bureau of Reclamation. The marketing and transmission of electric power produced at federal hydroelectric projects and reservoirs is carried out by the DOE's five Power Administrations. The DOE's organization also includes the Federal Energy Regulatory Commission, which is an independent regulatory organization within the Department.

(9a) Federal Energy Regulatory Commission (FERC) The FERC, an independent, five-member commission within the DOE, has retained many of the functions of the Federal Power Commission, such as set-

ting rates and charges for the transportation and sale of natural gas and for the transmission and sale of electricity and the licensing of hydroelectric power projects. In addition, the Commission establishes rates or charges for the transportation of oil by pipeline, as well as the valuation of such pipelines.

(9b) Bonneville Power Administration (BPA) Through a regionwide, interconnecting transmission system, the BPA markets electric power and energy from federal hydroelectric projects in the Pacific Northwest constructed and operated by the U.S. Army Corps of Engineers and the Department of the Interior's Bureau of Reclamation. It sells surplus power to areas outside the Pacific Northwest region and participates in exchanges of power. In addition, the BPA is responsible for energy conservation, renewable resource development, and fish and wildlife enhancement under the provisions of the Pacific Northwest Electric Power Planning and Conservation Act of 1980.

(9c) Southeastern Power Administration (SPA) The SPA was created by the Secretary of the Interior in 1950 to carry out functions assigned to the Secretary by the Flood Control Act of 1944, which pertain to the transmission and disposition of surplus electric power and energy generated at reservoir projects that are or may be under the control of the Department of the Army in the states of West Virginia, Virginia, North Carolina, Georgia, Florida, Alabama, Mississippi, Tennessee, and Kentucky. The SPA's program includes, among other things, the provision—by construction, contract, or otherwise—of transmission and related facilities to interconnect reservoir projects and to serve contractual loads; it also includes activities pertaining to the planning and operation of power facilities.

(9d) Alaska Power Administration (APA) The APA is responsible for operating and marketing power for two federal hydroelectric projects in Alaska. Power operations and marketing functions involving the Eklutna and Snettisham Hydroelectric Projects include the projects' transmission systems serving the Anchorage and Juneau areas.

(9e) Southwestern Power Administration (SWPA) The SWPA was created by the Secretary of the Interior in 1943 to carry out the

Secretary's responsibility for the sale and disposition of electric power and energy generated at certain projects constructed and operated by the Department of the Army. For these projects, the SWPA carries out the functions assigned to the Secretary by the Flood Control Act of 1944 in the states of Arkansas, Kansas, Louisiana, Missouri, Oklahoma, and Texas.

(9f) Western Area Power Administration (WAPA) The WAPA was established in 1977 by the Department of Energy Organization Act. The Administration is responsible for the federal electric power-marketing and transmission functions in 15 central and western states, encompassing 1.3 million square miles. The Administration sells power to cooperatives, municipalities, public utility districts, private utilities, federal and state agencies, and irrigation districts. The wholesale power customers, in turn, provide service to millions of retail customers in Arizona, California, Colorado, Iowa, Kansas, Minnesota, Montana, Nebraska, Nevada, New Mexico, North Dakota, South Dakota, Texas, Utah, and Wyoming. The WAPA is responsible for the operation and maintenance of 16,178 miles of transmission lines, 228 substations, and various auxiliary power facilities and also for planning, construction, and operation and maintenance of additional federal transmission facilities that may be authorized in the future.

(10) Department of Housing and Urban Development
(HUD) The HUD provides a variety of programs including: Federal Housing Administration mortgage insurance programs that help families become homeowners and facilitate the construction and rehabilitation of rental units; rental assistance programs for lower-income families; and programs that aid community and neighborhood development and preservation. The Department also takes steps to (a) encourage a strong private sector housing industry that can produce affordable housing and (b) stimulate private sector initiatives, public/private sector partnerships, and public entrepreneurship.

(11) Environmental Protection Agency (EPA) The EPA was
created to permit coordinated and effective governmental action on behalf of the environment. It endeavors to abate and control pollution systematically, by integration of a variety of research, monitoring, stan-

dard setting, and enforcement activities. The EPA coordinates and supports research and antipollution activities by state and local governments, private and public groups, individuals, and educational institutions. It also reinforces efforts among other federal agencies with respect to the impact of their operations on the environment, and it is specifically charged with publishing its determinations when those hold that a proposal is unsatisfactory from the standpoint of public health or welfare or environmental quality. Also, see Chapter 2 for additional descriptions of the EPA's roles and responsibilities in the EIS process.

(12) United States International Development Cooperation Agency (IDCA)   The IDCA's function is policy planning, policymaking, and policy coordination on international economic issues affecting developing countries. The agency's mission is twofold: first, to ensure that development goals are taken fully into account in all executive branch decisionmaking on trade, financing and monetary affairs, technology, and other economic policy issues affecting the less-developed nations; and second, to provide strong direction for U.S. economic policies toward the developing world and a coherent development strategy through the effective use of U.S. bilateral development assistance programs and U.S. participation in multilateral development organizations.

(12a) United States Agency for International Development (USAID)   The USAID administers U.S. foreign economic and humanitarian assistance programs in more than 100 countries in the developing world, Central and Eastern Europe, and the New Independent States of the former Soviet Union. To meet the challenges of the post–Cold War era, the USAID redefined its mission and developed a strategy for achieving sustainable development in developing countries. The agency supports programs in four areas: population and health, broad-based economic growth, environment, and democracy. USAID environmental programs support two strategic goals: reducing long-term threats to the global environment, particularly loss of biodiversity and climate change; and promoting sustainable economic growth locally, nationally, and regionally by addressing environmental, economic, and developmental practices that impede development. Globally, USAID programs focus on reducing sources and enhancing sinks of greenhouse

gas emissions and on promoting innovative approaches to the conservation and sustainable use of the planet's biological diversity.

(13) Advisory Council on Historic Preservation (ACHP) The ACHP is an independent federal agency established by the National Historic Preservation Act (NHPA). The Council advises the President and Congress on historic preservation matters, including issuing annual reports on preservation issues and activities, producing special reports and policy recommendations on preservation topics, and providing advice, technical assistance, and testimony on legislative proposals. The Council also carries out Section 106 (of NHPA) review, and it reviews federal agency historic preservation programs and policies. Membership of the ACHP comprises the following: four members of the general public (including the chairman); four historic preservation experts; Secretary of the Interior; Secretary of Agriculture; Architect of the Capitol; four federal agency heads (the Secretaries of the Treasury, Housing and Urban Development, Transportation, and the Director, Office of Administration, The White House); one governor and one mayor; President, National Conference of State Historic Preservation Officers, and Chairman, National Trust for Historic Preservation.[46]

(14) Endangered Species Committee (ESC) The ESC is an independent agency empowered to grant exemptions to requirements of the Endangered Species Act.

THE EIS CATCH-22

or how to conduct "meaningful analysis" early in the planning process when project information is only preliminary

The CEQ regulations state "Agencies *shall* integrate the NEPA process with other planning *at the earliest possible time* to ensure that *planning and decisions* reflect environmental values . . ." [emphasis added]. State regulations that implement SEPAs have similar guidance to integrate environmental review into the planning process as

---

[46]Advisory Council on Historic Preservation and the General Services Administration Interagency Training Center, *Introduction to Federal Projects and Historic Preservation Law, Participants Course Book*, 1993.

early as possible. The purpose for having early environmental review and integrating it with the planning process is to increase the probability that environmental concerns will be considered while a proposed project is still in its preliminary development. Once a project has gone through a great deal of planning and design, where much time and money have been invested, it becomes more difficult to make fundamental changes because to accommodate changes at that stage the planning and design work may have to be redone.

However, if a project is so preliminary that "meaningful analysis" cannot be done, the EIS process cannot begin. There is no standard of how well-defined a project must be to be "adequately" defined; the decision of whether a proposal is ready for analyses is made on a case-by-case basis. Generally, the more thoroughly a proposed project is defined, the easier it is to prepare an EIS. If a proposal is not well-defined, and questions remain regarding the who, what, when, where and why of a project, the ID team will have difficulty making conclusions regarding the potential environmental effects of a project. For any unanswered questions, the analyses would be based on assumptions or be left unanswered. If a proposed project has potential for significant impacts but the lack of project details prevents analyses of those impacts, a decisionmaker would not be able to balance the cost to the environment with the benefits of a proposal. Also, if an agency's decisions are based on an analyses of a poorly defined project, other reviewing agencies and the courts may consider the decisions "arbitrary and capricious" because the analyses will not support the agency's conclusions.

Project details, such as engineering drawings, are at the other end of the spectrum of defining a project. Detailed project information normally is not necessary to evaluate the environmental impacts of a proposed project. Detailed information should be provided if the information has relevance on a truly significant environmental issue. It may be worth delaying the process to obtain such information or to have the proponent's designers work out more of the detail than was initially thought necessary. Time spent up front defining a proposed project saves time in the long run [also see "The Project Planning Process" (later in this chapter) and Chapter 8 for discussions on defining the proposed project].

Program and Land Use Actions

If a proposed action is not for a specific project, such as an action to implement a large program, a phased or tiered EIS process may be appropriate. This is because the details of a broad agency action that covers a large area may not be known. Or a wide variety of potential activities could follow implementation of the program; however, the specific features of those activities (e.g., their locations, size, methods of implementation) are unknown at the time the program is being contemplated. Tiering allows the information that is available regarding a proposed action to be analyzed in a broad-level or programmatic EIS. This approach provides a decisionmaker with information necessary to decide whether and how to proceed with a proposed program.

Specific actions that may take place under a program may require further environmental analyses in additional EISs or EAs when information is available regarding those specific actions. Federal agencies tend to use programmatic EISs and the tiered process more than state or local agencies, even if state regulations permit tiering. A city's comprehensive plan, for example, would be a candidate for a programmatic EIS. At the federal level, the U.S. Forest Service and Bureau of Land Management frequently prepare programmatic EISs for their land use or natural resource management plans and programs. Tiering is also discussed in Chapter 13.

THE PERMIT PROCESS

Although many regulatory processes result in the issuance (or denial) of a permit, an EIS process has no permit associated with it. An EIS, when completed, is not a permit and the mitigation measures contained in an EIS are not enforceable conditions on a proposed project. An extra step is required to make mitigation measures enforceable, such as making the mitigation measures conditions of an agency's permit or part of a lead agency's record of decision.

Because EISs are required to evaluate a proposal's compliance with other environmental regulations, an EIS should identify all environmental permits applicable to a proposed action. The process of obtaining permits, however, normally is not part of an EIS process. Generally, there is no single process for obtaining permits for a given project.

The project proponent must contact all of the agencies with jurisdiction over a project and request information on the agencies' process for the necessary permits. One permit may require approval from several agency departments, and each department may have numerous requirements that must be met before it will give its approval. It is not unusual for departments within an agency (see Figure 3.1) to give conflicting requirements to a project proponent. He or she must then negotiate the departments' requirements to resolve the conflicts. As an example of permitting requirements for a proposed project, a city planning department required that trees be planted along a project roadway. But the city public works department prohibited planting trees because the slope along the roadway exceeded the department's guidelines. The angle of the slope could not be reduced because of the city planning department's requirement for a minimum buffer width between the road and wetlands, and the road couldn't be made narrower because of the width and design requirements specified by the city transportation engineers. Sixteen permits (with review and approval of several city departments for each permit) were obtained for this project.

State Permit Assistance Centers

Some states have established or are establishing permit assistance centers that provide information about permit requirements, coordinate permits for complex projects, and conduct preapplication meetings. The centers coordinate permits but do not issue permits. Center staff may mediate any problems between permit staff and applicants. States that provide permit assistance include California, Maryland, Massachusetts, New Jersey, New York, Oklahoma, and Washington. Most of these permit centers are relatively new and were established within the last three years. The centers were established to improve service to permit applicants, with a secondary purpose to attract businesses to the state.[47]

The goal of some state permit centers (e.g., Massachusetts and Washington) is to provide assistance with all types and levels of governmental permitting (federal, state, and local); however, this would

[47]Washington State Department of Ecology, *Improving the Permit Process: Integration, Tracking, and Assistance*, Publication #95-253, 1995.

THE PERMIT PROCESS ■ 91

be a very large undertaking and will take some time to develop such a capability. Other states, such as New York and New Jersey, coordinate permit processing and information with other agencies through their SEPAs[48] (New Jersey does not have a SEPA but houses its "Office of Permit Coordination" in the Department of Environmental Protection).

Washington's newly formed permit assistance center provides the following services:

■ *Permit Information and Assistance.* Serve as a clearinghouse and first point of contact for information on the state Department of Ecology's environmental permits, certifications, approvals, applications, and regulatory requirements. It plans to provide similar information for other Washington agencies as well as local and federal permitting assistance.
■ *Permit/Project Coordination.* Ensure that more complex projects (e.g., those involving multiple environmental permits and multiple regulatory personnel) are appropriately coordinated.
■ *Permit Tracking.* Track projects and information requests that come through the center to provide status updates and activity reports.
■ *Education and Training.* Provide training of potential project managers through the use of consultants and other agency regulatory personnel.[49]

In 1977 the California legislature passed the California Permit Streamlining Act and established the Office of Permit Assistance (OPA) which is within the Trade and Commerce Agency. The act requires the development permit process to be coordinated with the California Environmental Quality Act (CEQA) process. The OPA was created to help project applicants, localities, and the public to understand CEQA and the permitting process and to provide assistance and information to parties interested in the permit process. The OPA provides the following:

■ A single point of contact for state agency permits to answer questions about the permit process.

---

[48]*idem.*
[49]*idem.*

■ Identification of all state or local permits required for a project. The OPA can convene all state agencies at one time to identify and explain which permits are required for a project.

■ The OPA can arrange scoping meetings. The Office convenes meetings of the environmental staff of state and local agencies who will be involved in the CEQA review of projects. These meetings provide developer-applicants and environmental consultants with a chance to discuss all environmental issues and concerns early in the process in order to avoid wasted effort in unwarranted surprises in the EIR process.

■ The OPA has the authority to convene meetings to resolve questions or mediate disputes. When uncertainties or disagreements among agencies stall the permit process, the Office may be called upon to provide a forum for resolving the problem.[50]

While the goals of the permit assistance centers appear similar, the main area of difference between the California and Washington centers is that California's center focuses on the integration of *development* permits into the environmental review process; whereas Washington's permit assistance focuses on obtaining *environmental* permits.

Record of Decision

A record of decision (ROD) is a legal document under NEPA (as well as other environmental laws) that is a federal agency's decision on a proposed action following completion of an EIS. Although it is not a permit, a ROD includes the mitigation measures that an agency is stating it will implement or that it is making conditions of its approval for another entity's proposal. A ROD

■ States what the agency's decision was
■ Identifies all alternatives considered in reaching its decision, specifying the environmentally preferable alternative
■ Discusses the agency's preference of alternatives based on factors including economic and technical considerations and agency statutory missions

---

[50]California Office of Planning and Research, *California Environmental Quality Act Statutes and Guidelines*, 1995.

■ Identifies and discusses all factors, including national policy considerations, which were balanced by the agency in makings its decision and states how those considerations entered into its decision

■ States whether all practicable means to avoid or minimize environmental harm from the selected alternative have been adopted, and, if not, why they were not

■ Discusses monitoring and enforcement programs where applicable for any mitigation[51]

THE PROJECT PLANNING PROCESS

From a project proponent's perspective, a project that is the subject of environmental review has its own project planning and design process. A project, or proposed action, starts as an idea that results from the identification of a need or an opportunity. At this stage, an idea is usually not sufficiently defined to begin public or environmental review. To adequately define a proposed project, the project's operational, as well as its physical, characteristics must be determined. If construction is involved in the proposal, the method of construction and any phasing of construction would also be defined, at least preliminarily.

If a project is an agency's project, the agency will follow a prescribed procedure for development of a project from the conceptual stage to, for example, 35% design, 75% design, and so forth, until the details of the project have been determined. A proposed project is also subject to other internal agency reviews such as budgetary review, and it is reviewed by internal departments such as engineering, real estate, and planning.

A private individual or group's project will evolve also, but the steps for defining the project will vary from simple to complex depending on the type of project and the number of people sponsoring the project. Planners, architects, engineers, and others are frequently the specialists involved in a project that take it through its planning stages.

If a project is designed to minimize environmental impacts so it would not result in significant impacts, a lead agency may not require an EIS. But how does one know what the environmental concerns and

---

[51]40 CFR §1505.2.

constraints might be in order to effectively design the project until the proposed project receives rigorous environmental evaluation? With experience, a person can anticipate many of the environmental concerns associated with a specific project, particularly if he or she has familiarity with the area and with the agencies who have jurisdiction over the action. Design of a project to minimize or avoid impacts for known environmental concerns will go far toward streamlining the process. However, in most cases involving large or complex proposals, there is no certainty that all of the environmental issues have been considered until after working closely on the proposed action with agencies and the public.

Effects of the EIS Process

If an agency does not use the EIS process to make better decisions, the process is just another "hoop" to go through to get a project approved. Some individuals and companies that develop or design projects (public and private) do not include environmental considerations in their daily business. Architects, engineers, lawyers, and public relations experts might be considered necessary for a project's success. But environmental expertise is rarely sought unless a regulatory agency identifies the necessity. If time for environmental review was not included in a project schedule, it often will cause the original project schedule to slip by a year or more. Not many companies, including large development companies, have environmental staff. Many agencies do not have environmental staff either. Companies and agencies hire environmental consultants when they need them. However, this approach is reactionary rather than proactive because it is difficult to recognize when there is a need for environmental expertise unless someone who is knowledgeable about environmental requirements is involved in an agency's or company's daily business.

Whether projects are public or private, to work with the EIS process, make it a part of doing business. Identify and plan for environmental requirements, because delays and extra costs from environmental surprises are most often the cause of negative experiences with environmental requirements. Find ways to use the process to the organization's advantage in designing or developing better projects and possibly getting public relations acknowledgment, or advertising, that shows what your company or agency is doing for the environment (some companies already do this).

STREAMLINING THE PLANNING AND ENVIRONMENTAL PROCESSES

Roles and responsibilities of federal, state, and local agencies often conflict. The mandates of many federal and state agencies focus on limited uses, environmental protection, and regulations. Federal and state agencies have land and natural resource management responsibilities on public land. The mandates of local agencies, however, tend to focus on the orderly development and servicing of private land. Developers and agencies alike have been struggling with regulations or other agency requirements that overlap, duplicate, and conflict with other regulations. The situation results in unnecessary delays, burdens on agencies and their employees, and costs to taxpayers as well as developers. While regulations such as the CEQ regulations encourage agencies to combine their planning and EIS processes, few agencies have the knowledge or latitude to change their procedures (much less combine their procedures with other agencies' procedures) and attempt radically different approaches. However, federal, state, and local agencies in many areas of the country are identifying ways to reduce duplicative regulatory requirements and streamline the process by combining some planning and environmental processes.

Federal

The first step toward streamlining regulatory processes is the development of cooperative relationships and complementary goals among agencies that have similar or overlapping programs. For example, federal and state agencies have identified ecosystem planning as a cooperative method of controlling land use actions and management of natural resources. Federal agencies in the Department of the Interior, particularly research units of the U.S. Fish and Wildlife Service (USFWS), have developed computer-based gap analysis to measure elements of biodiversity as an initiative in ecosystem planning. In 1991, the agency conducted gap-analysis projects in 14 states in cooperation with the Bureau of Land Management, the U.S. Forest Service, and private groups such as The Nature Conservancy. Working together, these groups have been expanding gap analysis into a nationwide assessment[52] (ecosystem planning and gap analysis are also discussed below, under state initiatives in California).

Another example is the commitment made by the U.S. Department of Transportation, the U.S. Army Corps of Engineers, and the U.S. Environmental Protection Agency to integrate NEPA and Section 404 of the Clean Water Act in the transportation planning, programming, and implementation stages. The agencies agree to ensure the earliest possible consideration of environmental concerns pertaining to waters of the United States, including wetlands, at each of the above stages. A high priority is placed on the avoidance of adverse impacts to waters of the United States and associated sensitive species, including threatened and endangered species. Whenever avoidance of waters of the United States is not practicable, impacts will be minimized, and unavoidable impacts will be mitigated to the extent reasonable and practicable.[53]

The Department of the Interior and Department of Defense are working together at the strategic level to manage the Mojave Desert Ecosystem under the National Program Review (NPR). The planning area includes parts of four states: California, Arizona, Nevada, and Utah.[54]

California

In California, initiatives to streamline planning and permitting processes are cooperative efforts at federal, state, and local levels for specific environmental concerns. By working together and coming to agreements on goals and procedures, agencies are attempting to reduce some conflicts and redundancies in agency procedures. These initiatives focus on specific regions with environmental resources of concern to several agencies such as the management and preservation of endangered species' habitats.

Through the Statewide Agreement on Biodiversity, a federal, state, and local consortium in California introduced the concept of bio-

---

[52]Council on Environmental Quality, *Twenty-second Annual Report*, 1991.

[53]Southern California Association of Governments, *Regional Comprehensive Plan and Guide*, April 1995.

[54]Richard Crowe, Bureau of Land Management, Riverside, California, personal communication, September 1995.

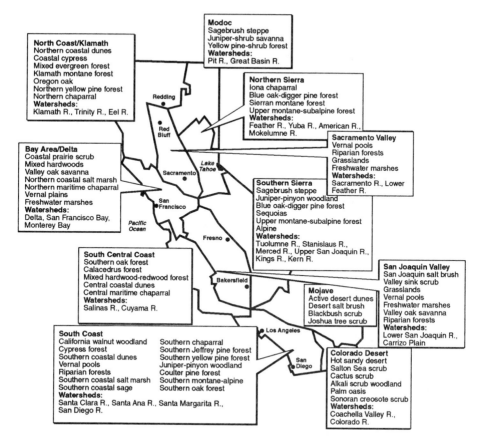

NOT TO SCALE

■ **FIGURE 3.5.** California bioregions with natural areas and watersheds. *Source*: Council on Environmental Quality, *Twenty-second Annual Report*, 1991.

regions—large, contiguous, geographic areas in which biological and physical components have similar structure and function—to the management of its public lands and waters. After identifying the bioregions (see Figure 3.5), the state entered into agreements with major federal and state land management agencies to cooperate on such concerns as the protection of endangered species and water quality. Signatories included the California departments of Fish and Game, Forestry, Fire Protection, Parks and Recreation, and Lands Commission; the University of California Division of Agriculture and Natural Resources; and the U.S. Fish and Wildlife Service, Bureau of Land Management, and U.S. Forest Service.[55]

Federal and state agencies maintain lists of rare, threatened, and endangered species and require agencies that control land use activities to develop policies for the species' protection. In Southern California, the federal listing includes approximately 50 species of plants and animals. Until recently, affected agencies developed plans for the protection of individual species on either federal or state lists. Planning for individual species has not proven to be effective in protecting species, and attention has shifted from planning for individual species to planning for multiple habitats and ecosystems. This approach is perceived to be more effective. It better handles the known and still poorly understood dynamics of all species in relationship with each other and their physical environment, it reduces or eliminates the need for piecemealing planning and streamlines project permitting processes, and it is more conducive to bringing all land use considerations and jurisdictions into the planning and management picture. Figure 3.6 shows areas and agencies that have or are preparing ecosystem plans (implementation of the plans, however, is another matter because of lack of funds and staff).

The U.S. Fish and Wildlife Service (USFWS) and California Department of Fish and Game (CDFG) regulate and monitor activities with potential to affect federal and state plant and animal species of concern. The analysis of the conditions of bioregional systems by these agencies in Southern California is through the application of gap analysis. The analysis is used to identify ecosystems at risk of dysfunction. Gap analysis involves computer overlays of mapped plant communities with private and public land ownership. Ownerships are assigned one of three protection levels depending on the degree the land is dedicated to biodiversity protection. Gap analysis has some limitations: it is coarse-scale and nonspecific in the location of species, does not recognize resource management problems, cannot "see" microhabitats, and cannot be used for project-scale planning. However, it is futuristic and very revealing at a regional scale given the trend in southern California of wholesale conversions of private land natural areas to develop-

---

[55]Council on Environmental Quality, *Twenty-second Annual Report*, 1991.

[56]Southern California Association of Governments, *Regional Comprehensive Plan and Guide*, April 1995.

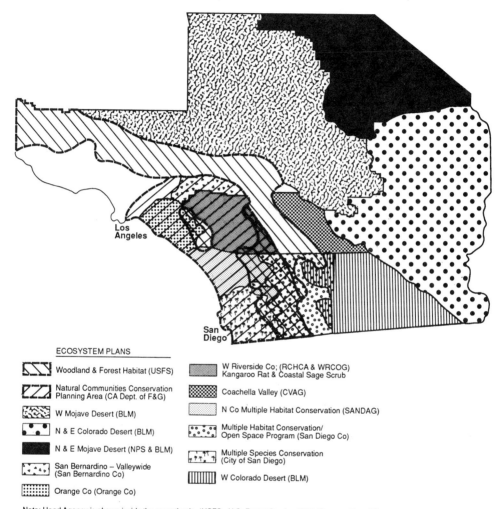

ECOSYSTEM PLANS

Woodland & Forest Habitat (USFS)

Natural Communities Conservation
Planning Area (CA Dept. of F&G)

W Mojave Desert (BLM)

N & E Colorado Desert (BLM)

N & E Mojave Desert (NPS & BLM)

San Bernardino – Valleywide
(San Bernardino Co)

Orange Co (Orange Co)

W Riverside Co; (RCHCA & WRCOG)
Kangaroo Rat & Coastal Sage Scrub

Coachella Valley (CVAG)

N Co Multiple Habitat Conservation (SANDAG)

Multiple Habitat Conservation/
Open Space Program (San Diego Co)

Multiple Species Conservation
(City of San Diego)

W Colorado Desert (BLM)

Note: Head Agency is shown inside the parenthesis. (USFS= U.S. Forest Service; BLM=Bureau of Land Management; NPS=National Park Service; RCHCA=Riverside County Habitat Conservation Agency; WRCOG=Western Riverside Council of Governments; CVAG=Coachella Valley Association of Governments; SANDAG=San Diego Association of Governments).

Plan names and boundaries not exact.

Source: Southern California Association of Governments, December 1994.   NOT TO SCALE

■ **FIGURE 3.6.**   Bioregional planning in Southern California.
Source: Southern California Association of Governments, December 1994.

ment.[56] The analysis is a predictive tool of where ecosystem planning should be headed. Large amounts of private land are determined to be at risk because of the pressure for development in the region. For example, the coastal sage scrub habitat has numerous endangered species and is 71 percent privately owned. Therefore, this habitat is a subject of ecosystem planning.

An example of an ecosystem plan is the West Mojave Coordinated Management Plan which includes the cooperative efforts of the USFWS, CDFG, counties, and cities working together to manage 9.5 million acres of public and private land with 103 plant and animal species of concern. A unique plan goal is to obtain USFWS and CDFG authorization to issue permits at the plan level for activities that are in accordance with the plan. The Bureau of Land Management is preparing a programmatic EIS for the plan. Thus, specific actions evaluated in the plan's programmatic EIS would not be required to go through project-by-project federal EIS or state EIR processes.[57]

Natural Communities Conservation Planning (NCCP) plans are required by a state of California by-law on ecosystem management. NCCP plans are for certain plant communities that have threatened or endangered species. Currently, the target plant community for NCCPs is Coastal Sage Scrub. Local governments are the lead agencies for NCCPs. San Diego County, Orange County, and Western Riverside County are preparing NCCP plans, also with the goal of obtaining plan-level permits.[58]

The above examples demonstrate that planning and problem solving is taking place through partnerships of landowners, federal, state and local agencies as well as private organizations. Although the initiatives for streamlining environmental and planning processes are encouraging, timing could be better. The cooperative planning initiatives came as ecosystems reached a crisis situation from the pressures of development in Southern California. At the same time, agency budget and staff cuts are inhibiting implementation of bioregional planning. Local governments and private landowners are expecting federal and state agencies to find the planning resources and provide the bulk of the lands that would be protected in open space reserve systems. The roles of cities and counties have historically been for orderly development, not protection of the environment. They are not prepared, with funds or staff, to implement bioregional or ecosystem planning.[59]

---

[57]Alden Sievers, Bureau of Land Management, Barstow, California, personal communication, September 1995.

[58]Richard Crowe, Bureau of Land Management, Riverside, California, personal communication, September 1995.

[59]*idem.*

Washington

Washington's House Bill (HB) 1724 attempts to streamline the requirements for comprehensive plans under the state's Growth Management Act (GMA) and EIS requirements under the State Environmental Policy Act (SEPA). Prior to the bill, EISs were required for comprehensive plans, but private or public proposals were also required to undergo site-specific environmental analyses. The legislators reasoned that the planning agencies should decide whether actions were environmentally appropriate for a given area at the time the agencies prepared their comprehensive plans, not when someone submitted a request for approval on his or her project. Under this procedure, if environmental information was not sufficient for a particular proposal, supplemental environmental analyses would still be required. However, the intent is to substantially reduce the amount of project-specific environmental documentation required under SEPA. As of this writing, the SEPA regulations have not been revised to reflect the changes resulting from HB 1724.

In spite of its goal to reduce project-specific analysis, the bill does not appear to make any real changes in the SEPA process. Under SEPA (and NEPA), tiered EISs (see Chapter 7) allow environmental documentation to take place in phases from a broad or programmatic level (when details of specific projects are not known), such as comprehensive plans, to more project-specific level SEPA documentation. If a proposal's actions were sufficiently reviewed under a programmatic EIS, further documentation would not be required. Agencies may not have been using the tiered approach for environmental review of comprehensive plans, and it would appear that the legislators did not consider the provision when they decided to revise SEPA. The bill essentially requires the same tiered approach to environmental analysis that is already in SEPA.

A potential problem with the bill is its assumption that agency programs and long-range plans can predict many of the projects that could be proposed, and, therefore, future projects within a given designation would receive evaluation of their environmental impacts in a comprehensive plan EIS.[60] Even though a proposed project is the same type as that contemplated in a comprehensive plan, the specific design

---

[60]Particularly jurisdictions (e.g., counties) that have large areas of undeveloped land and many environmental issues.

and operational features associated with a project can affect environmental impacts in a manner that was not known at the time a comprehensive plan and its EIS were prepared. The level of information in a comprehensive plan on potential projects would have to be sufficiently detailed to provide the appropriate level of environmental analyses so that further project-specific SEPA documentation will not be required. If not, the process will not have been streamlined because agencies reviewing specific projects will still be obliged to conduct additional environmental analyses. Alternatively, because of the bill, agencies would receive pressure (politically, and from the development community) to approve projects without additional analyses, and environmental impacts that were unknown at the time of the comprehensive plan may not be mitigated or otherwise considered by the agencies. Also, under this scenario, the public would not be afforded the opportunity to provide input on decisions for specific projects.

Effectiveness of Environmental Legislation

Many environmental professionals feel that the attempt to solve environmental problems through laws and regulations has not been altogether effective. The nature of regulations is that the focus tends to be on the legal process, rather than on the environmental problems. Some environmental improvements have been made, but disputes over the appropriate standards used in measuring pollution, the costs, the reporting requirements, the difficulty of working with bureaucracies, disagreements and conflicting requirements of regulatory agencies, and other problems have caused much of the effort and money for environmental purposes to be wasted on administrative matters. And, because the process deals with regulations and the possibility of lawsuits, much of the funds for environmental causes has gone to lawyers' fees rather than action toward improving the environment.

The current Congress is considering elimination or reduction of environmental regulations in response to criticisms that the regulations are stifling the economy and placing a burden on business and industry. The "easiest" response is to eliminate or reduce environmental regulations, an action with long-term costs to society. A more difficult, but wiser, course of action would be to eliminate the conflicts and duplication of effort. We need to plan our actions, consider potential effects on the environment, curb pollution, and clean up existing contaminated sites. We don't need multiple layers of bureaucracy, vague and conflicting regulations, and unnecessary bureaucratic paperwork. But Congress apparently is not to decide what agency functions are unnecessary, which agency responsibilities are redundant with another agency, and what paperwork to eliminate. The conflicting or redundant laws, after all, were passed by Congress (or more appropriately, by various Congresses).

Another related issue is that ours is a litigious society and that agency actions for compliance with NEPA or a SEPA (and other environmental statutes) are subject to lawsuits. As long as agencies have the specter of potential lawsuits hanging over their decisions, agencies will tend to be conservative and require more work, more information, and more analyses, resulting in more cost and delays to agencies, developers, and taxpayers than might actually be necessary. Despite the CEQ regulations' direction to agencies to prepare EISs that are not encyclopedic and that focus only on significant issues and eliminate from the study those issues that are not significant, agencies will try to "cover all the bases" to avoid litigation. Even if an agency decision is upheld by the courts, litigation is costly.

Solutions

Having said that the current regulatory system is not efficient, how do we find solutions for problems in the environmental system? Before solving the problems, we must recognize the nature of the problems and their complexities. The problems are complex because

- The environmental field is still evolving and contains many unknown factors
- Environmental information is incomplete
- Few experts have reached consensus on environmental issues
- Environmental regulations are primarily written by lawyers, few of whom have experience writing or managing EISs or other environmental "field" experience but have a natural tendency to focus on legal and administrative procedure
- How a problem is defined will affect the types of solutions; therefore, a "legal" problem will have a different solution than a physical (environmental) problem
- Environmental problems tend to have unique characteristics; therefore, cookbook solutions may not work

That environmental problems have unique characteristics is perhaps the most important point when considering solutions. The most effective and efficient solutions will be made at a local or regional level (depending on the level of the problem) where those who are most familiar with a problem are more likely to be able to develop creative or unique solutions for it.

As discussed under "Streamlining the Planning and Environmental Processes," above, agencies are identifying ways to solve some of the problems regionally and locally. They are doing so by working with each other and with private interests. When they have to, individuals are capable of finding creative solutions to problems. Bureaucracies, however, are not set up to deal with creative or unique situations on a large or national scale. In fact, the nature of government is to treat all things, including problems, as uniformly as possible, regardless of the ineffectiveness of the treatment. On a local scale, however, innovative environmental solutions are quietly being identified on a daily basis. Successes include dead lakes and rivers (e.g., the Potomac) that have fish inhabiting them again; overgrazed and eroded meadows that have been restored at minimal cost; companies that are operating at lower cost by identifying ways to be more efficient in the use of water, fuel, or other resources while also reducing pollution; and agencies (actually people within agencies) that have found ways to work cooperatively despite conflicting regulations and mandates.

Doing Away with Business as Usual

Some companies find solutions or ways to meet their environmental mandates, while others claim that the effort is too expensive or technically infeasible. Some agencies say "it can't be done" while others do it. Cost is frequently raised as a reason to not change an existing procedure. Is cost a prohibitive factor, or just an excuse? If a company or industry were providing a valuable public service or fulfilling a public need and found a technological solution to its environmental problem, but the solution was too expensive, would it be appropriate for taxpayers to subsidize part of the cost for the benefit of the common good? If agencies are not empowered to make decisions on these sorts of questions, we need to find and remove the roadblocks. If Congress is going to make sweeping changes, the sweeping changes should not be in reducing the effectiveness of environmental laws themselves, but in the way government agencies are allowed to operate. Perhaps the Office of Management and Budget (OMB), in consultation with those who have been involved in various environmental programs, should be given a mandate to find and remove impediments for environmental solutions in agency procedures.

From a private sector and public (government) partnership perspective, if information could be exchanged between those who have found a solution and those who have not, one obvious solution is to use or build upon the solutions of others. The question is how those who need solutions can find those who have dealt with a similar problem and found a solution. A private/public information exchange service could be created to not just provide a database, but consult and advise clients. Although consultants and agencies currently provide services that are solution-oriented, they do so independently, they are limited to specific segments of the environmental field, and they are not as all-encompassing as would be needed for, as an example, a national or global environmental solutions network. Such a service would require access to private sector and agency-derived environmental solutions, for everything from a technical pollution problem to long-range land use or natural resource management solutions covering all geographic areas and environments.

CHAPTER 4

PROPOSED ACTIONS

. . . it is proposed that directorate level OPR review and deletion or recertification of essentiality be accomplished on all Air Force publications.

From a directive titled
*Air Force Program for
Making Departmental Publications
Understandable by Users*

INTRODUCTION

A proposed action is the action under evaluation in an environmental impact statement (EIS) process. A proposed action is also called a *proposal* or a *proposed project*. If an agency is considering whether to approve, fund, or permit a private action, the agency should clarify which is the agency action and which is the proponent's proposed action in the narrative of an EIS. For example, a private company may have a proposed action to build houses. An *agency action* would be to decide whether to issue permits to allow the *proposed action* (the building of houses) to take place. For this example, an EIS process provides the agency decisionmaker with the information necessary to take the following basic types of actions:

107

- Permit or approve the proposed action as presented by the proponent
- Permit or approve the proposed action with conditions or changes to the proposed action
- Deny the permit or other approval

Some agency EIS procedures allow an agency to deny approval of a proposal if the impacts of the action are significant and cannot be adequately mitigated. Most agency regulations, however, allow agencies to choose to approve or proceed with a proposed action even if environmental impacts will remain significant after implementing mitigation measures. The agency decisionmaker is required to explain his or her reasons (e.g., economic and technical considerations, and the agency's statutory mission) for choosing to proceed with an environmentally damaging project or for not choosing the least environmentally damaging alternative.

Actions and Environmental Effects

In an EIS process, an action has environmental effects if it affects the physical environment. In reviewing federal EISs or an agency's decision to not prepare an EIS, courts have concluded that if the effect of an action does not affect the environment, the effect is not an environmental impact and does not need to be addressed in an EIS or require an EIS to be prepared, even if the effect is significant. As an example, if a socioeconomic effect of a proposed action is significant but the effect is not interrelated with physical impacts on the environment, an EIS would not be required to address the socioeconomic effect, and an EIS would not be required for a project that has only significant socioeconomic effects.[61]

AGENCY ACTIONS AND PRIVATE ACTIONS

As discussed in previous chapters, proposed actions may be those of an agency, organization, or individual. A decision on whether to proceed with an agency's proposed action is made within the agency's hierarchy or higher, such as at the departmental level, depending on the

---

[61]40 CFR §1508.14.

type of proposed action and the amount of decisionmaking authority granted to an agency.

Private actions on land owned or managed by an agency—or private actions on private land that must receive approval, permits, or funds from an agency in order to proceed—require an agency to decide whether to grant the requested agency action (the approval, permit, or funds). If an agency's action enables a private party to affect the environment, the action is subject to the National Environmental Policy Act (NEPA) or a state environmental policy act (SEPA).

A local or state agency action that requires approval, permits, or funds from a federal agency may be subject to federal EIS requirements (NEPA) if the federal part (federal action) in the overall action is more than de minimus.[62] Some federal agencies will conduct NEPA review of only the part of an action (whether private or another agency's) that is subject to the federal action. For example, if a proposed action requires a permit to fill wetlands from the U.S. Army Corps of Engineers (COE), the COE might prepare an environmental assessment for its part of the action (the wetland permit), rather than evaluating the entire action, if the environmental impacts of issuing the wetland permit were not significant. Chapter 8 includes a discussion of the COE's process for determining the scope of analysis of permit actions.

Piecemealing Actions

Breaking a proposed action into smaller actions, which individually do not result in significant impacts, is called "piecemealing." Project proponents will sometimes attempt to avoid a requirement to prepare an EIS by dividing a large project, with potentially significant environmental impacts, into smaller projects. Using a proposed housing project as an example again, an agency might have as one of its categorical exclusions (actions that don't require EISs) housing projects of 30 units or less. Therefore, a project proponent might submit a proposed action of 30 units of housing. The following year he might propose another 30 units, and another 30 the year after. The environmental review should be for the total development of 90 units, or the

---

[62]Legalese for being so minor as to be inconsequential.

development at build-out even though it might actually take 3 years to complete the development. Otherwise, an agency would be approving parts of a project without knowing the potential environmental impacts of the entire project after it is completed. The smaller projects are related actions, and they are part of an overall project that would take place in phases.

RELATED ACTIONS

Courts have used a number of tests to determine whether actions were related and therefore needed to be addressed in one EIS. An example is a U.S. Forest Service EIS that was the subject of a lawsuit. The court ruled that the Forest Service's proposed action to build a road was related to the subsequent action of harvesting timber in the area to which the road was proposed to be built. The purpose for building the road was clearly to allow access to remove the timber. Therefore, the two actions, of building a road and harvesting timber, were related and should have been evaluated as parts of one proposed action in the EIS. The court's evaluation was based on the "irrational or unwise test." The purpose of the test is to determine if a proposed action is so dependent on subsequent phases "that it would be irrational, or at least unwise, to undertake the first phase if subsequent phases were not also undertaken." The court considered it irrational for the Forest Service to construct a road to access timber and then not sell the timber accessed by the road.

Actions are not considered to be related if they are not interdependent. In other words, if an action could proceed on its own, and does not require another action to take place in order for either action to function, the actions are not related and are not considered parts of the same proposed action. This is called the "independent utility test." The focus of the test is whether one project will serve a significant purpose even if a second related project is not built.

In another example, the court ordered the Atomic Energy Commission to prepare an EIS for its liquid metal fast breeder reactor research and development program because the long-term commitment of resources to the program had the effect of foreclosing later alternative energy options. This test is the "irretrievable commitment test," and it focuses on whether completion of one action by the agency inevitably involves an "irreversible and irretrievable commitment of resources" to a following action or actions. If an agency irretrievably commits federal funds for closely related actions, the actions are related and must be considered together in one EIS.

MAJOR FEDERAL ACTIONS

The CEQ regulations define a "major federal action" as

. . . actions with effects that may be major and which are potentially subject to Federal control and responsibility. Major reinforces but does not have a meaning independent of significantly. . . . Actions include the circumstance where the responsible officials fail to act and that failure to act is reviewable by courts or administrative tribunals. . . .

(a) Actions include new and continuing activities, including projects and programs entirely or partly financed, assisted, conducted, regulated or approved by federal agencies; new or revised agency rules, regulations, plans, policies, or procedures; and legislative proposals. . . . Actions do not include funding assistance solely in the form of general revenue sharing funds, distributed under the State and Local Fiscal Assistance Act of 1972 . . . , with no Federal agency control over the subsequent use of such funds. Actions do not include bringing judicial or administrative civil or criminal enforcement actions.

(b) Federal actions tend to fall within one of the following categories:

(1) Adoption of official policy, such as rules, regulations, and interpretations adopted pursuant to the Administrative Procedure Act . . . ; treaties and international conventions or agreements; formal documents establishing an agency's policies which will result in or substantially alter agency programs.

(2) Adoption of formal plans, such as official documents prepared or approved by federal agencies which guide or prescribe alternative uses of federal resources, upon which future agency actions will be based.

(3) Adoption of programs, such as a group of concerted actions to implement a specific policy or plan; systematic and connected agency decisions allocating agency resources to implement a specific statutory program or executive directive.

(4) Approval of specific projects, such as construction or management activities located in a defined geographic area. Projects include actions

approved by permit or other regulatory decision as well as federal and federally assisted activities.[63]

CONTINUING ACTIONS

As stated in paragraph (a) above, actions include continuing actions as well as new actions. Most actions that are addressed in an EIS process are new actions, proposed actions with impacts that would be added to the existing environment. When evaluating continuing actions, unlike a new action—where one evaluates the change in the environment that would result from a proposed action—there is no change unless the continuing action is halted. Therefore, some mental gymnastics are necessary to reverse the approach from one normally used for a new action to one for a continuing action. If the action being evaluated in an EIS is to continue an existing action, an alternative to that action would be to stop the existing action. For the alternative of stopping the continuing action, evaluate what the environmental impacts would be without the effects of the continuing action. One would not evaluate what the *potential* environmental impacts of an action might be (they already exist) or whether to approve a proposed action. In fact, there isn't a *proposed action*; the activity under consideration is an existing activity. The decision to be made by an agency, or the *agency action*, is whether to allow the *existing action* to continue.

When a continuing action is the subject of an EIS process, the status quo or existing environment is the environment with the existing action in it. In other words, the no-action alternative is to allow the action to continue, not to stop the action. If an agency is proposing to modify a continuing action, that modification is a new or proposed action to the degree that it is different from the action which has been taking place. The status quo or no-action alternative would be for the existing action to continue without any modification.

As subjects of EISs, continuing actions are not as common as new actions. One reason is that the forward-looking EIS process is more conducive to evaluating new or proposed actions. The regulations refer almost exclusively to "proposals" or "proposed" actions, and some adjustments in an EIS's analysis of impacts must be made for continu-

---

[63]40 CFR §1508.18.

ing actions. Continuing actions are more or less forced to fit into the process. It is not clear when a continuing action would require an EIS if the continuing action is not being modified. Some legal experts state that NEPA does not apply if no significant changes are made to an existing project.[64] Also, since EISs are required to evaluate existing actions and impacts when evaluating a new action (in establishing the environmental baseline or cumulative impacts, or both), any continuing actions could thus be included in the evaluation of a new action. Lastly, if a continuing action of an agency were to become an environmental concern (e.g., previously unknown environmental impacts were discovered), an EIS for that existing action might be necessary. In evaluating continuing actions, instead of looking forward, an EIS appears to be retroactive.

NO ACTION

When evaluating alternative actions in an EIS, the no-action alternative must be one of the alternatives unless maintaining the status quo is prohibited, for example, by Congress.[65] The Council on Environmental Quality, however, states in its *Forty Most Asked Questions Concerning CEQ's National Environmental Policy Act Regulations* that the no-action alternative should always be included. On the other hand, some courts reason that the effect of the no-action alternative would be obvious: The proposed project would not go forward, its purposes would not be met, and the environmental impacts that would have resulted from the proposed action would not occur. The courts' reasoning overlooks the no-action evaluation's usefulness for disclosing environmental impacts from other actions that would result if the no-action alternative were selected. For example, road construction and increased traffic might result if a proposed railroad were not built.

The no-action alternative represents the existing environment, including existing environmental impacts. If the action under evaluation in an EIS is a new action, the no-action alternative means that

---

[64]League of Women Voters of Tulsa, Inc. v. United States Corps of Engineers (1984); Center for Nuclear Responsibility, Inc. v. United States Nuclear Regulatory Commission (1984) (Mandelker, 1995).

[65]Kilroy v. Ruckelshaus (1984).

the new action would not take place. If the action under evaluation in an EIS is a continuing action, the no-action alternative means that the existing action would continue.

> **If the action under evaluation in an EIS is a continuing action, the no-action alternative means that the existing action would continue.**

ENVIRONMENTAL REVIEW OF FEDERAL ACTIONS

Federal agencies have internal procedures for compliance with NEPA. The types of routine actions for which an agency has responsibility may be such that few actions have the potential to affect the environment, while other agencies deal with actions that affect the environment on a regular basis. Therefore the details of agency environmental procedures will vary. Some agencies have manuals or handbooks which provide agency policies, responsibilities, and extensive procedures for the agencies' environmental review of their actions. Many of the agencies whose actions are regularly subject to environmental review have responsibilities for managing the Nation's public lands, lands withdrawn from public use, and natural resources, as well as regulating activities of private parties. Public lands are managed by agencies. Public lands may be available for private uses including recreation (skiing, hunting, camping, fishing), cattle grazing, mining, and timber harvesting. Generally, commercial land uses require permits or leases, or other instruments that convey rights for private use of public land or resources. Agencies who manage these lands are subject to NEPA or a SEPA when evaluating agency actions such as granting a permit, or when proposing an agency-sponsored project or management activity such as a land use management plan or construction of a government facility.

Lands withdrawn from public use (the public is excluded from private use of the land) include some military and American Indian reservations. Federal agencies that manage withdrawn lands must also comply with NEPA for agency actions or projects required during the course of carrying out the agency's mandates and managing the natural resources of the land and water.

Agencies' procedures implementing NEPA must be consistent with the CEQ regulations. The procedures include listing agency actions normally requiring categorical exclusions, environmental assessments (EAs) and EISs. An agency would consult these lists to determine the appropriate environmental documentation, from none to an EIS, that might be required for a proposed action.

The Range of Actions Requiring Environmental Review

Because agencies were created to fulfill certain responsibilities or meet certain needs, the agencies' actions have been carried out on numerous occasions over the course of the agencies' history. Therefore, agencies should know (1) the types of actions for which they are responsible and (2) the level of environmental impact usually expected to occur from those actions. Based on this knowledge, federal agencies (and some state agencies with SEPAs) created lists of actions that fall into categories of (1) not requiring environmental documentation, (2) requiring environmental assessments to determine whether an EIS is required, and (3) requiring EISs.

Categorically Excluded Actions   Categorically excluded actions are those that normally do not individually or cumulatively cause significant environmental impacts and therefore do not require further environmental review. Some agencies have internal procedures to document the conclusion (e.g., in the form of a memo to the file) that an action is excluded from further environmental review. An action that is listed as normally being excluded from environmental review will not be excluded if, for a specific situation, there is potential to violate another environmental regulation. For example, if there is the possibility that an endangered species or its habitat could be adversely affected, the categorically excluded action is no longer categorically excluded. The following are examples of the U.S. Navy's actions that are normally categorically excluded and require EAs and EISs. Other federal agencies have similar categorically excluded actions, but not all federal agencies list the actions that normally require EAs and EISs. State agency examples are provided later in this section.

Examples of categorically excluded Navy actions include

- Routine personnel, fiscal, and administrative actions (e.g., recruiting, processing, and recordkeeping).
- Studies, data, and information-gathering that involve no physical changes to the environment (e.g., topographic surveys, bird counts, wetland mapping, forest inventories, timber cruising).
- Routine repair and maintenance of facilities and equipment, including maintenance of improved grounds such as landscaping and lawn care and minor erosion control measures.
- Decisions to close facilities (not base closures) and decommission equipment (where such equipment is not used to prevent/control environmental impacts) or temporarily discontinue use of facilities.
- Routine movement, handling, and distribution of materials, including hazardous materials/wastes that when moved, handled, or distributed are under applicable regulations.
- Demolition, disposal, or improvements involving buildings or structures neither on nor eligible for listing in the National Register of Historic Places and when under applicable regulations (e.g., removal of asbestos, PCBs, and other hazardous materials).
- Renewals and/or initial real estate ingrants and outgrants involving existing facilities and land wherein use does not change significantly. That includes, but is not limited to, existing or federally owned or privately owned housing, office, storage, warehouse, laboratory, and other special-purpose space.
- Grants of license, easement, or similar arrangements for the use of existing rights-of-way or incidental easements complementing the use of existing rights-of-way for use by vehicles (excluding significant increase in vehicles); electrical, telephone, and other transmission and communication lines; water, waste water, storm water, and irrigation pipelines, pumping stations, and facilities; and for similar utility and transportation uses.
- Transfer of real property from the military to another military department or to another federal agency, and the granting of leases (including leases granted pursuant to the agricultural outleasing program where soil conservation plans are incorporated), permits, and easements where there is no substantial change in land use or where subsequent land use would otherwise be categorically excluded.

- Renewals and minor amendments of existing real estate grants for use of government-owned real property where no significant change in land use is anticipated.
- Pre-lease exploration activities for oil, gas, or geothermal reserves (e.g., geophysical surveys).
- Return of public domain lands to the Department of the Interior.
- Land withdrawal continuances or extensions which merely establish time periods and where there is no significant change in land use.
- Temporary closure of public access to military property to protect human or animal life.
- Actions which require the concurrence or approval of another federal agency where the action is a categorical exclusion of the other federal agency.
- Maintenance dredging and debris disposal where no new depths are required, applicable permits are secured, and disposal will be at an approved disposal site.
- Approval of recreational activities that do not involve significant physical alteration of the environment or increase human disturbance in sensitive natural habitats and that do not occur in or adjacent to areas inhabited by endangered or threatened species.
- Routine maintenance of timber stands, including issuance of downwood firewood permits, hazardous tree removal, and sanitation salvage.
- Reintroduction of endemic or native species (other than endangered or threatened species) into their historic habitat when no substantial site preparation is involved.

State agencies such as California and Washington have very detailed and lengthy lists of categorically exempted actions in their CEQA and SEPA guidelines. Examples of categorically exempted actions include

- Construction of single family dwellings
- Maintenance and repair of existing roads
- Acquisition of property through eminent domain
- Environmentally protective orders adopted by a public agency
- Ongoing projects if there is no material change in project operations

State guidelines also list agencies whose actions are exempt from CEQA or SEPA review, actions that are ministerial[66] in nature and therefore do not require environmental review, actions that are statutorily exempt, and actions determined to be exempt by the courts from environmental review.

Actions Which Normally Require EAs   Actions which are not categorically excluded, and may not require EISs, are environmentally reviewed to determine whether significant environmental impacts would result from the action. The analysis is called an *environmental assessment* (EA). The amount of analyses varies depending on the agency and the type of proposed action. It may be brief—the CEQ regulations recommend a maximum of 15 pages—but is often 30 pages or more. Agencies prepare more EAs than EISs. At the conclusion of an EA, an agency decisionmaker will issue either a finding of no significant impact (FONSI), if the EA determines that an EIS is not necessary, or a notice of intent (NOI) to prepare an EIS if the EA determines that environmental impacts from the proposed action would be significant. Examples of Navy actions that normally require EAs include

■ Training exercises on or over (airspace) nonmilitary property
■ Dredging projects that increase water depth over previously dredged or natural depths
■ Proposed use of tidal and nontidal wetlands that would require a permit
■ Real estate acquisitions or outleases of land involving
   ■ New ingrants/outgrants only—that is, not renewals or continuances wherein land usage remains the same
   ■ Fifty acres or more where existing land use will change and will not be categorically excluded
   ■ Renewals of agricultural/grazing leases when changes in animal stocking rates, season of use, or conversions to or from crop land are involved

---

[66]A government decision involving little or no personal judgment by the public official as to the wisdom or manner of carrying out a project. The public official merely applies the law to the facts, and the decision involves only the use of fixed standards or objective measurements.

- Family housing projects when resident population changes substantially
- New target ranges or range mission changes which would increase environmental impacts
- Exercises conducted at the request of state or territorial governments (e.g., ship sinking for artificial reefs) wherein environmental impact might be expected
- Irreversible conversion of "prime or unique farmland" to other uses

Actions Which Normally Require EISs Some actions are of such a magnitude, or the activities associated with the actions are sufficiently disruptive of the physical environment, that the action clearly would result in significant environmental impacts. When there is no question that certain actions would have significant environmental impacts, those actions would require EISs. Examples of Navy actions that normally require EISs include

- Large dredging projects or dredging projects where dredged material disposal may result in significant impacts
- Proposed major construction and filling in tidelands/wetlands
- Establishment of major new installations
- Major land acquisitions that will result in changed use of the property
- New sanitary landfills[67]

Most agencies published their NEPA procedures (e.g., in the *Federal Register*), allowing the public and the CEQ to review and comment on the procedures and lists of categorically excluded actions prior to their final adoption by the agencies.

Limitations on Actions

Until a record of decision has been issued, the CEQ regulations prohibit federal agencies from taking actions concerning a proposed action which would (1) have an adverse environmental impact or (2) limit the choice of reasonable alternatives.[68] While a program (or pro-

---

[67]U.S. Navy, Chief of Naval Operations, "Procedures for Implementing the National Environmental Policy Act (NEPA)," *Environmental and Natural Resources Program Manual (OPNAVINST 5090.1B)*, 1994.
   [68]40 CFR §1506.1(a).

grammatic) EIS is in progress, and if the action is not covered by an existing program statement, agencies may not take a major federal action that may result in significant environmental impacts unless the action

- Is justified independently of the program
- Is itself accompanied by an adequate environmental impact statement
- Will not prejudice the ultimate decision on the program[69]

An action prejudices the ultimate decision when it tends to determine subsequent development or limit alternatives. For example, if an agency were to purchase property expressly for a proposed project, that action may limit or influence the agency's choice of sites for the project. However, if the property was not purchased expressly for the proposed project or program, and the land could be used for other purposes, it would not necessarily prejudice the agency's choice of alternatives.

The limitation on agency action does not mean that agencies or applicants could not continue the project planning process. Applicants or project proponents are allowed to develop plans or designs or perform other work necessary to support their applications.[70]

Effectiveness of an Agency's EIS Process

Despite agency manuals that have detailed EIS procedures and describe lists of actions that require various levels of environmental documentation, regulations are only effective if the individuals who implement regulations are knowledgeable of their duties. Duties to comply with environmental regulations are sometimes relegated to someone as additional duties—for example, to a civil engineer in the public works department of an organization.[71] If environmental duties

---

[69]40 CFR §1506.1(c).

[70]40 CFR §1506.1(d).

[71]It is not clear why environmental duties were most often given to engineers in government agencies. Perhaps, at least historically, agencies that had to construct facilities (an activity that requires environmental review) had engineers, so it was a matter of giving the duties to those who were available. And perhaps, for pollution prevention or hazardous waste issues, engineers are the appropriate technical people. But for environmental planning, the required technical training is more appropriately in the sciences and planning disciplines. Of course, ultimately, it isn't the job title, but the individuals' background and training, that matters.

are added to a person's job and that person does not know what the purpose or intent of the extra duty might be, he or she will not know how to carry out those responsibilities. For example, when filling out environmental checklists with questions regarding the presence of endangered species on a site or whether any historic properties would be affected, most people would not recognize endangered species or historic properties, or know how to identify them. An airplane hangar could be eligible for listing as a historic property, a weed could be an endangered species, and wetlands don't always have cattails or look wet. The technical subjects and laws and regulations are too specialized to be treated as additional duties. Ensuring that tasks are given to qualified people has been the responsibility of agency managers. But in a climate of budget cuts and increasing regulatory requirements, civil servants have been given the duties of several people. "Doing more with less" and having a "lean mean government" has reduced staff in some agencies but not responsibilities. As discussed in Chapter 3, agencies are attempting to streamline environmental and planning processes, in part, because of severe reductions in staff and budgets.

CHAPTER 5

OVERVIEW OF FEDERAL, STATE, AND OTHER NATIONS' EISs

No nation has a monopoly on good things.
Each one has something that the others could well afford to adopt.
Will Rogers

INTRODUCTION

This chapter describes some basic similarities and differences between state and federal environmental impact statements (EISs), and it discusses the worldwide interest in environmental impact assessment. Understanding similarities and differences should be helpful in reviewing EISs if you are familiar with state EISs, but not federal, or vice versa. Also, if a joint federal and state EIS is to be prepared, the differences are worth noting so that adjustments in the process can be made to accommodate both state and federal requirements. Some local agencies, such as cities and counties, have implementing regulations for state environmental policy acts (SEPAs) which may have requirements in addition to the SEPA, just as federal agencies have National Environmental Policy Act (NEPA) implementing regulations that have requirements in addition to the Council on Environmental Quality (CEQ) regulations. There are many hundreds of NEPA and SEPA agency implementing regulations. Therefore, comparisons between regulations can be made here for only a few of them. When preparing EISs, you will need to obtain the NEPA or SEPA implementing regulations of the particular agency or agencies with jurisdiction over a specific proposed action.

Other countries have adopted similar EIS requirements (most often called Environmental Impact Assessments) patterned after NEPA, the United States' federal law. These countries have environmental concerns similar to those of the United States, and there is growing recognition that the actions of countries around the world are having global environmental impacts. The actions of one country, which result in environmental impacts, are being felt by other countries in practical terms including economic impacts and impacts to public health from pollution or deterioration of the ozone layer. International environmental concerns include diverse but interrelated issues such as climate change, stratospheric ozone depletion, food security, health of the oceans, water supply, population change, conservation of biological diversity and natural resources, and pollution prevention and control. The environmental impact assessment requirements of other nations will not be discussed in detail in this chapter, but they are presented to provide a worldwide context for environmental concerns and the evaluation of environmental impacts.

FEDERAL AND STATE EISs

Because NEPA was the first law passed in the Nation that requires federal agencies to include consideration of environmental impacts in their decisionmaking process, and the CEQ regulations provide detailed guidance on the implementation of NEPA, the states, when passing their SEPAs, patterned their laws substantially after NEPA and the CEQ regulations. State laws do have their differences, such as the California Environmental Quality Act (CEQA), which has more detailed regulations than the CEQ regulations, or the Michigan Environmental Protection Act (MEPA), which emphasizes substantive rather than procedural requirements.

Sixteen states, the District of Columbia, and Puerto Rico (a U.S. territory) have environmental policy acts. An additional 11 states have limited environmental review requirements established by statute, executive order, or other administrative directives, for a total of 29 states and other governmental entities that require some form of environmental review for proposed actions. The states that have EIS and other similar environmental review requirements are shown in Figure 5.1. These states have "little NEPAs" or SEPAs that have general application to proposed actions or projects. Table 5.1 lists citations of state SEPA statutes. California, Minnesota, Washington, and Wisconsin have comprehensive environmental review requirements. Some have procedures that are roughly equivalent to NEPA, while others require environmental analysis only for certain limited, specific activities (e.g., power plant siting) or for state-initiated actions but not for actions that require a state permit or license. Table 5.2 lists the states that have limited environmental review requirements.

NEPA and SEPAs have similarities in the roles of the participants, the required steps in the processes, and the types of documents and legal notices. The timing of actions within the EIS processes are also similar, although the federal process generally allows more time for public review and takes more time to complete. Contrary to the belief of some EIS preparers, the level of detail required in EISs is not necessarily greater for federal EISs than for state EISs. The level of detail or amount of information is a function of the requirements of the lead agency and the range of issues, alternatives, and the study area that are determined during the scoping process.

Similarities of Roles

State and federal EIS processes have a project proponent, lead agency, and the public including other federal, state, and local agencies, affected individuals and American Indian tribes, environmental groups, and citizen groups. In both EIS processes:

■ A project proponent can be an agency or private party,[72] and the proponent's role is to provide all project-related information.

■ A lead agency is responsible for the EIS process and adequacy of an EIS or other document prepared pursuant to NEPA or a SEPA.

■ The public is an integral part of the process by identifying issues of significance and identifying alternatives (helping to define the scope of the EIS) and providing comments on the Draft EIS.

---

[72]Not all SEPAs apply to private actions.

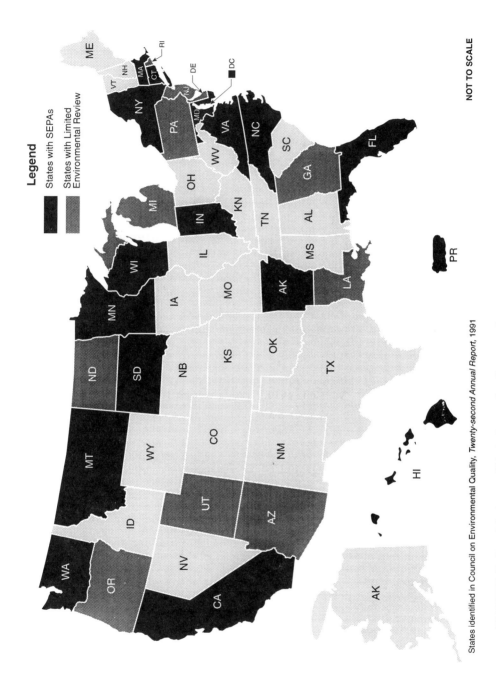

Legend

States with SEPAs

States with Limited
Environmental Review

NOT TO SCALE

States identified in Council on Environmental Quality, *Twenty-second Annual Report*, 1991

■ **FIGURE 5.1.** States with EIS or similar environmental review requirements.

126

■ **TABLE 5.1** Citations of States[a] with Environmental Policy Acts
(SEPAs)

State	Citation
Arkansas	Ark. Stat. Ann. §8-1-101 (1987)
California	Cal. Pub. Res. Code §§21000 et seq. (West 1982)
Connecticut	Conn. Gen. Stat. Ann. §§22a-14 to 22a-20 (West Supp. 1974-75)
District of Columbia	D.C. Code Ann. 1981 §6-981 et seq.
Florida	Fla. Stat. §§380.92 et seq.
Hawaii	Hawaii Rev. Stat. §§343-1 to 343-8 (1985)
Indiana	Ind. Code Ann. §§13-1-10-1 to 13-1-10-8 (West 1987)
Maryland	Md. Nat. Res. Code Ann. §§1-301 to 1-305 (1983 and Supp. 1987)
Massachusetts	Mass. Gen. Laws Ann. ch 30, §§61-62H
Minnesota	Minn. Stat. Ann. §§116D.01 et seq. (West 1977 and Supp. 1981)
Montana	Mont. Code Ann. §§75-1-101 to 105; §75-1-201 (1981)
New York	N.Y. Envtl. Conserv. Law §§8-0101 to 8-0117 (McKinney 1984)
North Carolina	N.C. Gen. Stat. §§113A-1 to 10 (1978)
Puerto Rico	P.R. Laws Ann. tit. 12, §§1121-1127
South Dakota	S.D. Codified Laws Ann. §§34A-9-1 to 34A-9-12
Virginia	Va. Code §§10.1-1200 through 10.1-1212
Washington	Wash. Rev. Code §§43.21C.010-43.21C.910 (1974); Wash. Admin. Code R. 197-11
Wisconsin	Wis. Stat. §1.11 et seq.; Department of Natural Resources WEPA rules are found in Wis. Admin. Code NR 150.01-40

[a]Sixteen states, the District of Columbia, and Puerto Rico have environmental policy acts or "little NEPAs."
Source: Council on Environmental Quality, *Twenty-second Annual Report*, 1991.

Similarities of Processes

Figure 5.2 illustrates the similarities between federal and California and Washington EIS processes. The processes follow the same steps as the flow diagram in Figure 1.1. The points at which decisions are made, the legal notices, documents, and filing requirements have different names in some cases but are essentially the same or serve the same purposes. States that have a SEPA call them various names

■ **TABLE 5.2** States with Limited Environmental Review Requirements[a]

State	Requirement
Arizona	An Executive Order mandates that the Governor's Commission on Arizonia Environment evaluate environmental problems, make recommendations to the Governor, and establish a clearinghouse for the exchange of information relating to environmental problems and their solutions.
Arkansas	In addition to a "little NEPA," Ark. Code Ann. §15-41-108 provides that the Arkansas Game and Fish Commission must prepare an EIS for cutting timber on Commission land.
California	In addition to a "little NEPA," the following California Codes require environmental impact reports: • Cal. Food & Agric. Code §33487 (new construction or repairs of dairy farms) • Cal. Gov. Code §7075 (establishment of enterprise zones) • Cal. Gov. Code §7087.5 (initial study and notice of preparation under Employment and Economic Incentive Act) • Cal. Gov. Code §51119 (timberland production zones; exemption) • Cal. Gov. Code §65950.1 (extension of time for EIR for planning and zoning of development projects) • Cal. Health & Safety Code §33333.5 (preparation and adoption of community redevelopment plans) • Cal. Health & Safety Code §56040 (implementation and administration for large scale urban development) • Cal. Pub. Res. Code §6873.2 (oil and gas leases on tide and submerged lands and beds of navigable rivers and lakes) • Cal. Pub. Res. Code §25540.4 (power facility and site certification) • Cal. Pub. Res. Code §30718 (implementation of port development under California Coastal Act) • Cal. Str. & H. Code §199.9 (mass transit guideway system) • Cal. Water Code §13389 (applicability of EISs to Clean Water Act)
Delaware	In the Del. Code Ann. tit. 7, Chapter 66 concerns wetlands permits, and Chapter 20, coastal zone permits.
District of Columbia	In addition to a "little NEPA," D.C. Code Ann. §43-1903 involves public utility environmental impact statements.
Georgia	The Code of Georgia provides that on certain types of actions on a case by case basis, the state may require that an environmental assessment be prepared; EAs would be reviewed by the state Department of Natural Resources, Environmental Protection Division.
Louisiana	La. Rev. Stat. Ann. §30.2021 (West 1991) covers interstate compacts on environmental control, for which the Louisiana Department

■ TABLE 5.2 (Continued)

State	Requirement
	of Environmental Quality serves as a clearinghouse for all statements of environmental impact to be prepared or reviewed by state agencies (other than Department of Transportation and Development), in accordance with NEPA. The Department of Wildlife and Fisheries is responsible for review and comment on any EIS regarding fish and wildlife resources or their habitat, as well as the discharge of dredge and fill material into state waters. The Department of Health and Human Resources is responsible for EISs regarding public health.
Massachusetts	In addition to a "little NEPA," Mass. Gen. Laws Ann. ch. 111H, §30 concern low-level radioactive waste facility licensing.
Michigan	Executive Order 1974-4 requires each state agency to prepare a formal environmental assessment for all major activities of the agency having a possible significant impact on the environment or human life. Mich. Comp. Laws Ann. §281.655 et seq. covers EISs for sand dune mining and model zoning plans under the Lakes and Rivers Sand Dune Protection and Management Act.
New Jersey	Executive Order No. 53 (1973) requires all state agencies and departments to submit to the Department of Environmental Protection a description of the environmental impact of all major construction projects. • N.J. Rev. Stat. §13:19-7 provides for EISs for such projects as coastal protection, N.J. Rev. Stat. §27:23-23.5 covers EISs for the New Jersey Turnpike authority (highways and turnpikes). • N.J. Rev. Stat. 52:13F-4 refers to environmental impact statements on specific legislative bills.
North Carolina	In addition to a "little NEPA," N.C. Gen. Stat. §104G-11 concerns low-level radioactive waste management technology licensing.
North Dakota	N.D. Century Code §§54-01-05.4 provides that the Governor may require EISs of a limited nature prior to the transfer of any interest in state-owned land to federal agencies.
Oregon	Although Oregon Rev. Stat. ch 46 does not require EISs for major actions having a significant impact on the environment, permit applications for siting of major energy generating facilities must include a background report addressing various anticipated environmental impacts.
Pennsylvania	Executive Order requires Environmental Assessments for all transportation projects. Various state regulations require EAs for other state actions.
Rhode Island	R.I. Gen. Laws §23-63-3 provides that a facility which plans to export tires for burning must submit an EIS conforming to the EPA standards.
South Dakota	In addition to a "little NEPA," S.D. Codified laws Ann. §49-41B-21 et seq. concern energy conversion and transmission facilities.

■ **TABLE 5.2** (Continued)

State	Requirement
Utah	State of Utah Exec. Order (Aug. 27, 1974)
Washington	In addition to a "little NEPA,"
	• Wash. Rev. Code Ann. §78.52.125 (oil and gas conservation drilling that affects surface waters)
	• Wash. Rev. Code Ann. §70.95.700 (solid waste incineration or energy recovery facility)
Wisconsin	In addition to a "little NEPA,"
	• Wis. Stat. Ann. §144.68 (solid waste, hazardous waste, and refuse)

ªEighteen states and the District of Columbia have limited environmental review requirements established by statute, executive order, or other administrative directives. Some of these states also have SEPAs (see Table 5.1).
Source: Council on Environmental Quality, *Twenty-second Annual Report*, 1991.

such as California Environmental Quality Act (CEQA), Washington State Environmental Policy Act (SEPA), Michigan Environmental Protection Act (MEPA), and New York's State Environmental Quality Review Act (SEQRA).

The federal or state EIS process begins with a proposed action or project, often in the form of an application to a federal, state, or local agency if it is a private proposal. An agency that has a proposed action (a public proposal) will submit the request in accordance with agency procedures and through the various agency departments. An EIS process begins at the time an agency is considering a proposal; or said another way, an EIS process cannot begin until there is a proposal or a proposed action. Therefore, an EIS would not be required for something that an agency *might* do or has the power to do, but has not proposed to do.

Identification of the Lead Agency Generally, for federal and state projects, if a proposed action is an agency's project, that agency is the lead agency for the proposed action. However, if a proposed action requires substantial approval from another agency (for example, a state agency action requires federal agency approval in order to proceed), the federal agency making the decision to approve or not approve the action would be the lead agency. For that scenario, the EIS process would be under NEPA or it would be a joint NEPA and SEPA process.

Federal	California	Washington
Environmental Protection Agency (EPA) (EIS filings) **Council on Environmental Quality (CEQA)** (NEPA implementing regulations and lead agency disputes)	**Office of Planning & Research (OPR)** (Prepares CEQA guidelines, resolves lead agency disputes) **State Resources Agency** (Adopts CEQA Guidelines)	**Department of Ecology (DOE)** (Does everything)
Lead Agency: Federal (some state agencies)	Lead Agency: State or Local	Lead Agency: State or Local
Categorical Exclusions	Categorical Exemptions (Notice of Exemption)	Categorical Exemptions
Environmental Assessment (EA)	Initial Study	Environmental Checklist
Finding of No Significant Impact (FONSI)	Negative Declaration/ Notice of Determination	Determination of Nonsignificance (DNS)
or	or	or
Notice of Intent (NOI)	Notice of Preparation (NOP)	Determination of Significance (DS) and Scoping Notice(DS/SN)
NOI Published in *Federal Register*	NOP Published in State Clearinghouse Newsletter	DS/SN Published in *SEPA Register*
Public Scoping Meeting Optional	Public Meetings Optional	Public Scoping Meeting Optional
Prepare Draft EIS	Prepare Draft EIR	Prepare Draft EIS
Draft EIS Notice of Availability Published in *Federal Register*	Notice of Completion Filed with OPR and Public Notice Distributed	Draft EIS Notice Published in *State Register*
Public Review 45–60 days	Public Review 30–90 days	Public Review 30–45 days
Prepare Final EIS	Prepare Final EIR	Prepare Final EIS
FEIS Notice of Availability Published in *Federal Register*	Optional Review of Final EIR	Final EIS Notice Published in *State Register*
—	Certification of Final EIR	—
30 Day Waiting Period	—	7 Day Waiting Period
Record of Decision (ROD) Published in *Federal Register*	Notice of Determination Filed with OPR (lead state agency), or County Clerk (lead local agency)	Notice of Action Taken (NAT) (optional)
Agency Action May Proceed	Agency Action May Proceed	Agency Action May Proceed
Project Planning Continues	Project Planning Continues	Project Planning Continues
Supplemental EIS*	Subsequent EIR*	Supplemental EIS*

* If project changes or new circumstances

■ **FIGURE 5.2.** Federal, California, and Washington EIS Processes.

Threshold Determinations For federal and state EIS processes, a threshold determination must be made by a lead agency on the appropriate level of environmental review for a proposed action. A threshold determination in a federal or state EIS process is the determination that the environmental impacts of a proposed action will not result in significant environmental impacts and, therefore, will not require the preparation of an EIS; or that the proposed action will result in significant environmental impacts and, therefore, will require the preparation of an EIS.

Scoping and Public Notices Scoping, the process that determines what actions, alternatives, environmental effects, and mitigation measures will be addressed (the scope) in an EIS, is required in federal and state EIS processes.[73] Public scoping is the process by which the public gets involved in determining the scope of an EIS. The amount of scoping varies by agency and by the type of environmental review (e.g., whether for an EA or an EIS). Scoping is accomplished informally, by a lead agency and by consultation with other agencies with jurisdiction by law or expertise; it is also accomplished formally, by publishing notices or sending letters to interested parties (public scoping). A formal public scoping process begins by a lead agency publishing a notice about a proposed action and requesting comments and concerns to help define the significant issues, range of alternatives, and mitigation measures to be addressed in the EIS. Public scoping also helps an agency make its threshold determination if it has not already decided that an EIS is required for a proposed action.

A notice of intent (NOI) to prepare an EIS is published in the *Federal Register* to start formal public scoping in the federal EIS process. States also have a public notification process. California's notice of preparation (NOP) is published in the State Clearinghouse newsletter.[74] Washington's notice, a determination of significance and scoping notice (DS/SN), is published in the Washington Department of Ecology's

---

[73]California's CEQA guidelines refer to "consultation" rather than "scoping," although the processes (to obtain public input and provide public notification) are similar.

[74]The EIR Monitor discussed in the CEQA guidelines is not in use (Terry Rivasplata, Chief of the California State Clearinghouse, Office of Planning and Research, personal communication, October 1995).

SEPA Register. Additional and optional forms of public notification for scoping include mailing letters, publication of notices in local or national newspapers, and posting notices on the proposed project site. If a federal or state lead agency decides that an EIS will not be required, public notices of that decision are also provided. A federal agency's notice to not prepare an EIS is a finding of no significant impact (FONSI), California's is a negative declaration ("Negdec"), and Washington's notice is a determination of nonsignificance (DNS).

Public scoping is encouraged, but many agencies in the federal and state EIS processes do not require it for environmental documents such as environmental assessments or checklists (i.e., anything less than an EIS). In some cases, limited public scoping—such as notification of some of the potentially affected or interested parties and allowing an abbreviated public scoping period—may be conducted.

Public scoping meetings are generally optional in both the federal and state EIS processes. Specific agencies may have different requirements for scoping meetings depending on the nature of the proposed action (e.g., the type of action and its likelihood for public controversy regarding potential environmental effects).

Public Review of EISs Federal and state EIS processes provide for public review and comment on EISs. The minimum amount of time allowed to review a federal Draft EIS is 45 days, while California and Washington provide a minimum of 30 days for review. Federal and Washington processes allow for additional time (a minimum of 15 days) if requested by a reviewer and if the lead agency decides to grant the additional time. CEQA guidelines do not specify an extension period, but agencies may grant extensions of review periods.[75] Responses to public comments are provided in a Final EIS or EIR. Before finalizing their responses, California agencies send draft responses to agencies who commented on a Draft EIR. Other public commentors may or may not receive responses to their comments (other than in the Final EIR) depending on the lead agency's policy.[76]

The federal EIS process requires a 30-day waiting period from the time a Final EIS is published to the time the lead agency can act

---

[75]*idem.*
[76]*idem.*

on its decision and publish a record of decision (ROD). Washington state requires a 7-day waiting period after a Final EIS is published, and publication of a notice of action taken (NAT) is optional. California's CEQA guidelines do not require a formal "waiting period" after issuance of a Final EIR. Federal and state lead agencies generally forward any public comments on a Final EIS to the decisionmaker, along with any responses to comments for the decisionmaker's information and consideration; but the regulations do not require responses to those who comment on a Final EIS. In other words, those who provided comments on the Final EIS may or may not get responses from the lead agency, depending on the agency's policies.

The requirement for public involvement, as well as the level of analyses, is not as rigorous for EAs as it is for EISs presumably because the impacts under evaluation are not significant. Regulations encourage scoping and public involvement, and the courts feel that public involvement is important when making a threshold determination (particularly if the determination is that an EIS will not be prepared).

Similarities of Documents

EISs and their basic contents are similar for federal and state EIS processes; the format, however, may be different. Documents other than EISs probably vary the most in format and content, although the purposes of the documents are similar (see below). Categories of actions are for three basic types of environmental review: those that (1) normally do not require environmental review, (2) normally require environmental review to determine whether an EIS would be required, and (3) normally require EISs. These categories of actions were described in Chapter 4.

Figure 5.3 summarizes the environmental review process (shown as a flow diagram in Figure 1.1) as represented by the possible "tracks" for environmental documentation. The tracks are described for the NEPA process, but the state processes are basically the same.

■ **FIGURE 5.3.** Federal environmental documentation tracks. *Source*: Modified from Joseph A. Wellington, A Primer on Environmental Law for the Naval Services, *Naval Law Review*, 1989.

Track	CLASS		EFFECT			Scoping	DOCUMENTS						
	Minor	Major	Unknown	Insignificant	Significant	Scoping	EA	FONSI	DEIS	FEIS	ROD	SEIS	Initiate
1	■												■
2		■	■			■	■	■					■
3		■	■			■	■		■	■	■		■
4		■		■									■
5		■			■				■	■	■		■
6		■			■				■	■	■	■	■

Notes:

Track 1.–This track describes a minor project. By definition a minor project is: (1) Limited in scope and (2) without significant environmental effects. The sponsoring agency is required only to undertake an in-house environmental analysis to ensure that this is the case and may then proceed to initiation. The in-house analysis need not be made available for comment by anyone outside the organization. Should the in-house analysis predict significant impacts, the project becomes "major," and proceeds down one of the other tracks.

Track 2.—This track represents the situation where it is initially unknown if the environmental impacts of a federal project will significantly affect the environment. Here, the federal agency may solicit public and other agency comment on the project through scoping to assist in identifying all relevant issues by allowing the public and other agencies to review and comment on the project before alternatives are developed. The lead agency will then compile and record the results of this review in an environmental assessment (EA). If the project and its consequences do *not* significantly affect the environment, the agency prepares a finding of no significant impact (FONSI).

Track 3.—Like Track 2, Track 3 depicts a major project whose effects are initially unknown. It differs from Track 2 in that, once the EA is complete, the agency determines that the project *will* have significant impact on the environment. The agency then continues with the preparation of a Draft and Final environmental impact statement (DEIS and FEIS), as well as a record of decision (ROD).

Track 4.—Track 4 describes a major federal project which does *not* have significant environmental impact. An example of a Track 4 project might be a decision by the Federal Highway Administration to change the shapes or colors of all the road signs in the federal insterstate highway system. While such a project would be a major undertaking, it is clear from the outset that there would be no environmental impact and, thus, no reason to prepare any of the documents.

Track 5.—Track 5 represents the situation where a major federal project is determined from the start to involve significant environmental impacts. A Draft and Final EIS and ROD would then be prepared.

Track 6.—Track 6 is similar to Track 5, except that it provides for the use of a supplemental environmental impact statement (SEIS) in the event of new evidence, or significant change in circumstances during the period (1) between preparation of the DEIS and the FEIS, or (2) between preparation of the FEIS and the execution of the ROD, or (3) between issuance of the ROD and the initiation of the project. In these cases, the SEIS ensures that the environmental documentation remains accurate as the process evolves. An SEIS may also be used to correct subsequently identified deficiencies in a previously published FEIS.

Categorical Exclusions and Exemptions NEPA, CEQA, and SEPA processes require the identification of categorically excluded or exempted actions, actions that normally do not individually or cumulatively result in significant environmental impact. Federal and state EIS regulations nullify an action's categorical exclusion if there is potential for a proposed action to affect endangered species, wetlands, or historic properties. Not all agencies have a requirement to document the rationale for determining that an action is categorically excluded or exempted or for documenting that the potential for impacts to endangered species, wetlands, or historic properties were considered.[77] Categorically excluded actions are those which fall in Track 1 of Figure 5.3.

Environmental Assessments, Initial Studies, and Environmental Checklists A document prepared to determine whether a proposed action has the potential for significant environmental impacts (the threshold determination) is an environmental assessment (EA) in the federal EIS process, an initial study in California, and an environmental checklist in Washington. Although the purposes of the EA, the initial study, and the checklist are the same, the content and format vary considerably. The CEQ regulations provide broad guidelines for the content of an EA, stating that an EA shall include brief discussions of the need for the proposal, the alternatives, the environmental impacts, and a listing of agencies and persons consulted. Therefore, the format of one federal agency's EA may vary widely from another's. The initial study and environmental checklist, on the other hand, are standardized forms with lists of questions related to each element of the environment. The questions are designed to identify the potential impacts of a proposed action. A state or local agency reviews the applicant's answers on the form to determine if the responses are sufficient to make a threshold determination. If not, the agency may request the applicant to provide the required additional information.

If a proposed action is a private proposal, the federal and state EIS processes allow project proponents or consultants to prepare an EA, initial study, or environmental checklist. The federal, state, or lo-

---

[77]CEQA provides for public notices of exemption.

cal agency provides independent review and is responsible for the adequacy and accuracy of the documents. If the proposed action is a public proposal, the agency with responsibility for the proposal, or a consultant to the agency, would normally prepare the EA or comparable state document.

Finding of No Significant Impact, Negative Declaration, and Determination of Nonsignificance After an EA, initial study, or checklist is completed, the lead agency makes a threshold determination which determines whether the process will continue on Track 2 or 3 of Figure 5.3. Federal and state EIS processes require the preparation and publication of a public notice on the lead agency's threshold determination. If the threshold determination is that a proposed action will not result in environmental impacts, and an EIS is not required, a federal agency would issue a finding of no significant impact (FONSI), a state or local agency in California would issue a negative declaration, and a Washington state or local agency would issue a determination of nonsignificance (DNS) (Track 2). In both processes, the methods of publication or issuance of a FONSI, "Negdec," or DNS are similar. Publication is in the *Federal Register*, State Clearinghouse Newsletter, or *SEPA Register* and can be in local, regional, or national newspapers or other publications depending on the areas of public interest. Notices can also be mailed to those who are interested or potentially affected by a proposed action.

Environmental Impact Statements and Reports If a federal or state agency makes a threshold determination that an EIS or EIR is required for a proposed action, a public scoping notice is published as described under "Scoping," above. The federal notice is a notice of intent (NOI), California's notice is a notice of preparation (NOP), and Washington state's notice is a determination of significance and scoping notice (DS/SN). EISs are prepared in Tracks 3 and 5 of Figure 5.3.

Most states call the document that addresses significant environmental impacts an EIS, but California calls it an environmental impact report (EIR). The basic content of a federal EIS is similar to that of a state EIS. Federal and Washington EISs are required to discuss the "affected environment" (or existing conditions) while California calls the discussion the "environmental setting." A federal EIS also requires the discussion of "environmental consequences," while California and Washington EISs call it the "environmental impacts."

Federal agencies tend to follow the format of EISs recommended by the CEQ regulations. NEPA and CEQA regulations describe similar formats for EISs and EIRs (although the regulations allow flexibility in formats of the documents). The main difference between the NEPA/CEQA format and Washington's SEPA format is that in NEPA and CEQA documents, the affected environment and environmental consequences are presented in separate sections, whereas SEPA EISs present the impacts to a particular resource in the same section as the discussion of the existing environment for that resource (see Figure 5.4).[78] In terms of logical flow, it is easier for the reviewer to follow the discussion in the SEPA format than in the NEPA/CEQA format. If, for example, in the federal EIS format a reviewer is reading about *impacts* to water quality, he or she must go back to a previous section or chapter to find the discussion on *existing* water quality conditions. In SEPA's EIS format, the reader does not have to search as hard or as far for information related to a particular topic. Because EIS formats are not mandatory, federal agencies have, in some cases, prepared EISs with the more reader-friendly format of presenting information regarding a particular topic in the same section. The following compares the two basic formats:

NEPA	*SEPA*
Affected environment	Animals
Animals	Affected environment
Water	Environmental impacts
Traffic	Mitigation measures
Cultural resources	Unavoidable adverse impacts
Environmental impacts	Water
Animals	Affected environment
Water	Environmental impacts
Traffic	Mitigation measures
Cultural resources	Unavoidable adverse impacts
and so forth	and so forth

---

[78]CEQA also prefers to have the discussion of growth-inducing impacts and mitigation measures in separate sections.

Federal (NEPA)

California (CEQA)

* Note: Only required for agency plans, policies, or ordinances; or resolution of local agency formation commission.

Washington (SEPA)

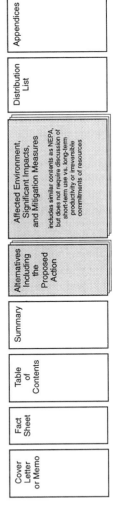

* Note: The most important sections of an EIS or EIR are shaded for comparison. Note that SEPA combines the discussion of the affected environment, proposal impacts, and mitigation measures in one section.

■ **FIGURE 5.4.** EIS/EIR contents (regulations allow flexibility in EIS and EIR formats).

Federal and state EIS processes include the preparation of a Draft EIS and a Final EIS as two distinct documents. In NEPA and Washington's SEPA processes, Final EISs are comprised of responses to comments on the Draft EIS. If responses to comments are such that any changes are minor, the Final EIS may consist of an errata sheet, the public comments, and the agency's responses to comments. The Draft EIS need not be reissued as a Final EIS. In this case an EIS would consist of the Draft and Final EIS—both documents would be necessary to have all of the information that resulted from the EIS process. In preparing Final EISs, however, federal, state, and local agencies do revise and reissue the contents of a Draft EIS even if changes are minor, particularly if the agency is not paying for the EIS. In California, CEQA guidelines require the Draft EIR or a revised Draft EIR to be included in the Final EIR, although a Draft EIR does not have to be distributed to those who have already received it.

In Washington, SEPA regulations do not require alternative sites for a private proposal to be identified and evaluated unless the proposal would require the site to be rezoned. However, if the rezone is for a use allowed in an existing comprehensive plan that was adopted after review under SEPA, an analysis of alternative sites is not required. The federal regulations do not have such a limitation on the review of alternative sites for private or public proposals.

Under NEPA, CEQA, and SEPA regulations, in the absence of physical environmental impacts, economic or social effects do not require the preparation of an EIS. A federal Final EIS must identify the preferred alternative, whereas the CEQA and SEPA regulations do not have such a requirement. In most cases, an agency's proposed action is the agency's preferred alternative.

Supplemental EISs and Subsequent EIRs Supplemental EISs (SEISs), under NEPA and SEPA, and Subsequent EIRs (SEIRs), under CEQA, are required under similar circumstances by federal and state processes (reflected in Track 6 of Figure 5.3). An SEIS or SEIR is required if an agency or project proponent makes substantial changes in a proposed action or if significant new circumstances or information, relevant to the environmental concerns of the proposed action, become known. A new circumstance must present a seriously different picture of the environmental impact of the proposed action than was presented in the original EIS. Both Draft and Final EISs may be supplemented

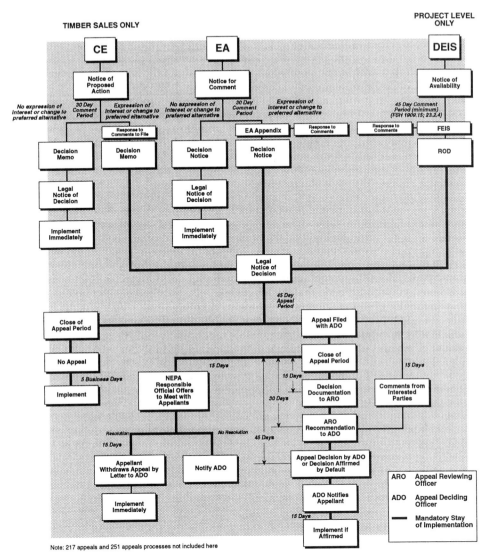

■ **FIGURE 5.5.** Example of a U.S. Forest Service NEPA appeal process: 36 CFR Part 215 (for site-specific projects).
Source: Modified from Tongass National Forest, Chatham Area 12/20/93.

because of new or changed circumstances or to correct deficiencies in a previous Draft or Final EIS.

NEPA and Washington state's SEPA do not require public scoping for supplemental EISs, and NEPA does not require scoping for legislative EISs. CEQA, however, requires the entire EIR process to be started from the beginning for a subsequent EIR.

      Administrative Appeals For those who disagree with an agency decision, some agencies provide an appeal process whereby aggrieved parties take their grievance to a higher level in the agency bureaucracy. If an agency does not have an administrative appeal process, and someone disagrees with an agency decision, the only recourse is judicial appeal or a lawsuit. NEPA does not specifically provide for appeals; however, some federal agencies such as the U.S. Forest Service (see Figure 5.5) and Bureau of Land Management have an administrative appeal process. State and local jurisdictions may also have appeal processes. Washington's SEPA regulations have provisions for appeals (both administrative and judicial) and provide guidance to agencies who implement administrative appeal procedures by rule, ordinance, or resolution (see Figure 5.6 for an example of a local agency appeal process). With a few exceptions, if an agency provides an administrative appeal procedure, that procedure must be used before a judicial review may be initiated (see "Exhaustion of Administrative Remedies" in Chapter 6 for additional information). California's CEQA guidelines set strict statutes of limitations on appeals.

APPLICATION OF NEPA IN OTHER COUNTRIES

Courts have held that NEPA does not apply to activities of the United States federal agencies in other nations. It does, however, apply to the United States trust territories. Presidential Executive Order 12114 furthers the purposes of NEPA and requires agencies to implement procedures for considering the environmental effects of major federal actions that occur outside the boundaries of the United States. The type of action specified as requiring an EIS in the executive order is a major federal action that significantly affects the global commons outside the jurisdiction of any nation (e.g., the oceans, upper atmosphere, or Antarctica) in which nations have common interests. For other federal actions, agencies may prepare EAs or EISs, or bilateral (participation by two nations) or multilateral environmental studies with other nations or other types of environmental reviews.[79]

---

[79]EO 12114, *Environmental Effects Abroad of Major Federal Actions*, 1979.

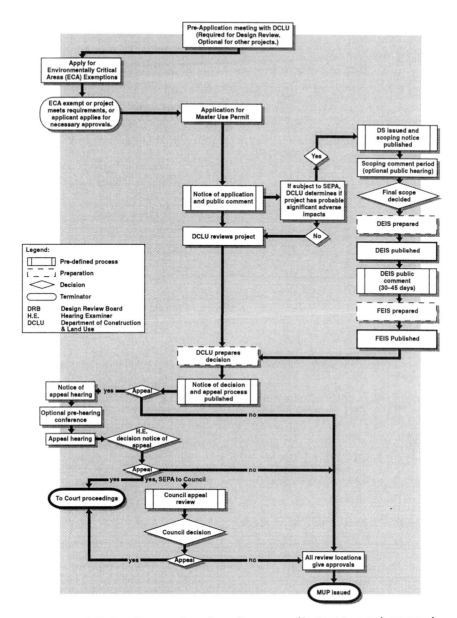

■ **FIGURE 5.6.** Example of a city permit, SEPA, and appeal process flow chart. *source*: Seattle Department of Construction and Land Use, 1995.

THE INTERNATIONAL ENVIRONMENT

The United States and other nations have established cooperative relationships regarding environmental concerns. Global warming, acid rain, pollution, and the rapid extermination of plant and animal species are worldwide issues. All of these concerns are caused, or at least accelerated, by human activities. As human populations explode, other species are diminishing either from being killed (a direct impact) or as a result of losing habitat and food (indirect impacts). Slashing and burning of rain forests, while providing developable land and farmland, is eliminating plants and animals and their habitats and threatening to alter the global climate and weather patterns. Tropical forests cover 6 percent of the Earth's surface but are home to perhaps half of the planet's plant and animal species. The loss of tropical forests and other habitat may be causing a worldwide rate of extinction of species that is 1000 times the natural rate.[80] In addition, exportation of endangered species or their parts (e.g., pelts, horns, and tusks) is a lucrative source of income in some countries, and it contributes to the species' decline.

Maintaining biodiversity, which refers to the diversity of life—species, their genetic structural, behavioral, and functional variability and the communities and ecosystems of which they are comprised—is an international issue. Human interaction, biodiversity, and ecosystem dynamics are highly integrated.[81] Scientists and citizens of countries around the world believe that maintaining a diversity of species is important to the survival of ecosystems, of which humans are a part.

ENVIRONMENTAL IMPACT ASSESSMENTS OF OTHER NATIONS

NEPA has not only been copied by many states in the United States, it has served as a model for environmental impact assessment regulations throughout the world. The European Council of Ministers approved a directive on June 27, 1985 requiring its "member states" (member nations) to adopt environmental impact assessment (EIA) procedures. Similar to NEPA, the EIA process includes a threshold de-

---

[80]Council on Environmental Quality, *Twenty-second Annual Report*, 1991.
[81]Council on Environmental Quality, *Twenty-fourth Annual Report*, 1993.

termination of whether a proposed project would result in significant environmental effects. The EIA procedures are applicable to government actions and private projects. The Directive establishes basic assessment principles and procedural requirements, and it lists types of projects for which environmental assessments are mandatory and projects for which environmental assessments may be prepared. The member states have considerable discretion in developing the details of their national legislation and may determine which of the projects that are not on the mandatory list will require EIAs. Figure 5.7 is a flow diagram of the basic EIA process. Member states that have implemented EIA regulations include Austria, Belgium, Denmark, Finland, France, Germany, Greece, Ireland, Italy, Luxembourg, The Netherlands, Portugal, Spain, Sweden, and the United Kingdom.

The types of environmental impacts evaluated in the EIAs, including direct and indirect impacts, are similar to those evaluated in a NEPA EIS. The types of projects subject to environmental assessment appear to be primarily related to development. National defense projects and projects authorized by prior legislation are exempt from environmental review. Member states are required to inform other member states whose environments are likely to be significantly affected by a proposed project.[82]

The Canadian Environmental Assessment Act, enacted in 1992, is also similar to NEPA in many respects. Federal agencies are required to evaluate the environmental impacts of their actions, and an EIA is not required if an action is on an exclusion list. A screening process is used to determine whether an action may have significant environmental effects. Unlike NEPA, the Canadian EIA process includes a mediator and review panel to review actions that have questionable adverse environmental effects or other concerns, along with follow-up programs to verify the accuracy of an EIA and determine the effectiveness of any mitigation measures.[83]

The United States works with the World Bank and other multilateral development banks to integrate environmental considerations in lending practices. Efforts focus on environmental impact assessment procedures, energy efficiency and conservation programs, sensitive or

---

[82]Daniel R. Mandelker, *NEPA Law and Litigation* (release #3), 1995.
[83]*idem.*

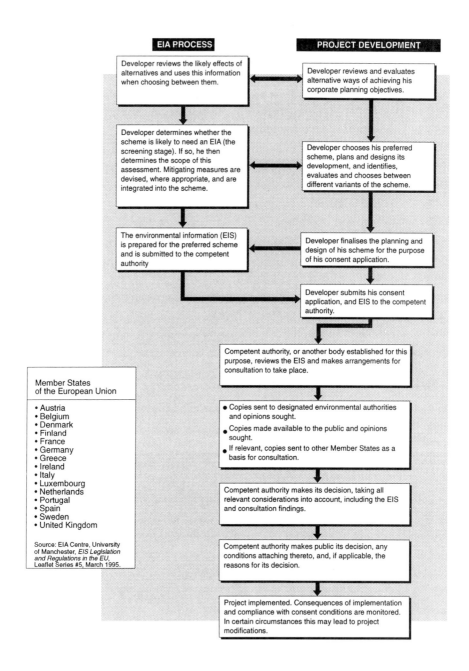

EIA PROCESS	PROJECT DEVELOPMENT
Developer reviews the likely effects of alternatives and uses this information when choosing between them.	Developer reviews and evaluates alternative ways of achieving his corporate planning objectives.
Developer determines whether the scheme is likely to need an EIA (the screening stage). If so, he then determines the scope of this assessment. Mitigating measures are devised, where appropriate, and are integrated into the scheme.	Developer chooses his preferred scheme, plans and designs its development, and identifies, evaluates and chooses between different variants of the scheme.
The environmental information (EIS) is prepared for the preferred scheme and is submitted to the competent authority	Developer finalises the planning and design of his scheme for the purpose of his consent application.

Developer submits his consent application, and EIS to the competent authority.

Competent authority, or another body established for this purpose, reviews the EIS and makes arrangements for consultation to take place.

Member States of the European Union

- Austria
- Belgium
- Denmark
- Finland
- France
- Germany
- Greece
- Ireland
- Italy
- Luxembourg
- Netherlands
- Portugal
- Spain
- Sweden
- United Kingdom

Source: EIA Centre, University of Manchester, *EIS Legislation and Regulations in the EU*, Leaflet Series #5, March 1995.

- Copies sent to designated environmental authorities and opinions sought.
- Copies made available to the public and opinions sought.
- If relevant, copies sent to other Member States as a basis for consultation.

Competent authority makes its decision, taking all relevant considerations into account, including the EIS and consultation findings.

Competent authority makes public its decision, any conditions attaching thereto, and, if applicable, the reasons for its decision.

Project implemented. Consequences of implementation and compliance with consent conditions are monitored. In certain circumstances this may lead to project modifications.

■ **FIGURE 5.7.** European Union simplified flow chart of the EIA process and its relationship to project appraisal, authorization, and implementation. *Source*: By permission, EIA Centre, University of Manchester, *Five Year Review of the Implementation of the EIA Directive*, Leaflet Series #14, Manchester, UK, August 1993.

endangered ecosystems, and public participation. Countries, national organizations, and lending institutions that have adopted EIA requirements, and other environmental regulations are shown in Tables 5.3 and 5.4. The tables are based on a survey by the International Institute for Environment and Development (IIED), and thus they represent information submitted by those countries and agencies that responded to the survey. The IIED does not consider the information authoritative; however, it is very interesting in terms of the number of countries and areas purported to require environmental review. Also, the EIA Centre at the University of Manchester reports having 88 countries in its EIA network.

While many of NEPA's attributes were adopted by other nations, none of the sources consulted for the above discussion on EIA requirements of other nations noted a requirement to analyze the environmental impacts of alternatives to a proposed action. This would seem to be a large area of difference between EIS requirements in the U.S. and those of EIAs because the alternatives analyses is the "heart" of a NEPA EIS. The purpose of discussing the environmental impacts of alternative ways to accomplish a project's objectives is to allow an agency decisionmaker to make a choice among alternatives while taking into consideration their relative environmental benefits and costs. EIAs seem to focus only on the impacts of a proposed project, making alternatives analysis a concern unique to the United States.

■ TABLE 5.3 National Environmental Guidelines by Sector

	General EIA	Health	Social	Agriculture	Aquaculture/Fisheries	Coastal Zone	Energy/Power	Forestry	Human Settlements	Industry	Mining	Risks/Hazards	Tourism	Transport	Waste	Water	Wetlands	Other
AFRICA																		
Africa Regional	●															●		●
Botswana	●																	●
Ghana	●									●						●		
Mozambique	●																	●
Namibia	●																	●
Nigeria	●	●								●								
South Africa										●								●
Tanzania	●																	●
Zimbabwe																		●
ASIA/PACIFIC/MID. EAST																		
Asia and Pacific Regional	●	●	●	●		●	●			●		●	●			●	●	
Bangladesh															●			
Bhutan	●																	
China	●						●											
Hong Kong	●																	
India	●				●	●	●	●	●			●	●	●		●		●
Indonesia	●					●											●	
Israel	●																	
Japan	●																	
Korea	●																	
Kuwait	●																	
Malaysia	●																	
Nepal	●																	
Oman	●																	
Pakistan	●					●												
Philippines	●																	
Singapore	●																	
Sri Lanka	●					●	●											
Thailand	●					●										●		
AUSTRALASIA																		
Australia	●						●											
New Zealand	●																	
CARIBBEAN																		
Caribbean Regional	●			●									●					
Bermuda	●																	
St. Lucia	●																	
Trinidad and Tobago	●																	
Turks and Caicos Islands	●																	
EUROPE*																		
European Regional	●																	
Belgium	●																	
Cyprus	●																	
Denmark	●																	●
Finland	●						●											
France	●						●			●			●	●				
Germany	●																	
Netherlands	●																	●
Norway	●																	
Portugal	●						●											
Spain	●												●					
Sweden	●												●					
United Kingdom	●									●			●					●
Eastern Europe Regional	●																	
Estonia	●																	
Russian Federation											●							
LATIN AMERICA																		
Latin American Regional	●					●				●			●					
Bolivia							●	●	●				●					
Brazil	●																	
Chile	●																	
Costa Rica												●	●					
Ecuador	●																	
Honduras	●																	
NORTH AMERICA																		
Canada	●															●		●
United States	●		●				●									●		

Source: Dilys Roe, Barry Dalal-Clayton, and Ross Hughes, *A Directory of Impact Assessment Guidelines*, International Institute for Environment and Development, London, UK, 1995, p. 6.

*Note: Austria enacted an EIA law on July 1, 1994. Greece passed an EIA law in 1986 (implementation was not until 1990). Ireland passed a number of regulations on EIA (1989, 1993, and 1994). Italy passed laws on EIA in 1986 and 1988. Luxembourg passed an EIA law on May 9, 1990. (Source: EIA Centre, University of Manchester, UK, *EIS Legislation and Regulations in the EU*, Leaflet Series #5, March 1995).

■ TABLE 5.4 Agency Environmental Guidelines by Sector

	General EIA	Health	Social	Agriculture	Aquaculture/Fisheries	Coastal Zone	Energy/Power	Forestry	Human Settlements	Industry	Mining	Risks/Hazards	Tourism	Transport	Waste	Water	Wetlands	Other
MULTILATERAL DEVELOPMENT BANKS																		
ADB	●																	
AsDB	●	●	●	●		●	●			●		●		●				●
CDB	●																	
EBRD	●																	
IADB	●																	
World Bank	●	●	●			●	●		●	●		●			●	●		
BILATERAL DONORS																		
AIDAB	●								●	●								
BMZ	●		●			●			●	●								
CIDA	●																	
DANIDA	●																	
DGIS	●												●					
FINNIDA	●																	
JICA	●			●	●		●	●	●	●								
NORAD	●			●	●		●	●	●	●				●	●	●		
SIDA	●																	
UK ODA	●	●																
USAID	●			●	●						●							
UNITED NATIONS AGENCIES																		
UNDP	●						●				●							
UNEP	●			●	●	●		●	●	●	●	●	●			●		
UNESCO																●		
UNIDO										●						●		
UN-ESCAP				●						●				●		●		
FAO		●	●	●	●													
IFAD	●		●	●														
WHO	●	●		●					●	●		●			●	●		
INTER-GOVERNMENTAL ORGANISATIONS																		
European Community	●																	
IUCN	●					●	●	●			●							●
OAS	●											●						
OECD	●	●	●									●						
OECS	●																	
SPREP	●					●												

List of acronyms used in the table

ADB	African Development Bank
AIDAB	Australian International Development Assistance Bureau
AsDB	Asian Development Bank
BMZ	Bundesministerium fur Wirtschaftliche Zusammenarbeit
CDB	Caribbean Development Bank
CIDA	Canadian International Development Agency
DANIDA	Danish International Development Agency
DGIS	Directorate General for International Cooperation- The Netherlands
EBRD	European Bank for Reconstruction and Development
FAO	Food and Agriculture Organisation of the United Nations
FINNIDA	Finnish International Development Agency
IADB	Inter-American Development Bank
IFAD	International Fund for Agricultural Development
IUCN	The World Conservation Union (International Union for Conservation of Nature and Natural Resources)
JICA	Japanese International Cooperation Agency
NORAD	Norwegian Agency for Development Cooperation
OAS	Organisation of American States
OECD	Organisation for Economic Cooperation and Development
OECS	Organisation of Eastern Caribbean States
SIDA	Swedish International Development Authority
SPREP	South Pacific Regional Environment Programme
UK ODA	United Kingdom Overseas Development Administration
UNDP	United Nations Development Programme
UNEP	United Nations Environment Programme
UNESCO	United Nations Educational, Scientific and Cultural Organisation
UNIDO	United Nations Industrial Development Organisation
UN-ESCAP	United Nations Economic and Social Commission for Asia and the Pacific
USAID	United States Agency for International Development
WHO	World Health Organisation

Source: Dilys Roe, Barry Dalal-Clayton, and Ross Hughes, *A Directory of Impact Assessment Guidelines*, International Institute for Environment and Development, London, UK, 1995, p. 7.

CHAPTER 6

EFFECTS OF COURT DECISIONS

When a law is passed, the only thing agreed upon is the words.

and

The law is whatever the courts say it is.

Anonymous

INTRODUCTION

The environmental impact statement (EIS) process exists because of laws passed by Congress and by some states. Federal and state agencies interpret National Environmental Policy Act (NEPA) and state environmental policy act (SEPA) requirements with the aid of agencies such as the Council on Environmental Quality (CEQ) and state counterparts. When there is a disagreement about an agency's implementation or interpretation of the laws and regulations, and a lawsuit is filed regarding an agency's action, the final decision on what the law says or requires is made by the courts. Decisions, however, are not written in stone. The courts' interpretations of the law change or are revised, and new interpretations are continually being added to the case law.

NEPA does not address judicial review of agency actions by citizens or parties in disagreement with an agency's action under the law.

SEPAs (e.g., California, Michigan, Washington), however, do provide for judicial or administrative review. Under NEPA, judicial review is through the Administrative Procedure Act. This chapter discusses court decisions that have affected the way agencies view EISs and the EIS process, but it does not discuss the legal process or how decisions were derived. Sources of information on NEPA case law are listed in the Bibliography for further reference.

One should consider past court decisions because the courts have the final say on the adequacy of an EIS or the appropriateness of an agency's decision. This doesn't mean that the decisions of the courts are always logical, or that a court would agree with an agency's actions because they were in accordance with a past court's decision, but it helps understand what may be required to comply with regulations when the regulations themselves are vague.

The discussion in this chapter is primarily regarding federal court decisions regarding NEPA. Thousands of NEPA cases were reviewed by federal courts since 1970. State courts have considered federal court decisions when evaluating similar issues and have come to many of the same conclusions. The federal Supreme Court, however, has not ruled against a federal agency regarding the adequacy of an EIS. In general the federal Supreme Court gives more deference to the decisions of an agency than do the lower courts or the state courts for state agency actions (also, state judicial review standards differ from federal standards).

The courts interpret the meaning and intent of the law. Some courts take a narrow reading of the law, limiting scope and applicability, while others take a broader reading and expand the law's requirements. Some courts favor private causes, others favor agencies. Thus the courts have reached some inconsistent conclusions on similar issues.

ROLE OF THE COURTS

In some early cases, plaintiffs expected the courts to exercise powers (e.g., to change an agency's decision or action) that the courts essentially said the law did not grant. The extent of the courts' authority, or the courts' judicial role, was interpreted differently by different courts. In some cases, the federal Supreme Court stated that its proper judicial role was

. . . to ensure that the agency has taken a "hard look" at environmental consequences; it is not to second-guess the correctness of the agency's decision.[84] The courts have recognized that NEPA does not impose a substantive duty on federal agencies to protect the environment. Rather, NEPA is essentially procedural; it does not require the federal agency to adopt the environmentally preferred alternative or to mitigate significant impacts. Once an agency has made a decision subject to NEPA's procedural requirements, the only role for a court is to insure that the agency has considered the environmental consequences.[85]

Another court's interpretation of its judicial role was as follows:

The role of the courts is simply to ensure that the agency has adequately considered and disclosed the environmental impact of its actions and that its decision is not arbitrary and capricious.[86]

The above interpretation of judicial role goes beyond the first interpretation by saying that the court should review not just whether the agency followed procedures and "considered" the environmental consequences, but whether the consideration and disclosure of environmental impacts were "adequate," and whether the agency's decision was "arbitrary and capricious."

When it comes to matters that are within an agency's purview and an agency's technical expertise, the courts have stated that they will not substitute their judgment for that of the agency or its technical experts.

The Court is not empowered to substitute its judgment for that of the agency.[87] The court does not determine whether an effect was significant or whether an alternative was reasonable.[88]

NEPA LITIGATION HISTORY

By reviewing the history of NEPA litigation over the past 25 years, one notices recurring complaints in the lawsuits and that some agencies are defendants of their NEPA decisions more often than others. The obvious reasons that some agencies receive more challenges in court

---

[84]Kleppe v. Sierra Club (1976).
[85]Strycker's Bay Neighborhood Council Inc. v. Karlen (1980).
[86]Baltimore Gas & Electric Co. v. Natural Resources Defense Council, Inc. (1983).
[87]Citizens to Preserve Overton Park v. Volpe (1971).
[88]Citizens for Better Henderson v. Hodel (1985).

than others is that those agencies are regularly involved in actions that have the potential to affect the environment, and the actions are of concern to the public. Actions that affect land and natural resources are under the control of, and subject to the mandates of, agencies in the Departments of Agriculture, Interior, Defense (including the COE), and Transportation, among others. Agency responsibilities are described in Chapter 3, where it is noted that the Departments of Agriculture and Interior manage millions of acres of public lands for multiple public uses. Public views on the appropriateness of agency policies are often in disagreement with other public views and with agency views. Other actions that have the potential to affect a large number of people include transportation projects or construction of highways, which are under the purview of the Department of Transportation. The Department of Defense is also responsible for actions that have environmental consequences that are of public concern. Thus, these agencies are frequently in court for review of their actions. One could argue that if these agencies made the "correct" decisions in the first place, they would not be in court. And in some cases, agencies were derelict in their NEPA duties. However, an agency action will rarely be perceived as correct by all the public. The public's motives are "partly political, partly philosophical[89];" therefore, an agency action that is major enough and environmentally controversial enough to require an EIS will not please everyone.

Table 6.1 lists the number of NEPA cases by year, the most frequent defendant and plaintiff, and the most common complaint. The most frequent defendant is the Department of Transportation, and the most frequent plaintiffs are citizen and environmental groups. The most common complaint is that an agency should have prepared an EIS for an action (when it didn't), and the second most common complaint is that an agency's EIS was inadequate. After hearing many cases, the courts decided which standards were appropriate to use for reviewing allegations that an agency made inappropriate decisions. These standards include the "arbitrary and capricious" standard and the "rule of reason."

---

[89]Cabinet Mountains Wilderness v. Peterson (1981).

■ TABLE 6.1 Cumulative NEPA Litigation Survey, 1970–1992

Year	Cases Filed	Injunctions	Most Frequent Defendants	Most Frequent Plaintiff	Most Common Complaint	Second Most Common Complaint
1970–1975ᵃ	654	50	DOT HUD COE DOA	Citizen and environmental groups	No EIS	Inadequate EIS
1976	119		DOT DOI HUD	Citizen and environmental groups	No EIS	Inadequate EIS
1977	108		DOT DOD DOI	Citizen and environmental groups	No EIS	Inadequate EIS
1978	114		DOT DOI DOD HUD EPA	Citizen and environmental groups, individuals	No EIS	Inadequate EIS
1979	139	12	DOT HUD DOI USDA DOD	Citizen and environmental groups, individuals	No EIS	Inadequate EIS
1980	140	17	DOT DOI DOD HUD EPA	Individuals or citizen groups, environmental groups	No EIS	Inadequate EIS
1981	114	12	DOD DOT DOI USDA HUD NRC	Environmental groups, individuals or citizen groups	No EIS	Inadequate EIS
1982	17	19	DOI COE DOT USDA HUD EPA	Individuals or citizen groups, environmental groups	Inadequate EIS	Inadequate EIS
1983	146	21	DOI DOT USDA FERC NRC	Individuals or citizen groups, environmental groups	No EIS	Inadequate EIS

■ **TABLE 6.1** (Continued)

Year	Cases Filed	Injunctions	Most Frequent Defendants	Most Frequent Plaintiff	Most Common Complaint	Second Most Common Complaint
1984	89	14	USDA DOT DOI	Environmental groups, individual or citizen groups	Inadequate EIS	No EIS
1985	77	8	DOI DOT COE FERC	Environmental groups, individual or citizen groups	Inadequate EIS	No EIS
1986	71	16	DOT DOI USDA	Citizen and environmental groups	No EIS	Inadequate EIS
1987	69	3	USDA DOD DOI	Citizen and environmental groups	No EIS	Inadequate EIS
1988	91	7	DOT DOI COE USDA	Citizen and environmental groups, individuals	No EIS	Inadequate EIS
1989	57	5	DOT USDA	Citizen groups, environmental groups, individuals	No EIS	Inadequate EIS
1990	85	11	DOT COE EPA	Individuals or citizen groups, environmental groups	No EIS	Inadequate EIS
1991	94	14	ARMY DOT DOI	Environmental groups, individuals or citizen groups	No EIS	Inadequate EIS
1992	81	5	DOI DOT USDA EPA	Individuals or citizen groups, environmental groups	No EIS	Inadequate EIS

[a]Data taken from Council on Environmental Quality, *Seventh Annual Report*, 1976.
Source: Council on Environmental Quality, *Twenty-fourth Annual Report*, 1993.

JUDICIAL STANDARDS OF REVIEW

The "hard-look doctrine" applies to judicial review of agency NEPA decisions:

> Assumptions must be spelled out, inconsistencies explained, methodologies disclosed, contradictory evidence rebutted, record references solidly grounded, guesswork eliminated, and conclusions supported in a "manner capable of judicial understanding."[90]

Chapters 8, 10, and 12 discuss methods of preparing EISs to withstand scrutiny under the hard look doctrine.

Agency Decisions to Not Prepare an EIS: Arbitrary and Capricious Standard

In 1989 the federal Supreme Court held that the "arbitrary and capricious" standard was the correct standard of review for agency decisions to not prepare an EIS or supplemental EIS.[91] Under the arbitrary and capricious standard, a reviewing court considers whether the agency's determination is based on a consideration of the relevant factors and whether there has been a clear error of judgment.

Supplemental EISs are required when there is substantial new information or change in a proposed action that affects environmental impacts. The most frequently litigated issue concerning significant new information is whether the information is new. If the subject matter of the information was discussed in the original EIS, the information is not new.[92]

Once an agency has complied with NEPA's procedural prerequisites, it would not violate NEPA if it proceeded with an unwise action that was environmentally destructive.[93] An agency's actions are not arbitrary or capricious even if the actions are directly contrary to NEPA's purposes as long as the agency complied with NEPA's procedures.

---

[90]E. I. duPont de Nemours & Co. v. Train (1976).
[91]Marsh v. Oregon Natural Resources Council (1989).
[92]Natural Resources Defense Council, Inc. v. City of New York (1981).
[93]Robertson v. Methow Valley Citizens Council (1989).

Adequacy of EISs: The Rule of Reason

The courts use the "rule of reason" when reviewing the adequacy of an EIS or a supplemental EIS. Under the rule of reason, reviewing courts determine whether an EIS or a supplemental EIS is adequate by considering whether the EIS provides sufficient detail so that reviewers may understand and consider the relevant effects; whether an EIS explains fully its methodology, analysis, and rationale; whether it contains enough scientific reasoning to alert specialists to problems in their fields of expertise; and whether the EIS is concise, clear, and written in plain language.

An EIS does not have to be perfect,[94] but it should be adequate, complete, and provide a good faith effort at full disclosure. Agencies are not required to elevate environmental concerns over other appropriate considerations.

Selecting Alternatives When evaluating the adequacy of an EIS, the range of alternatives and the discussion of alternatives is subject to a rule of reason.[95] An EIS is not inadequate because it did not include every possible alternative or consider remote and speculative alternatives, but the range of alternatives must be reasonably comprehensive. If remote and speculative alternatives are discussed, they do not have to be discussed in detail. The alternatives analysis must provide enough information to allow reviewers to understand the differences in the environmental effects of the alternatives and to make a choice among the alternatives.

If alternatives would not be consistent with the purpose of the proposed action, or is not feasible for some other reason, they do not have to be included in the analysis. An EIS should, however, briefly discuss why alternatives are not feasible. If, based on the evaluation of an alternative, the alternative is determined to be not feasible, or remote and speculative, the EIS is not required to continue analyzing the alternative.

---

[94]Sierra Club v. Morton (1975).

[95]Life of the Land v. Brinegar (1974); Natural Resources Defense Council, Inc. v. Hodel (1988).

EISs Procedural and Substantive Requirements

SEPAs and NEPA have two broad types of requirements: procedural and substantive, although the laws and regulations are not necessarily divided into those categories. A common view of these requirements is that the procedural requirements apply to the EIS process while the substantive requirements apply to the contents of the EIS. Procedural requirements include how an EIS should be prepared and circulated and how the public should be invited to participate. Substantive requirements include the adequacy of an EIS in terms of its disclosure of impacts, identification of alternatives and mitigation measures, appropriateness of methodology used in the analyses, and how it meets the spirit of the law (furthers the goals of the law) as well as the letter of the law (how it meets the law's specific requirements).

However, one legal expert's opinion is that from the perspective of the courts, an agency's "procedural" duty to comply with NEPA refers to the duty to prepare an adequate impact statement. An agency's "substantive" duty would be to reject or modify an action because of the proposed action's potentially negative environmental impacts identified in an EIS.[96] These are two different views on the meaning of procedural and substantive requirements. One is from the standpoint of EIS preparation, and the other is from a legal standpoint. Both views are valid facets of an EIS process.

The CEQ regulations intend that judicial review of an agency's compliance with NEPA take place after a final EIS has been filed or after a finding of no significant impact has been issued. Plaintiffs can sue an agency for not following the required procedures or for preparing an inadequate EIS. Generally, a minor violation of procedures will not be sufficient grounds for a court to rule against an agency.[97] As stated earlier, the most common complaint is that an agency neglected to prepare an EIS. Agencies are usually fairly careful to follow prescribed procedures, but a decision to not prepare an EIS is based on an agency's determination that a proposed project will not result in significant environmental impacts. The determination of nonsignificance

---

[96]Daniel R. Mandelker, *NEPA Law and Litigation* (release #3), 1995.
[97]40 CFR §1500.3.

can be a point of disagreement for a plaintiff who feels that the proposed action will result in significant impacts, and therefore the procedural disagreement is that an EIS should have been prepared.

Plaintiffs can allege that an EIS is inadequate for a number of reasons. Common allegations include the following: The impacts were not adequately analyzed, the range of alternatives was inadequate, mitigation measures were insufficient, the analysis was biased, the data were inadequate or flawed, and many others. It is very difficult to judge the sufficiency of the analysis in an EIS because no matter how much analysis, data, or detail is included, there can always be more. Thus, environmental groups that make a business of suing agencies for inadequate EISs can use the same complaints from one suit to the next. The wording of some complaints are so general and recycled so frequently as to become recognizable with a particular group or organization (e.g., "the EIS is fatally flawed and woefully inadequate"). Furthermore, on a given technical issue, experts can be found who will disagree with other experts. Thus, a court will normally not "substitute its judgment" for that of the agency's experts[98] unless the EIS is very obviously lacking in detail and does not substantiate or support an agency's decision.

Exhaustion of Administrative Remedies

Legal opinions vary regarding whether administrative remedies must be exhausted before someone can seek judicial review of an agency's compliance with NEPA. One opinion is that as a general rule, plaintiffs must exhaust applicable administrative remedies before seeking judicial review. In addition, if a plaintiff fails to comment on an agency's proposed action despite public notice of the action, or if a plaintiff fails to follow an agency's administrative review procedures, or maintains silence on NEPA issues at the administrative level and raises the issues for the first time in court, the plaintiff may not be entitled to judicial review of the agency's alleged noncompliance with NEPA. The doctrine of exhaustion of administrative remedies is flexible, however. Judicial review is permitted despite a plaintiff's failure to ex-

---

[98]An agency is entitled to rely on the reasonable opinions of its own qualified experts (Marsh v. Oregon Natural Resources Council, 1989).

haust administrative remedies when the remedy is inadequate, when the plaintiff would suffer irreparable harm if judicial review was not permitted, or when it would be futile for plaintiffs to pursue administrative remedies.[99]

In contrast, another opinion is that the exhaustion rule is appropriate only when an issue is expressly delegated to an agency that deals exclusively with and has expertise in the area. The exhaustion rule does not apply to NEPA because NEPA applies to all federal agencies and no agency has expertise in determining whether an impact statement is required in a particular instance. In addition, the Supreme Court has held that federal courts cannot require exhaustion of remedies under the Administrative Procedure Act (APA) unless the agency's statute or rules explicitly make exhaustion a prerequisite to judicial review. NEPA does not contain a judicial review provision. Because the APA applies to NEPA appeals, the court's decision also governs exhaustion of remedies in NEPA cases.[100]

Highlights of Significant Court Decisions

Court cases are referenced in several places throughout this book. The following are some of the more frequently cited cases, as well as decisions that were repeated in several cases, and therefore have some agreement among the courts:

Vermont Yankee Nuclear Power Corp. v. Natural Resources Defense Council, Inc.

■ NEPA, while establishing "significant substantive goals for the Nation," imposes upon agencies duties that are "essentially procedural."

■ An agency does not have to elevate environmental concerns over other appropriate considerations.

■ An EIS does not have to ferret out every conceivable alternative.

Marsh v. Oregon Natural Resources Council

■ The arbitrary and capricious standard applies to judicial review of whether to prepare a Supplemental EIS.

---

[99]Valerie Fogelman, *Guide to the National Environmental Policy Act: Interpretations, Applications, and Compliance*, 1990.

[100]Daniel R. Mandelker, *NEPA Law and Litigation* (release #3), 1995.

■ An agency is allowed to rely on the reasonable opinions of its own qualified experts.

■ A supplemental EIS is not required every time new information comes to light because an agency would always be awaiting updated information, only to find the new information was outdated by the time the decision was made.

Robertson v. Methow Valley Citizens Council

■ A mitigation plan is not required in an EIS.

■ NEPA does not require an agency to adopt the environmentally preferred alternative.

■ NEPA does not mandate particular results, but simply prescribes the necessary process. NEPA merely prohibits uninformed—rather than unwise—agency action.

Strycker's Bay Neighborhood Council, Inc. v. Karlen

■ Once an agency has made a decision subject to NEPA's procedural requirements, the only role for a court is to ensure that the agency has considered the environmental consequences. It cannot interject itself within the area of discretion of the executive as to the choice of the action to be taken.

■ Environmental factors do not have to be the determining element in a decision.

Kleppe v. Sierra Club

■ Agencies have broad discretion in defining the scope of their EISs (disagreeing with Sierra Club's assertion that the Department of the Interior had a regional coal program that required a programmatic EIS). A proposal exists when an agency has a goal and is actively preparing to make a decision on one or more alternative means of accomplishing it, and the effects can be meaningfully evaluated.

■ Ruling has been interpreted to mean that the discussion of cumulative actions in an EIS need only include those actions that are "proposals;" actions that are only planned or contemplated need not be included.[101]

■ Court cannot substitute its judgment for that of the agency.

---

[101]Daniel R. Mandelker, *NEPA Law and Litigation* (release #3), 1995.

Citizens to Preserve Overton Park, Inc. v. Volpe

- Supreme Court applied the arbitrary and capricious judicial review standard and the hard-look doctrine to evaluate whether an agency engaged in "reasoned decisionmaking."
- Court is not empowered to substitute its judgment for that of the agency.

Trout Unlimited v. Morton

- EIS did not have to contain a formal cost–benefit analysis.

Natural Resources Defense Council, Inc. v. Morton

- Applied "rule of reason" to consideration of alternatives even though they are outside the lead agency's jurisdiction or would require legislative changes. (However, most court decisions limit alternatives that must be discussed in an EIS to those that meet the purposes of the proposed project.)

Hanly v. Kleindienst (II)

- Mere public opposition to a project does not require an EIS. An action must be controversial on environmental grounds for an EIS.

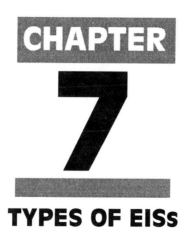

CHAPTER
7

TYPES OF EISs

. . . who can plan for the long term when survival is today?
Ian McHarg

INTRODUCTION

This chapter will address major types of environmental impact statements (EISs) and the appropriate use of those EISs. Regardless of the type of EIS, the required content of EISs are similar; differences are primarily in the EIS process. Agencies may use different types of EISs because EISs must be prepared early in the planning process and because EISs are prepared by agencies with very different responsibilities for a wide variety of actions. However, if an EIS is to be prepared early in the planning process, project information may be preliminary and lacking in detail, or the specific projects that follow an agency decision may not be known. Therefore, environmental review may take place in phases, and the type of EIS will depend on the type of action that is proposed and the amount of information available at the time an agency is considering whether to act on a proposal.

Agency responsibilities that result in actions to manage large tracts of land and resources (such as the preparation of plans and programs) or to respond to legislative mandates require preparation of EISs. The agency action may be to implement a program, policy, or plan, but the specific actions that result from the agency action are the actions that result in environmental impacts, and those specific actions sometimes are not known at the program, policy, or plan stage. In other words, an agency action to implement a program has no direct environmental impact. Environmental impacts result from subsequent actions because the agency's program will allow those actions. Nonetheless, an EIS must be prepared for the program, before specific actions or projects at specific locations are known.

Proposed actions range from the adoption of a large program with regional or national effects, to an action on a given site. Preparation of one EIS for a specific proposed project may be sufficient, while a large program that is lacking in details at the early planning stage may require two or more EISs, or an EIS followed by environmental assessments (EAs), for the specific actions or projects within a program. Types of EISs include

■ Tiered and programmatic
■ Project-specific
■ Legislative
■ Supplemental
■ Single-issue or limited-scope EIS

TIERED AND PROGRAMMATIC EISs

Tiering refers to the process of preparing environmental analyses from the broad program, plan, or policy level, to a site-specific analysis, or to a program, plan or policy analysis of lesser scope. "Tiering" is the process, and "tiered" or "programmatic" refers to the type of EIS prepared in the process. Each level in the process goes from the more general to the more specific. California Environmental Quality Act (CEQA) guidelines call this a "staged environmental impact report (EIR)" because the contemplated action would proceed in stages. For example, a programmatic EIS (also called a "program EIS") might be prepared for a proposed action to allow geothermal leasing on public land in a large region.[102] In this case, the broad action under review is

---

[102]The Department of the Interior prepared a programmatic EIS for the national geothermal leasing program in the 1970s.

the geothermal leasing program. Geothermal activities, however, take place in stages including the exploration, development, production, and utilization phases. The results of exploration activities such as geophysical surveys and shallow drilling determine where development of a geothermal resource, if found, might take place. Exploration could take many years, and the result of those activities may show that a geothermal resource is not present, thus concluding further activity. If a resource were discovered, the location and type of development of that resource is not known at the time an agency is considering the leasing plan or program within a region (e.g., in a national forest). Therefore, it may not be possible to predict where within the region specific activities would take place in order to evaluate site-specific impacts. Currently, the Bureau of Land Management (BLM) prepares resource management plans for various resources [with National Environmental Policy Act (NEPA) documentation], including geothermal energy.[103]

Tiering "helps the lead agency focus on the issues which are ripe for decision and exclude from consideration issues already decided or not yet ripe."[104] An EIS, or an EA, prepared subsequent to a programmatic EIS can reference analyses contained in the programmatic EIS to avoid repeating the same information. The more specific EIS or EA can then focus on the issues related to the site-specific analyses or on an action that has a narrower scope than was addressed in the programmatic EIS. A programmatic EIS is useful for evaluating a program's contribution to cumulative environmental effects in a study area because it is reviewing potential impacts from a total program rather than impacts that result project-by-project.

To determine whether proposed actions are related and should be included in one EIS, the Council on Environmental Quality (CEQ) regulations identify the following ways to consider the question:

■ Geographically, including actions occurring in the same general location such as body of water, region, or metropolitan area
■ Generically, including actions which have relevant similarities, such as common timing, impacts, alternatives, methods of implementation, media, or subject matter

---

[103]The BLM prepares a "reasonable development scenario," which makes assumptions on the expected geothermal activities.
[104]40 CFR §1508.28.

■ By stage of technological development including federal or federally assisted research, development, or demonstration programs for new technologies, which, if applied, could significantly affect the quality of the human environment.[105]

The above considerations are appropriate when evaluating whether to include actions in an EIS, even if an EIS is not for a proposed program. That is, actions that are not related as parts of an overall program still should be included in an EIS (e.g., to evaluate cumulative impacts).

Determining the scope of a programmatic EIS is often less obvious than determining the scope of a project-specific EIS because of the unknown elements in a program. An EIS team must decide which issues to evaluate, the level of analyses to provide in the programmatic EIS, and which issues to reserve for later analyses. The EIS should explain why certain issues or analyses are more appropriate for analyses in subsequent environmental documents. The analysis of impacts is necessarily general in a programmatic EIS, since specific impacts are related to location; and at this stage, the locations of future activities may not be known. Some assumptions must be made in the analyses, and it may not be possible to quantify impacts or discuss them with any degree of precision. For these reasons, reviewers of a programmatic EIS tend to feel that the EIS is not fully disclosing potential impacts or is not detailed enough to support an agency's decision to implement a program, and their most frequent comment is that the EIS needs more analyses and more project information.

PROJECT-SPECIFIC EISs

The most common type of EIS is prepared for a specific project. A project-specific EIS evaluates a proposed action with known characteristics in a specific location. Project-specific EISs frequently involve construction of structures or facilities. The project itself may be large, or the project's effects may be over a large area, but if specific features of the project's implementation, construction, and operation are known, and they are not related to other actions that are part of an overall program, they should be evaluated in a project-specific EIS.

---

[105]40 CFR §1502.4(c).

Let's continue with the geothermal example: If, based on the programmatic EIS, geothermal leasing were allowed, EISs or EAs may be prepared for specific geothermal activities that are not categorically excluded by the lead agency. In some cases, an EIS of narrower scope than the programmatic EIS might be the second tier of environmental review if potential environmental impacts of an action under a program would be significant and were not addressed in sufficient detail in the programmatic EIS. The second tier could be a project-specific EIS, or more specific (than the first tier) but not quite specific enough to end the environmental review of subsequent actions. Environmental assessments might provide the third tier of analyses for subsequent actions which, by then, should be fairly site-specific. The EIS process, including public involvement and review, is the same for project-specific and programmatic EISs.

LEGISLATIVE EISs

A legislative EIS is prepared for a bill or legislative proposal to Congress that would result in significant effects to the environment. A bill or legislative proposal must have the substantial cooperation and support of a federal agency so that the proposal is predominantly that of the agency. Legislative proposals do not include appropriation requests, authorizations for further study of a contemplated action, and agency responses to congressional inquiries. The legislative EIS is part of a formal transmittal to Congress, but it may be transmitted to Congress up to 30 days later if necessary to complete an accurate EIS. The EIS must be available in time for congressional hearings and deliberations.

The EIS process is abbreviated for legislative EISs. Scoping is not required and, with some exceptions, only a Draft EIS is required to be prepared and submitted to Congress. Both a Draft and Final EIS are required when

- A congressional committee with jurisdiction over the proposal has a rule requiring it
- The proposal results from a study process required by statute such as those required by the Wild and Scenic Rivers Act and the Wilderness Act
- Legislative approval is sought for federal or federally assisted construction or other projects which the agency recommends be at specific geographic locations

■ The agency decides to prepare Draft and Final EISs

A lead agency receives the comments on a legislative EIS, and it forwards them (with the agency's responses to comments) to the congressional committee with jurisdiction over the proposed legislation.[106]

SUPPLEMENTAL EISs

Supplemental EISs (SEISs) are prepared if substantial changes relevant to environmental concerns are made to a proposed action, or if significant new circumstances or information relevant to the environmental concerns of the proposed action become known.[107] Both Draft and Final EISs may be supplemented because of new or changed circumstances or to correct substantial deficiencies in a previous Draft or Final EIS. New information that would result in the preparation of an SEIS must be available before a lead agency makes its decision on the proposed action[108]; information made available after the agency decision has been made is not required to be evaluated in a SEIS (except as discussed below).

Some NEPA experts feel that an EIS should be supplemented whenever significant new information or changes to the project are discovered, even after a decision has been made on the action. Others feel that an EIS should not be supplemented after a decision is made because the purpose of an EIS, to aid a decisionmaker in making a decision, has been fulfilled. The first opinion would require an agency to review its decisions essentially in perpetuity. A more reasonable view is that if a record of decision has been issued, but some part of the decision remains to be made (and could still be modified based on new information), the agency should supplement its EIS. If there remains "major federal action" to occur, and if the new information will affect the quality of the human environment in a significant manner or to a significant extent that has not been already considered, a supplement must be prepared.[109] As with legislative EISs, scoping is not required for federal SEISs.

---

[106]40 CFR §1506.8.
[107]40 CFR §1502.9(c).
[108]Wisconsin v. Weinberger (1984).
[109]Marsh v. Oregon Natural Resources Council (1989).

SINGLE-ISSUE OR LIMITED-SCOPE EISs

The CEQ regulations for federal EISs encourage narrowing the scope of EISs to eliminate insignificant issues from study, but they do not specifically address limited-scope EISs. Some state agencies take the concept a step further by discussing single-issue or limited-scope EISs to further encourage the preparation of very focused documents. In most cases, preparers of EISs will at least explain why certain impacts were considered insignificant and eliminated from further study so that reviewers will not assume the impacts were overlooked.

An EIS that has a pointedly limited scope will raise questions and skepticism regarding an agency's good faith effort at full disclosure of potential environmental impacts. The CEQ and courts that review federal EISs express views that generally support a broad interpretation of the regulations and frown on excessively limited EISs. A balance must be struck when focusing the analysis on significant issues so that the analysis does not discard issues before evaluating their significance. If (1) a thorough scoping process was conducted whereby agencies, groups, and individuals all commented on a single issue or only a few issues and (2) the nature of the proposed project is such that it clearly has only one type or limited types of impacts, a limited-scope EIS may be appropriate.

CHAPTER 8

BUILDING THE EIS FRAMEWORK

If we do not learn from history, we shall be compelled to relive it.
True. But if we do not change the future, we shall be compelled to
endure it. And that could be worse.

Alvin Toffler

INTRODUCTION

While the concepts in this chapter must be understood by an environmental impact statement (EIS) project manager or team leader and by all team members, they would also be useful for anyone involved in the EIS process, including agencies and other public reviewers of EISs. Before beginning to write any part of an EIS, the team members need to define the premises, such as the purpose and need for a proposed action and any assumptions (the foundation), and the scope of the analyses and overall methodology (the framework) on which the analyses in an EIS are based. The premises, assumptions, scope, and methodology must be carried through the EIS by each team member in his or her methodology, analyses, and conclusions. The logic should flow and be consistent throughout the EIS from beginning to end, from author to author. The reason these concepts are also important to reviewers of EISs is that the logic of arguments presented in an EIS is one of the areas that should be analyzed and critiqued during public review of a Draft EIS.

Information that makes up the foundation and framework of an EIS is usually presented in an introductory chapter and in the chapter describing the proposed action, its purpose and need, and its goals or objectives. However, few EISs discuss the scope of analyses to be covered by an EIS, and fewer still describe an overall methodology for the EIS. The scope and methodology for analyses of individual environmental resources or subject areas are normally given in the beginning of the section that discusses the resource or subject. This is also necessary because scope and methodology will vary somewhat depending on the accepted methodology and professional standards for a particular subject. EIS team members must have a clear understanding of an EIS's overall scope of analyses and methodology as well as the methodologies for their specific analyses. The methodologies used also should be explained in an EIS for the public's understanding. For specific analyses, team members should explain any necessary deviations from the overall scope and methodology. By placing specific analyses within an overall context, EIS team members maintain consistency in their analyses and provide the "big picture" for EIS participants and reviewers to understand how the analyses fit, or relate to each other, within an EIS.

OBTAINING IN-HOUSE AGREEMENT

Anyone who will be reviewing in-house versions of the EIS and who has a say in whether the EIS is adequate should be in agreement with decisions made regarding the assumptions, scope, and methodology that will be used in the EIS. In other words, anyone who is in a position to stop the process because of a disagreement with the EIS's basic foundation and framework should be involved in their development. He or she should not get involved for the first time during review of an in-house draft EIS. Substantial delays and revisions occur when in-house counsel (the lead agency's lawyer) or in-house technical experts disagree with an EIS's basic approach, assumptions, premises, scope, or methodology, and, sometimes, terminology. The lead agency's EIS project manager should ensure in-house agreement on an EIS's foundation and framework before having the EIS team begin its analyses.

DEFINING THE PROPOSED ACTION AND SELECTING ALTERNATIVES

EISs usually have a chapter or section that describes the proposed action and alternatives. The importance of the thought and work that goes into this chapter cannot be overstated because it is a large part of an EIS's foundation and framework. Everyone on an EIS team, and all reviewers of an EIS, should understand the need for the proposal, its purpose, its objectives, and how this information determines the range of alternatives to be included in the EIS. All EIS participants should know what the scope of an EIS will be, what subjects it will cover and what subjects were omitted and why, and what the geographic study area will include and how the study area was identified. Without a clear understanding of the proposed action—its purpose and objectives—EIS team members will not have the proper focus in their analyses and may include inappropriate assumptions in their analyses. Reviewers of EISs frequently skip the description of the proposed action to save time; however, reviewers cannot fully understand the rationale and conclusions in an EIS if they don't understand the basic assumptions in the EIS, the scope of the EIS, the objectives of the proposed project, and how the alternatives were identified.

The chapter in an EIS that describes the proposed action and alternatives includes a description of the proposal's purpose and need, the proposal's goals and objectives, the alternatives to the proposed action that will be evaluated in the EIS, and the features of the proposal itself such as the methods that will be used in construction, any phasing or timing considerations, and how the proposed project will operate.

Defining a Proposal's Purpose and Need

The purpose-and-need statement is simply an explanation of why an action or project is being proposed. An action or project is usually proposed because there is a deficiency or problem to be remedied, a need to be filled, or a statutory mandate that requires an agency to take a specified action that will result in environmental impacts. The filling of a need is often accompanied by an opportunity as well, which can also be included in the purpose and need for an action. Opportunities can be a number of things from a window of opportunity to an opportunity for a return on investments. Federal EISs may describe a proposed project's purpose and need in a separate section (see Figure 5.4). All EISs should describe a proposal's purpose and need and whether federal, state, or local regulations mandate the need for the proposed action. If a project's need developed over a period of time and was influenced by various factors, explanations of how the project's purpose and need evolved can be presented, briefly, as background information.

A proposed action's purpose and need should be provided by the person, agency, or organization making the proposal. It's logical that whoever proposed an action must have seen a need and purpose for doing so. Some EIS lead agencies take the position that the proponent's statement of need and purpose for his project belongs to the proponent and the agency is not in a position to agree, disagree, or modify the proponent's purpose and need. If a project proponent says that he sees a need for more golf courses and, therefore, he is proposing to construct one, the lead agency assumes the need is valid for the purposes of conducting an EIS process. Many private development proposals require loans or financial backing which lending institutions or investors will not provide without assurances regarding the economic feasibility of a proposal. Thus, agencies can reasonably conclude that a proponent would not propose a project that is unnecessary or expected to fail. Other agencies feel that the purpose-and-need statement for a project is as much the lead agency's responsibility as any other part of the EIS, and therefore the agency has the authority to change the proponent's statement of purpose and need if the lead agency feels it is necessary to make the statement more accurate in the agency's estimation.

Defining Project Goals and Objectives

A project proponent or applicant for a proposed action or project will have certain goals and objectives for the action. Therefore, the sponsor of the action should provide a description of an action's purpose, need, and objectives. If a proposal is an agency proposal, the goals and objectives of a particular proposed action are usually part of an agency's mandates or actions that are required to carry out the agency's responsibilities. For example, the goal of a proposed low-level military training route may be to have fighter pilots capable of evading detection by radar. The objectives of the proposal may be (1) to provide certain types of practice opportunities along the route that require specific features such as certain types of terrain, (2) to be within a prescribed geographic area considering limitations of distance from the military airfield as a function of available fuel, and (3) to avoid densely populated areas and wildlife reserves or other designated noise-sensitive areas. These objectives become criteria for selecting alternatives and help define appropriate alternative training routes. The criteria also provide the basis for analyzing and selecting an alternative (the agency's preferred alternative) that best meets the objectives of the proposed project (including mitigation measures such as objective 3, above).

The objectives of a proposed project or action are also its requirements. If those requirements are not met, the objectives of the project would not be met or the project would not be viable. Some projects have different types of objectives which range from mandatory to discretionary. The mandatory objectives or requirements are those that are essential for a project to be viable. If even one of the requirements was not met, the action or project could not take place. Other types of project requirements, if not met, may not jeopardize a project, but the project may be less effective than if all of the requirements were met.

Defining Alternatives

Alternatives are alternative means of accomplishing a proposed *project's* goals, not alternative means of accomplishing an *agency's* or *proponent's* goals.[110] Therefore, how a proposed action or project's goals and objectives are defined is very important and can require some concentration to clearly describe the requirements or objectives of a project, rather than an agency's general mission objectives.

Since projects have (1) physical features such as where they are located and how they are designed and constructed and (2) operational features such as times and methods of operation, alternatives can be comprised of modifications of any of these project features. The purpose for modifying project features and creating alternatives is to avoid or reduce environmental impacts. If the design of a project does not result in a significant environmental impact, it is not necessary to look

---

[110]Van Abbema v. Fornell (1986) (Fogleman, 1990).

for alternative ways to design the project. On the other hand, if the design of a project or its location would fill wetlands, potential alternatives would be to modify the project's design, or relocate the project, to avoid or reduce the amount of wetland that might be filled.

The features of a proposed project depend on the type of project. Some proposals (such as airspace designations by the Federal Aviation Administration, or operational changes for an existing action) have no construction associated with them, so construction alternatives would not be applicable and would not be identified or evaluated.

For proposed projects that involve construction, the size of a project is commonly a factor for identifying alternatives. Since the decisionmaker should be presented with a range of alternatives, or alternatives that provide a range of environmental impacts, agencies often evaluate a proposed project in small, medium, and large scopes (e.g., variations in size and other characteristics), assuming the proposed project is for the alternative with the largest scope. An alternative that would result in similar or greater environmental harm than the proposed action does not have to be considered.[111] An EIS would identify, evaluate, and compare how well the objectives of the project are met by each alternative and would then compare the environmental impacts of the alternatives. The decisionmaker can then see that, for example, Alternative "A" would result in the least environmental impact compared to the other alternatives, but also would meet less of the project objectives than the other alternatives, and so forth for each alternative.

> **An alternative that would result in similar or greater environmental harm than the proposed action does not have to be considered.**

Alternatives can be described as alternatives to the proposed action (e.g., the proposed action and Alternatives "A" and "B") or as alternatives including the proposed action (e.g., the proposed action is alternative "A," and the other alternatives are Alternatives "B" and "C")—it is a matter of preference for the lead agency or the EIS team

---

[111]Northern Plains Resource Council v. Lujan (1989); Sierra Club v. Morton (1975); Iowa Citizens for Environmental Quality, Inc. v. Volpe (1973) (Fogleman, 1990).

leader.[112] The EIS team should be told which description to use in an EIS so that everyone is using the same approach to alternatives. Better still, the outline to be used for an EIS should show how alternatives are to be described, and it should be given to EIS team members before they begin their analyses.

Reviewers of alternatives should keep in mind the following:

■ Alternatives must be "reasonable," that is, they must meet the objectives of the proposed project. Therefore, a suggestion to have an elementary school in place of a proposed project for a firearms range is not a reasonable alternative.

■ Suggestions that the lead agency find another site are not as useful as suggestions of where such alternative sites might be.

■ A suggestion of an alternative site is not useful if it does not meet the criteria used in the EIS to evaluate alternative sites. If, however, you feel that the criteria used in the EIS were inappropriate, and your suggestion for an alternative site is reasonable because it will meet the proposed project's objectives, you should provide your rationale.

For detailed alternative site analyses, a project's objectives guide the identification of site selection criteria, or criteria that are used to evaluate how well alternative sites or locations for a proposal meet the objectives and goals of a proposal. An EIS team should identify site selection criteria early in an EIS process so that all team members have the alternatives and evaluation criteria when conducting their analyses. If a proposed project's location could be in an infinite number of locations or in numerous locations, an EIS does not have to evaluate every potential site. The evaluation of alternatives must be reasonably comprehensive, however; and the rationale for selecting alternatives and methodology for evaluating them must be clear and understandable.

The basic consideration, when deciding which and how many alternatives to include in an EIS, is that the document must present the full range of possible impacts and mitigation measures. The "range" will depend on the facts and circumstances of the situation, such as

---

[112]The CEQ regulations discuss "alternatives including the proposed action" (40 CFR §1502.10(e)).

the type of proposal, its purpose and objectives, and potential environmental impacts. To reduce alternatives to a manageable number when a proposal has many possible alternatives, those with similar impacts and mitigation measures can be represented by one alternative as a "type" of alternative. Thus, 30 possible alternatives might be reduced to five or six that represent the range of alternatives that will be evaluated in the EIS.

An EIS team evaluates alternatives for a proposal that has numerous possible alternatives in steps or phases. First, the EIS team develops the evaluation criteria based on the proposal's purpose and objectives. Second, the EIS team (with input from the public) identifies the possible alternatives which may include alternative sites, designs, operational features, and mitigation measures. Third, if the identified alternatives are a reasonable number, the EIS team can proceed with the evaluation of how well each alternative meets the requirements of the proposed action and the potential environmental impacts of each alternative. If an alternative does not meet the basic requirements of a proposal, one does not have to analyze its environmental impacts, and the alternative may be eliminated from further evaluation. The EIS team members must explain why alternatives do not meet the requirements of a proposed action or are unfeasible for technical or other reasons, and therefore the alternatives should not receive further consideration. The remaining alternatives are the "reasonable alternatives" that receive environmental evaluation.

> **If an alternative does not meet the basic requirements of a proposal . . . the alternative may be eliminated from further evaluation.**

If the EIS team and the public identify alternatives that are too numerous to be manageable, the team must identify alternatives with common features that represent types of alternatives with a range of possible environmental impacts and mitigation measures. This step takes place before the third step described in the previous paragraph. The process of developing a short list of reasonable alternatives from a much larger number of possible alternatives must be completed before an EIS team begins its analysis of environmental impacts. The number of alternatives that is considered reasonable will depend on the type of action and its scope of impacts. A proposal with primarily local impacts, such as construction of a building, may have fewer alternatives than a proposal with regional impacts, such as a major transportation project. Three alternatives may be reasonable for the former example, while ten alternatives may be reasonable for the latter. An EIS must describe its process of identifying

- The criteria for evaluating alternatives (based on the proposal's purpose and objectives)
- Potential alternatives
- The evaluation criteria used to develop a short list of alternatives

The following are general examples of alternative evaluation criteria:

- Site Criteria
  - Economically feasible (e.g., for construction)
  - Technically feasible (e.g., type of site)
  - Must be available
  - Must be a certain minimum size or configuration, or both
  - Must have infrastructure in place (e.g., utilities, access) or within a reasonable (specified) distance
  - Must not have environmental constraints that would preclude meeting the purpose and objectives of the proposal
- Operational criteria (identify specific requirements)
- Design criteria (identify specific requirements)

The specific criteria for a proposed action will depend on the circumstances and type of proposal. As an example, specific criteria used in developing a short list of alternatives were identified in an EIS for a proposal to acquire land and realign training areas on a Marine Corps base in North Carolina to meet the need for an increase in firearms and artillery training. Criteria were classified in three major categories:

Operational
- Fulfills firing range requirements
- Fulfills tank maneuver area requirements
- Supplements existing facilities
- Aids training
- Impacts existing training facility or area
- Represents more efficient land use

■ Provides ease of access

■ Soils suitable (traffickable) to support training

Environmental

■ Compatible with existing natural setting/resource management plan

■ Wetland impact

■ Endangered species impact

■ Degree of wildlife disturbance/habitat loss

Socioeconomic

■ Airspace requirements/conflicts

■ Existing land use

■ Number of residences displaced

■ Development potential/county tax base, losses

■ Road conflicts

■ Utility conflicts

■ Cultural resources conflicts

■ Noise impacts

■ Simplifies acquisition transaction (number of landholders)[113]

Some of the listed criteria do not express their intent (e.g., "wetland impact"). The intent was to minimize environmental impacts. All of the above criteria were described in more detail in the narrative of the EIS so that readers could understand the necessity for each criteria. A matrix was also prepared that showed the degree to which each alternative met the criteria.

Defining the Proposed Action

Similarly to the purpose and need, the description of a proposed action also comes from a project proponent or applicant. The person, agency, or organization proposing an action should have all of the information necessary to accurately and thoroughly describe a proposed action. The lead agency's responsibility is to ensure that a proposed action is adequately defined before starting an EIS process. If a project proponent, or applicant, does not have experience defining his or her proposal for a NEPA or a SEPA process, the description may need revision or additional detail to be suitable for an EIS. If necessary, the

---

[113]*Draft Environmental Impact Statement, Proposed Expansion and Realignment of the Marine Corps Base, Camp Lejeune, Onslow County, North Carolina*, 1989.

lead agency's EIS project manager, or consultant's EIS team leader, could assist the project proponent in defining a proposed project. Also, the EIS lead agency could provide the proponent with an outline of the information necessary for a description of the proposed action. The proposed action generally includes the name and type of proposal, who is proposing it, the proposed location, the proposed project schedule, and what the project's features will include such as how it will be designed, constructed, and operated. If some features of a proposed project mitigate environmental impacts, or features are built into the proposal for the express purpose of mitigating impacts, the description of the proposed action should identify those mitigation measures.

The following sections describe some of the basic components of a description of the proposed action. The topics do not have to be in the order shown here, but they should be logically presented. If a proposed project is in its preliminary stages of development, some of the information regarding the project's features may have to be estimates or approximations.

Type of Action

Proposed Action or Project The proposed action, project, or proposal can be an agency (public) action or private action. This discussion introduces the reader to the proposed project, gives the name of the proposal, and presents a general description of what the proposal entails.

Agency Action (If Different from the Proposed Action) If a proposal is a private project, the agency action required on that proposal should be identified. The agency action is the decision contemplated by the agency decisionmaker that would allow, or not allow, the proposed project to continue in its planning process toward eventual implementation of the proposal.

Project Location The proposed project location is usually given in several ways: a legal description, a physical description that would be recognizable to local residents, and a project location map.

Design A description of the preliminary design of a proposed project includes a description of the physical features of the project, its siting or layout, and preliminary design drawings. In particular the "footprint" of a proposed project (the physical area that would be affected by a project) may be relevant if there are natural constraints to development such as unstable slopes, wetlands, or sensitive wildlife habitat areas. This information is helpful when shown graphically on maps and other drawings of the preliminary project design.

Construction A description of the preliminary plans for construction of the proposed project include the schedule for construction such as how long it would take, whether it will be accomplished in phases, what actions would take place in each phase, where excavation or filling might take place, how much material would be excavated, how much fill would be imported to the site, and what the basic methods of construction would be.

Operation A description of the preliminary plans for operation of the proposed project include any seasonal considerations of the project's operation (e.g., year-round or seasonal operation), days of the week, and times of the day of operation. Many other operational features, besides schedules of operation, could be described depending on the type of proposed action.

Project Mitigation Measures Project mitigation measures do not have to be described in a separate section, but may be described under design, construction, or operation if the mitigation measure is part of that particular feature of the proposal. For example, erosion and sedimentation control plans are commonly required during construction. Special lighting and windows might be part of the proposed project's design to reduce the amount of light and glare that the project might produce.

Example Description of a Proposed Action
The following is an example of a detailed description of a proposed action. The example, from a federal EIS prepared for a proposed thoroughbred racetrack, consists of excerpts from a much longer proposed action description. The figures and appendices noted in the excerpt are not included with this example.

Project Description   The applicant, Auburn Racing, is proposing to construct a thoroughbred horse racing park on approximately 165 acres in Auburn, Washington. Although specific improvements would differ, the proposed facility would be similar in scope and operation to Longacres Park (located in Renton, Washington), which closed in 1992. . . . The applicant anticipates that live thoroughbred racing would occur 125 days per year between approximately April 1 and October 1 (similar to Longacres). . . . The facility is being designed to accommodate 7,500 people on average.

Location   The site is located in the City of Auburn, which is situated in the southwest portion of King County, Washington (see Figure 1-1). The property is bounded on the south by a sewer line approximately 1,300 feet north of 15th Street N.W., on the east by the Burlington Northern Railroad tracks, on the north by 37th Street N.W., and on the west by the Union Pacific Railroad tracks in the City of Auburn (see Figure 1-2). Four parcels within these boundaries are not part of the project. The site is located in Sections 1 and 12, Township 21 North, Range 4 East, and in Section 36, Township 22 North, Range 4 East. State Route 167 and Interstate 5 are located approximately 1,000 feet and 2.5 miles west of the site, respectively. . . .

Major Aspects of the Proposal (or Design Features)   Figure 1-2 shows the proposed site plan for the proposed facility. Further refinement of the depicted layout can be expected during the final design process. The following is a description of the project components.

Grandstand   The most prominent structure on the site would be the grandstand located just southwest of the center of the site. The grandstand area would occupy approximately 5 acres west of the racetrack. It would consist of the grandstand structure, the area between the front of the grandstand and the track (apron), a paddock (saddling area), and the grassy knoll and overflow tarmac (an area for standees on overflow days) located north of the grandstand structure. The grandstand structure would be a partially enclosed, six-level facility with approximately 240,000 square feet of floor area. The structure would be aligned in a north–south fashion, with the front of the structure facing east, toward the track (see Figure 1-2). The structure is oriented toward the east to prevent the afternoon sun from interfering with spectators' view. The grandstand would be organized as follows:

Basement:	Service Facilities
Level One:	Apron-Level Grandstand
Level Two:	Upper Grandstand
Level Three:	Mezzanine
Level Four:	Clubhouse/Turf Club
Level Five:	Officials/Press

Patrons would be accommodated on Levels One through Four. . . . The grandstand would house restaurants, food concession booths, bars, restrooms, and mutuel (betting) windows. The grandstand also would contain kitchens, mechanical and electrical rooms, storage, and administrative space.

The structure would be approximately 93 feet high. The grandstand exterior skin would consist of a masonry veneer at the first level only. From the second level to the roof, the structure would be finished with a prefinished corrugated metal siding. On the track side of the building, the clubhouse would be enclosed with a 22-foot-high glass curtain wall. . . .

Racetrack   The project would include a 1-mile (eight furlongs) oval racetrack. The racing surface would be 90 feet wide. The entire racetrack area, including the infield, would occupy approximately 49 acres of the site.

The racetrack consists of two turns with a radius of 419 feet, and two straightaways (the front stretch and backstretch) both 1,320 feet in length. . . . The track also includes chutes, or extensions of the front stretch and backstretch. The chutes provide a longer straight stretch so that during a race, the horses can establish position before entering the turns. . . .

The surface of the track would consist of 10 inches of a sand, silt, and clay mixture underlain by 6 inches of limestone screening material and 12 inches of compacted fill. . . . The track would be lighted for nighttime use. The lighting system would consist of approximately 45 light poles located around the inside rail of the track, approximately 120 feet apart. . . . The fixtures would produce a defined beam directed downward at the track surface. . . .

Barns   The proposed project would include 12 barns plus a pony barn, which would contain stalls for approximately 1,400 horses. The barns would be located north of the racetrack and would occupy approximately 20 acres of the site. Each barn would be 456 feet by 110 feet, one story in height, and contain approximately 110 stalls. . . . The barns also would include equipment rooms (tack rooms) and wash racks. The barns would be constructed of a concrete block exterior, wood stall system. . . . Each barn could store approximately 1,800 bales of hay and 900 bags of grain. Facilities for handling animals waste and trash, and areas for placement of circular electric horse walkers (hot walkers), would be located between the barns.

Approximately 150 dormitory rooms for 300 backstretch personnel would be located at the ends of the barn buildings. . . . Restroom and laundry facilities, as well as mechanical rooms, also would be included in this area. Dormitories would be in use only during the racing season.

Parking The proposal would provide parking for about 5,100 vehicles; parking areas would occupy approximately 43 acres of the site. Most of the parking would be located north and west of the racetrack; additional parking would be provided south of the grandstand. Parking areas would be surfaced with asphalt paving. Overflow parking for an estimated 2,000 additional vehicles would be necessary for one or two days of the year with implementation of a Transportation Management Program. . . .

Access Access to the site would be provided on the south from 15th Street N.W. via a public road ("H" Street N.W.) and on the north from 37th Street N.W. It is estimated that 60% of the traffic would access the site from the south. Twenty-ninth Street N.W. from the east and west would end with an emergency gate and turnaround constructed on the applicant's property. Access from this point would be allowed for emergency purposes only. Internal access would be provided by a north–south private road along the western border of the site. . . .

Accessory Structures Accessory structures would include a paddock, state holding and test barn, veterinary clinic, jockey building, administrative building, and a maintenance building (see Figure 1.2). Two 1,000-gallon fuel tanks would be located near the maintenance building.

Operation The proposed horse racing park would operate approximately 125 days per year during the April 1 to October 1 racing season. These dates are subject to approval by the Washington State Horse Racing Commission. The grandstand would be open for simulcast racing events during the off-season. Simulcast racing involves the satellite broadcasting of racing events taking place at other tracks. . . . A tentative schedule for the racing season would have the racetrack open during the following times:

April 1 to October

Wednesday	3:30 p.m.–7:30 p.m.
Thursday	3:30 p.m.–7:30 p.m.
Friday	6:30 p.m.–10:30 p.m.
Saturday	1 p.m.–5 p.m.
Sunday	1 p.m.–5 p.m.

On Wednesdays, Thursdays, and Fridays, there would be nine races per day. On Saturdays and Sundays, there would be 10 races per day. Each race lasts for about one to three minutes, depending on the distance of the race, with approximately 15 to 30 minutes between the races. During the off-season, simulcast racing would occur as follows:

October 1 to April 1

Wednesday	12 p.m.–4 p.m.
Friday	12 p.m.–4 p.m.
Saturday	12 p.m.–4 p.m.
Sunday	12 p.m.–4 p.m.

Simulcast activity is expected to draw about one-fourth as many people as live racing (on average 1,500 to 2,000 people per event). Other uses during the off-season could include community events. Based on experience at Longacres, these events are very small in comparison to racing events. . . .

Animal Waste Management Plan The proposed animal waste management plan would control animal waste generated on the site. . . . The majority (96%) of animals waste would be generated in the barn stalls. Stall flooring would consist of 2 feet of compacted clay or a comparable impervious surface, which would be covered with straw bedding. . . . Removal of animal waste deposited in stalls in the barns would be the responsibility of owners, trainers, and their employees. Track management would require that animal waste be removed from the stalls daily. . . . Grooms would transfer straw bedding material, along with the animal waste, onto a wheelbarrow or manure cart in the stalls, travel the short distance to the manure bins located at the end of each barn, and dump the waste into bins where it would be picked up daily. Separate removal procedures would be followed for conventional solid waste such as paper, metal, and other debris. Manure bins would be constructed with a concrete floor that would be sloped to the rear, concrete back and side walls 6 feet in height, and a cover. . . .

Animal waste would be hauled off site daily, including weekends and holidays. . . . During the racing season, an average of 28 loads would be hauled off site weekly, with each load representing approximately 185 cubic yards. It is anticipated that waste would be hauled to Albany, Oregon, and sold to a large mushroom-growing company. . . .

Grading, Drainage, Erosion, Sedimentation Control Plan A Grading, Drainage, Erosion, Sedimentation Control Plan (GDESCP) would be implemented during the grading and construction phases of the project as required by the City of Auburn (see conditional use permit condition 1 in Appendix G). The GDESCP would meet or exceed all requirements for issuance of a City of Auburn grading permit. GDESCPs incorporate Best Management Practices (BMPs) to minimize potential for erosion and control sediment when erosion does occur. BMPs that would be used in the GDESCP include the following:

- Stabilizing and protecting disturbed areas as soon as possible in accordance with the schedule presented in the City of Auburn GDESCP requirements
- Retaining sediment within the site area, either by filtering or detaining runoff. . . .
- Implementing a thorough maintenance program to ensure all control practices are working properly. . . .

The GDESCP temporarily would make use of the large pond that is proposed as part of the permanent racetrack facility. This pond, which would become part of Detention System No. 1 (described below), would be excavated first and would be used as a sedimentation pond for water runoff and sediment control. . . .

Storm Water Management System The proposed on-site storm water management system would have two primary functions: (1) to control the volume and rate of storm water runoff leaving the site and (2) to provide water quality treatment for all runoff leaving the site. A secondary function of the system would be to provide groundwater recharge. Surface water would be allowed to infiltrate and recharge groundwater through wet detention ponds and biofiltration swales, to the extent possible. The wet detention ponds would be multi-celled. Lining of some cells may be required to protect groundwater quality. Groundwater recharge would occur in the later cells after water quality treatment has occurred. . . . Appendix C, the Detailed Storm Drainage Narrative, provides additional information.

The storm water management system would be designed ultimately for the 100-year storm event with a multi-orifice control structure to release 50% of the 2-year flow through the lower orifice, the 10-year flow through the middle orifice, and the 25-year flow through the upper orifice. . . . Runoff releases would not exceed pre-development rates for all other storm events. Figures 2.2 and 2.3 show the conceptual storm drainage plan for the project.

The proposed onsite storm water management system would consist of catch basins equipped with oil/water separators, tight lines, and French drains, which would collect water from all surfaces, including paved and compacted clay roadways, walkways, and parking areas; from building roofs; and from below the track surface. . . .

Three separate detention systems are proposed which would discharge from the site at two different locations . . . [detailed descriptions of each detention system were provided but will not be repeated here].

Restoration of Mill Creek and Wetland Creation, Restoration, and Enhancement (Wetland Mitigation Plan) Approximately 56.6 acres of off-site property (Thormod site), which straddles the Valley Freeway (SR 167) north of Main Street in Auburn, Washington, would be used for off-site wetland mitigation. The Thormod site includes a quarter-mile reach of Mill Creek and approximately 45.8 acres of adjacent wetlands, as well as 10.8 acres of upland pastures. Currently, a portion of the site is used by livestock. Restoration of the creek would involve recontouring the banks and establishing an approximately 200-foot-wide riparian corridor. Open water areas also would be constructed to achieve maximum fish and wildlife habitat values. The riparian zone, which is now largely dominated by reed canarygrass, would be planted with willow, black cottonwood, Sitka spruce, and other wetlands trees, shrubs, and herbaceous plants.

Not all existing wetlands on the Thormod site would be enhanced, and all existing upland areas would be converted to wetland by creation or restoration. The proposed open water, aquatic bed emergent, scrub-shrub, and forested marsh habitats would be developed in areas that are now covered largely by heavily grazed pastures and by reed canarygrass, blackberry, and other weedy invasive plants. Removal of cattle from the eastern portion of the site also would help mitigate impacts on wetlands and water quality . . . for more detail see Appendix B3, Wetland Mitigation Plan.[114]

The above example is for a proposed project that has well-defined features of design, construction, and operation. Mitigation measures that are inherent features of the proposed project's operation are the frequency and timing of its operation. Because the racetrack would be in operation during a racing season (spring through early fall), and races are limited in duration, the EIS notes that noise, traffic, and other impacts to wildlife and human neighbors would be less than for an alternative of warehouses (a probable action under the no-action alternative) that would operate all year and 24 hours a day. In addition, several mitigation measures are built into the project itself, such as the animal waste management plan and storm water management plan. The plans were developed in consultation with many agencies. Although both state and federal EISs for this proposal were challenged, the courts upheld the adequacy of the EISs.

SCOPE OF THE EIS

NEPA and SEPA implementing regulations admonish preparers of EISs to narrow the scope of EISs to focus on significant issues, to avoid the collection of extraneous background information (avoid encyclopedic EISs), and to eliminate insignificant issues from detailed study. The scope of environmental analysis consists of two basic types: the physical scope, or the geographical study area, and the subjects or issues to be analyzed in the document. Both types of scopes are important parts of the EIS framework because they control the extent of analyses in terms of the area of impact to investigate, as well as what issues or impacts to evaluate within that area of impact.

Defining the Scope of the Study Area

Any environmental analysis, whether an EIS or some other document, should identify the physical area under analysis. An EIS normally has many authors. If they are not told the specific geographic area of the EIS's analysis, each author may analyze a smaller, larger,

---

[114]*Final Federal Environmental Impact Statement (NEPA), Auburn Thoroughbred Horse Racing Facility*, 1995.

or entirely different area. "Western Halifax County" or "the beach" is too vague, and it assumes that a person will know the boundaries or limits of the study area. An EIS team, including the lead agency, should decide what the geographic scope of the document will be and should draw a boundary around the area of the project's expected environmental impacts. A map of the study area is essential, especially if members of the EIS team will be doing any field work. Without knowing the specific boundaries for their analyses, EIS team members may be looking too far or not far enough. If reviewers of environmental documents do not know the study area under analysis, they cannot evaluate the relevance or accuracy of the information.

Defining a study area's scope requires some knowledge of the types of environmental impacts that are expected from a proposed project, as well as knowledge of how far those impacts are expected to be felt from a project. If a project is stationary, such as a structure or facility, the study area might be smaller than if a project is mobile and moves across a large area. A stationary project, however, may have mobile features that result in environmental impacts such as generation of traffic, or discharge of pollutants to air or water (these may have stationary sources, but the pollutant is mobile and the area of environmental impact may be some distance from the source). Therefore, proposals may require some brainstorming by EIS team members to adequately define the scope of the study area for the overall EIS. Generally, an EIS team will define a study area that is large enough to encompass most of the proposed project's expected impacts. If one environmental topic's impacts will be felt twice as far as the other subjects in the EIS, it may not be necessary to double the size of the study area. The author can explain the reason for the difference in his or her study area in the section of the EIS that evaluates the environmental topic.

If an EIS will include evaluation of alternative sites, the team leader and lead agency will have to decide how to define the scope of the study area for each site. In this case, there may be a regional study area as the overall study area for the EIS, and then specific study areas for each site. Whatever the decision, the study area should be defined for the EIS team members and should be explained for the understanding of the reviewers.

Defining the Scope of Analyses

The possible environmental resources and elements or topics for analyses in an EIS are numerous (environmental elements are listed in Chapter 12). However, most proposed projects would not result in significant impacts to all possible environmental elements. To define the scope of analyses for an EIS, the lead agency and ID team leader must identify a proposal's expected environmental impacts. The EIS team should be given a list of the environmental elements or subject areas that are to be analyzed in an EIS so that team members will know their individually assigned topics and the other topics that will be analyzed in the document. The level of effort each team member will expend in preparing his or her analysis is that person's scope of work. A scope of work for each individual normally is not discussed in an EIS; however, it is necessary for the EIS team members to know what their scopes of work will be. Team members' scopes of work provide guidelines on their assigned tasks, the time they should spend on their tasks, and what type or level of analysis to conduct. It may be useful for team members to include information regarding the scope of the analysis in an EIS when it is relevant to explaining the methodology used for analysis of a particular subject.

U.S. Army Corps of Engineers' Scope of Analysis Agencies may define the scope of analysis for environmental documents differently from other agencies in response to the agencies' regulatory responsibilities and mandates. The U.S. Army Corps of Engineers (COE) is an example of a federal agency that has regulatory and permitting responsibilities which require the agency to simultaneously meet the mandates of NEPA and other laws. When considering whether to issue permits ("permit actions") requested by project applicants, the COE must decide the scope of its NEPA documents. If an application for a COE permit covers only part of a larger project (e.g., a permit for a pier which is part of a larger, upland oil refinery project), the COE must determine the scope of analysis or the portion of the total project it will evaluate in its NEPA documentation.

The COE authorizes the discharge of dredged or fill material when it issues section 404 (of the Clean Water Act) permits. Therefore, the action the agency studies in its NEPA document is the discharge of dredged or fill material. Under section 10 of the Rivers and Harbors Act, the COE regulates work and structures in navigable waters of the United States. Actions evaluated in a NEPA document are those subject to section 10 of the act. The same applies to COE authorizations under section 103 of the Ocean Dumping Act.

NEPA does not expand the COE's authority to approval of activities outside waters of the United States (e.g., construction of the upland facility may proceed without a COE permit). The only activity legally dependent on COE action is the permitted activity. In some cases, however, the COE expands the scope of its analysis to include portions of the project outside the agency's jurisdiction, where the agency has sufficient control and responsibility for the activities, so that the environmental impacts are essentially a result of its action. Generally, however, the scope of analysis would be confined to environmental effects of only the activity requiring a COE permit. For example, when a pipeline or electric utility line requires a COE permit to cross a navigable water of the United States, only the crossing, including the structures in the immediate vicinity which affect the location and configuration of the crossing, would be analyzed to determine whether the crossing would result in significant impacts.

The determination of whether the COE has significant control over a project to expand the environmental analysis to include the entire project is made by the COE's District Commander. The point at which there is sufficient control or responsibility over a project is not defined in the agency's regulations. Typical factors to be considered in determining whether significant control and responsibility exists include

■ Whether the regulated activity comprises "merely a link" in a corridor-type project (e.g., transportation or utility transmission project)
■ Whether there are aspects of the upland facility in the immediate vicinity of the regulated activity which affect the location and configuration of the regulated activity
■ The extent to which the entire project will be within COE jurisdiction
■ The extent of cumulative federal control and responsibility[115]

---

[115]U.S. Army Corps of Engineers, *Environmental Quality; Procedures for Implementing the National Environmental Policy Act (NEPA); Final Rule,* 1988.

Public Input   The public has a significant role in determining an EIS's overall scope, and the level of detail to be provided for any topic that is included in an EIS. During a scoping process, the lead agency will invite the public to help define the scope of an EIS. The scope of an EIS is somewhat dynamic because an EIS process is one of discovery. As more information becomes available, an EIS's scope may be revised to include additional topics or to change the emphasis of analyses.

Controlling Scope

While changes in an EISs scope are normal, changes often cause disruptions in an EIS process. Whenever an EIS's scope changes, work may need to be revised and additional resources and expertise may be necessary. This is particularly true if an EIS is being prepared by consultants under contract because the contract's scope of work is tied to the EIS's scope, and any changes in one will affect the other.

An EIS's scope is less likely to change if a thorough public scoping process was conducted and if the proposed action or project was well-defined and clearly presented to the public. An experienced lead agency and EIS consultant can also make a difference in terms of how well the proposal is presented and the ability to anticipate many, if not all, of the significant environmental issues before the EIS process begins.

METHODOLOGY OF ANALYSIS

The methodology used in an EIS is important because the adequacy of the analysis is partly a function of how well it meets applicable scientific and professional standards. Reviewers are interested in the methodology used in an EIS, the rationale for using a particular methodology, and, when appropriate, the rationale for not using another methodology. The overall methodology for an EIS is the way the EIS was prepared in relation to guidelines or requirements of NEPA, a SEPA, or internal agency guidelines. If the subject of an EIS is regarding a highly technical or scientific discipline, it may be appropriate to provide an explanation of the EIS's overall technical methodology.

The more complex or unusual an EIS, the more important it is to explain the overall EIS methodology. For example, joint agency EISs prepared to meet the requirements of more than one agency, or to meet federal and state requirements, will need an explanation of how the various agencies' EIS contents, formats, and procedures will be incorporated into one methodology for one EIS. This will ensure that all EIS team members understand how the EIS will be prepared, including the sequence of events that must take place. Reviewers of an EIS should also understand an EIS's methodology to understand the document's logic and conclusions.

An EIS should provide a methodology for the technical analysis in each environmental section or subject. An overall methodology may also be useful for preparers of an EIS if the subjects or issues are closely related to each other. For example, a proposed project's environmental impacts might affect wetlands, which will in turn affect water quality, which then affects fish. Even though each topic has its own methodology of analysis, the team members responsible for the analyses of related topics should discuss and come to agreement on a methodology for the parts of the analyses that they have in common. Examples include the approach, areas of emphasis, regulatory and scientific standards to be used, units of measurement to be used, and data sources. Whenever there is more than one author, and choices to be made in methodology, there will be the potential for an author to use a methodology inconsistent with another author's methodology.

Choosing a Methodology for an EIS

The EIS team and lead agency must identify and agree upon the methodology used for an EIS. At a minimum, the overall EIS methodology would include (a) the basic process identified in EIS regulations and (b) a systematic, interdisciplinary approach that ensures the "integrated use of the natural and social sciences and the environmental design arts."[116] An interdisciplinary approach is one in which the EIS team members conduct their analyses in consultation with each other. Beyond the basic requirements, methodologies will vary depending on the nature of the proposal and its probable environmental impacts. The discussions in this book are substantially about methodologies for preparing EISs (there are probably as many methods as there are people who prepare EISs), particularly in Chapter 10 through Chapter 12.

---

[116]40 CFR §1501.2(a); §1502.6; NEPA Section 102(2)(A).

The CEQ regulations require the methodology used in EISs to have professional and scientific integrity and to include sources relied upon for conclusions in an EIS.[117] Several factors affect the choice of methodology for analyzing a specific topic:

■ How appropriate the methodology is for the topic under analysis
■ How well-accepted the methodology is in the particular field or scientific community
■ How much information is available about the proposed action or about the environmental topic in the area of study
■ How much time, money, or other resources are available to use the methodology under consideration

Most of the factors to consider in choosing a methodology involve having knowledge of the subject area and study area and an ability to identify cause-and-effect relationships (e.g., how features of a project will cause certain effects). However, some things, such as the availability of information, are unknown at the start of an EIS. Also, there are practical considerations when choosing a methodology. The EIS team member who will prepare the analyses of a technical subject is often not the person who decides on the amount of time or money available for his or her analyses. If computer modeling, for example, is determined to be the best methodology for a particular analysis, but obtaining the program would take additional time, and running the program would require a substantial increase in budget, the lead agency must decide whether to use a less-than-best methodology or to add more time to the schedule and cost to the budget. The courts have determined that the "best" methodology does not have to be used in an EIS. The decision should be based on whether the information is necessary to allow the decisionmaker to make an informed decision. If a second- or third-best methodology will still permit the decisionmaker to make an informed decision, the extra time and expense to use the "best" methodology may not be warranted. The purpose of EISs is not to generate scientific or technical data—even though the information might be useful to the scientific and regulatory community—it is to generate informed decisions by a lead agency for a proposed action.

---

[117]40 CFR §1502.24.

Explaining the Choice of Methodology

Most authors of EISs will go to the trouble of describing his or her methodology but will not explain why that methodology was used, what limitations affected the methodology, and why other methodologies were not used. For reviewers of EISs, the reasons for using or not using certain methodologies are useful in understanding how the author conducted his or her analysis. Reviewers of Draft EISs frequently ask why a particular method was used instead of another, or why a certain technical step or approach was not used. If an EIS author does not explain how the use of a methodology was affected or limited by some physical or technical circumstance, the reviewer will not understand the rationale for an author's analysis. Furthermore, the reviewers may disagree with the methodology because of their lack of understanding and, therefore, disagree with the conclusions of the EIS. While there will always be some disagreement regarding methodology, many of the concerns result from not understanding the author's approach or reasons for using certain methods in the analyses. If EIS authors provide thorough explanations or rationale for the appropriateness of one methodology over another, and how any limitations may have affected the methodology, the reviewers' concerns will be reduced.

USE OF DATA

Data, or the information presented in an EIS, receives the same public scrutiny for scientific and professional integrity as the methodology used in obtaining and analyzing the data. The source of data in an EIS should be referenced or noted in a manner appropriate for the standards of the professional or scientific discipline. Factors to evaluate the appropriateness of data for an EIS include the following:

- Is it relevant (to the scope—issues and area—being analyzed in the EIS)?
- Is it current? (If it's not current, is it still relevant?)
- Is it enough (to be useful)?
- Is it reputable? (Does the source have scientific or professional credibility?)
- Is it usable? (Is it in a format that you can use?)

Data must be critically evaluated before using it because the quality of the analyses will be affected by the quality of the data. Of the factors listed above, the most difficult can be obtaining usable information. Often, agencies or other members of the public will state that a "great deal" of data on an issue is available. However, data may not be what people thought it was; or, upon review of the data, you may find that no one has any confidence in the data because of the way they were collected, or the methodology used is incompatible with your methodology, or the data are simply not applicable for your purposes. The assessment of the data and reasons for not using them is relevant to the analysis and should be explained in the EIS.

Most of the data used in EISs come from existing sources. The most common sources for data are local, state, and federal natural resource and land use agencies; libraries; and colleagues at the office, at other agencies, or at universities. A proposed project may have a unique aspect for which there are little existing data. In those cases, the lead agency for the EIS must decide on the type and amount of data that must be generated (e.g., by field work) specifically for the EIS. Occasionally, a reviewing agency will hold an EIS's project "hostage," by stating that the reviewing agency doesn't have enough data[118] to agree or disagree with the conclusions of an EIS, and therefore the reviewing agency will state that the EIS is inadequate and withhold its approval (if any) on the proposed project. The lead agency or project proponent must then decide whether to spend the time and money to attempt to generate the data the reviewing agency claims it requires.

EISs are not required to have all data that any reviewer might say are necessary. Many agencies and other public reviewers feel that certain data are necessary for their own programs or personal interest, but the EIS is only required to have data relevant to the decision to be made on a proposed action or project. On the other hand, an EIS is required to (a) disclose any data gaps or missing information relevant to the proposed action and (b) discuss what affect the lack of data has on the analyses. Any limitations on the EIS's analyses from the lack of data should be discussed. Also, if it is likely that EIS reviewers will know of data and expect to see it in the analyses, but the data were not used for some reason, explain why the data were not appropriate and were excluded from the analyses. Otherwise, review-

---

[118]The data may be something the reviewing agency has lacked funds to obtain on its own, but it sees the opportunity to have another party incur the cost to obtain the data if that party is seeking approval from the reviewing agency.

ers may assume that the data were overlooked and that the analysis is incomplete or incorrect without it.

> **EISs are not required to have all data that any reviewer might say are necessary.**

Documentation of Data Sources

Data sources are from written sources such as books, reports, government publications, and unpublished records, and they are also from verbal sources such as personal communications on the telephone or in meetings. All sources of information should be documented in an EIS. The method of documentation, whether in parentheses after the topic or as a footnote, depends on the style used for the EIS. Sources of data obtained verbally can be overlooked because it's easy for authors of EISs to forget to record conversations. EIS team members should have a method for writing the name of the person (the source), date, and topic of conversation that provided the data for the EIS. Possible methods for documenting verbal sources include writing notes on a calendar, in a project notebook, or on telephone conversation forms.

DISCUSSING CONTROVERSY REGARDING ENVIRONMENTAL EFFECTS

Environmental effects or impacts are sometimes controversial, not only in terms of the public's reaction to a proposed project's environmental effects, but also in terms of the scientific or technical community. Topics such as the effects of electromagnetic frequencies (EMFs) have been the subject of numerous studies with inconclusive or contradictory conclusions. Some studies link EMF with higher incidences of cancer, while other studies claim there is no such evidence. The author of an EIS that is addressing this issue must identify the issue and present both sides of the controversy. An EIS that discusses one side of an environmental issue or controversy will present a biased analysis. The methodology for preparing the analysis of a controversial issue is not necessarily different from that of any other analysis, but the author must ensure that he or she presents all relevant information regarding the issue for the public's and the decisionmaker's consideration.

CHAPTER SUMMARY OF TIPS

■ For the benefit of reviewers, explain any deviations in methodology of specific analyses from the EIS's overall methodology.

■ To avoid delays and changes in the EIS process, before work on an EIS is begun:

Ensure in-house agreement on the elements of an EIS's foundation and framework.

Identify and provide to the EIS team the scope of an EIS, what subjects it will cover, what subjects will be omitted and why; what the geographic study area will include and how the study area was identified.

Identify the geographic scope of the document, and draw a boundary around the area of the project's expected environmental impacts.

Provide EIS team members an outline to be used for an EIS that shows, among other things, how alternatives are to be described.

Identify and provide to the EIS team site selection criteria including the alternatives and evaluation criteria.

■ To reduce alternatives to a manageable number, identify alternatives with similar impacts and mitigation measures that can be represented by one alternative as a "type" of alternative.

■ To assist a project proponent in defining his or her proposal, provide an outline (and example) of the type of necessary information for a description of the proposed action.

■ To assist in making a threshold determination, and as part of any environmental document, include any features of a proposed project that mitigate environmental impacts in the description of the proposed action.

■ For EIS team members' and the public's understanding, describe an EIS's methodology, particularly if it is unusual. For example, for joint EISs, explain how it will meet federal and state requirements and how the contents, formats, and procedures will be incorporated into one methodology for one EIS.

■ To avoid inconsistencies in the analyses, all authors of an EIS should use compatible methodologies for the analyses of related topics. Examples include the approach, areas of emphasis, regulatory and scientific standards to be used, units of measurement to be used, and data sources.

■ For reviewers to understand the analyses, explain why a particular methodology was used, what limitations affected the methodology, and why other methodologies were not used.

CHAPTER 9

EIS PROJECT MANAGEMENT

FAST. GOOD. CHEAP. Pick Two.
Anonymous

INTRODUCTION

Project management, as discussed in this chapter, is the management of staff, budget, and schedule during the preparation of an environmental impact statement (EIS). The primary role of a project manager is to make sure that (1) the EIS team members understand their roles, the tasks they are to complete, the time and the budget they have available to complete their tasks and (2) the EIS meets the procedural and substantive requirements of regulations. This may or may not involve managing a contract, depending on whether the service of a consultant is obtained. If an agency is preparing an EIS with in-house staff, a contract is not necessary; however, a schedule and scope of work are still necessary to effectively manage an EIS team.

Project management of EISs requires attention to detail, an ability to identify and organize tasks and develop methodical approaches for the tasks, and an ability to plan and orchestrate the myriad of internal and external events that must take place. For this responsibility, a good memory is only secondary to an ability to communicate with everyone involved in the process, to lead, to make decisions, and to negotiate when necessary.

The EIS team leader is also the project manager for an EIS. If a consultant is preparing an EIS for a lead agency, the consultant and lead agency may both designate project managers. Their roles, however, are different because the consultant's project manager is involved in the details of EIS preparation and the lead agency's project manager is responsible for meeting requirements of the agency's EIS process and overseeing the consultant's work. Agency project management responsibilities include placing public notices in the required publications (e.g., federal or state registers, newspapers) at various points in the EIS process, deciding whether to have public hearings or meetings (and if so, when and where to have them), and carrying out any other requirements of the agency's EIS process that are not delegated to an EIS consultant.

EISs are prepared by individuals who search for information, find answers to questions, and develop solutions (such as mitigation measures) to problems. An EIS project manager sometimes relies on intuition as well as technical knowledge. EIS team members are self-directed and motivated but comfortable working as a team. The task of management, then, is primarily of managing this group or team of people who each have a specific role or job that is different from the others, to ensure that everyone is going in the same direction armed with a common understanding of the proposed project and the EIS methodology. Regardless of whether a contract is used in an EIS process, the EIS team members need to have the same type of guidelines normally found in contracts on the assignment of responsibilities, expected level of effort, and the amount of time available to accomplish tasks. Project managers are responsible for providing these guidelines to team members and for preparing contracts when necessary.

CONTRACTS

If an agency plans to prepare an EIS with in-house staff, a contract is obviously unnecessary. When an agency decides to have a consultant prepare its EIS, or a private party proposes a project that requires an EIS, an EIS consultant is retained under contract with the lead agency or project proponent to prepare the EIS. A lead agency is responsible for an EIS's preparation and for conducting the EIS process regardless of whether the EIS is prepared by a consultant; however, the cost of an EIS for a private project is usually paid by the project proponent.

Contract Terminology

Basic contract terminology used in this chapter are the following:

Client   The client is the agency, person, or company that wishes to have an EIS prepared and is usually the party that pays for an EIS (the paying client). If the paying client is a private party, the lead agency is also a client from the consultant's perspective because the lead agency must be satisfied with the EIS. In this case, the consultant is trying to satisfy two clients, the paying client and the lead agency, even though the lead agency makes the final decisions on the EIS.

Consultant   The consultant, also called the "contractor," is the person or company retained to prepare an EIS. More than one consultant or subconsultant (consultant to the consultant) may be involved in preparing an EIS. The client may have contracts with each consultant; or a prime consultant (the lead consultant), with overall responsibility for meeting the requirements of a contract, may have contracts with its subconsultants.

Scope of Work   The scope of work is usually an attachment to a contract that prescribes the tasks the consultant agrees to perform in preparing an EIS. Those who sign a contract agree to the issues to be addressed, the alternatives to be analyzed, the level of effort for the analyses, the method of preparation, the number of reviews, the number of copies, the number of meetings, and other tasks associated with preparation of an EIS.

Budget   The budget, also an attachment to a contract, is the expected cost associated with preparation of an EIS in accordance with the contract scope of work. If the scope of work were to change, the budget may also change. Those who sign a contract agree that the budget is acceptable for carrying out the scope of work.

Schedule A schedule for accomplishing the scope of work may be a separate attachment to the contract or shown as part of the scope of work. The schedule shows when various tasks and events are expected to be completed or to take place. If events related to the EIS do not take place, the consultant may not be able to meet the schedule for completion of tasks. For example, if information were to be made available to the consultant by a particular date, but the information is delayed, the consultant may have to delay completion of a task until the information becomes available. Those who sign the contract agree that the schedule is acceptable and that they are committing to the schedule if they have responsibilities to meet by certain dates.

Contractual Roles

The major participants in a contract to prepare an EIS are a lead agency and an EIS consultant. If the proposed project that will be evaluated in the EIS is the lead agency's, the agency and consultant are the only ones with a contractual relationship in the EIS process. If the proposed project is a private proposal, the lead agency, EIS consultant, and project proponent will have various roles depending on the type of contract.

Agency Project For an agency proposal, the lead agency and EIS consultant negotiate and sign the contract. This is the most straightforward relationship because a consultant is given direction on an EIS by the "paying client" who is also the lead agency. The scope of work, budget, and schedule are negotiated between the two parties, and the contract is a two-party contract. Information regarding the proposed project, contractual matters, and EIS preparation are all given to the consultant by the lead agency.

Private Project For a private proposal, the lead agency, EIS consultant, and project proponent negotiate and sign a contract (a three-party contract); or the lead agency and EIS consultant negotiate and sign a contract with input from the project proponent[119]; or the project propo-

---

[119]The project proponent can express preferences that the lead agency will consider, but the proponent does not sign the contract.

nent and EIS consultant negotiate and sign a contract with input from the lead agency. Federal lead agencies select the EIS consultant, even though the contract may be signed by the consultant and project proponent. Regardless of who signs the contracts, the consultant is concerned with the satisfaction of all parties. An agency's regulations or policies will dictate the type of contract an agency may use. If a project proponent is actively involved in the preparation of an EIS, the lead agency may require the proponent to give all information, comments, or concerns to the agency, who evaluates the information before giving it to the consultant. In other cases, the consultant receives comments and information from the proponent directly. In the first case, the interaction between the consultant and proponent is controlled by the lead agency, and it takes more time for information to be communicated to the consultant. In the latter case, information is communicated directly from the proponent to the consultant who may use the information in the EIS. But if, upon review of the EIS, the lead agency disagrees with the proponent or wants the consultant to change something provided by the proponent, the consultant must spend additional time and effort to make the agency's changes.

For environmental assessments (EAs), a lead agency does not have to provide direct control over a consultant, although a consultant usually coordinates with the lead agency during preparation of an EA. A lead agency reviews the EA for adequacy to ensure that it meets the agency's requirements.

Project proponents who do not have prior experience with EISs often assume that the role of an EIS consultant is to prepare the "minimum" EIS possible (in terms of cost and effort) that will "satisfy" the lead agency. The consultant, who may have prepared dozens of EISs and has experience with the lead agency, is expected to know how to prepare a "minimally acceptable EIS." In reality, agency requirements for EISs are not identical from project to project, and consultants cannot choose to prepare an inexpensive versus expensive EIS in the same manner as someone might choose an economy car over a luxury car (see Figure 9.1). Requirements for an EIS will vary, even within the same agency, depending on the agency project manager and the type of project being proposed. The level of public controversy and political interest will also affect the scope and cost of an EIS. These factors tend to be unpredictable to some degree.

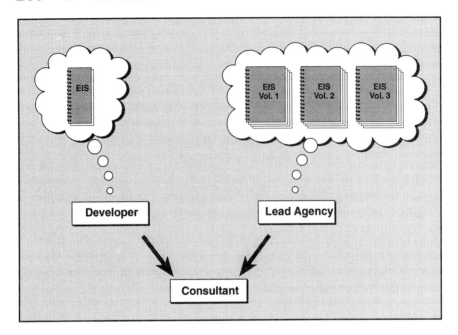

■ **FIGURE 9.1.** Three-party contract for private projects.

To prepare agreements or contracts for EISs, project managers must identify the goals of those who are signing the contract and the factors that will affect the contract. It is not enough to say the goal is to prepare an EIS. What are major factors that affect the type of EIS to be prepared, the methodologies to be used, the expertise that will be required to develop the EIS team, the schedule, and the cost? Answers to the questions will depend on the type of proposed project, the lead agency's EIS process, and the anticipated environmental issues. Without these answers, the contract and, consequently, the EIS may not meet the needs or goals of the participants in the process.

Those who are regularly involved with EISs will agree that an EIS process is dynamic. The process starts early in planning for a proposed action and requires public involvement to define the scope of the EIS. Changes are inevitable, flexibility is a must in the EIS team, and that is how the process was designed to be. Yet, the contract parties in an EIS process insist on a static or rigid approach to the EIS contract. The problem is in the nature of contracts. Contracts are legally binding instruments that primarily specify the expectations of a client (e.g., a lead agency) while simultaneously setting limits on what a contractor is obligated to do. A "flexible contract" is most likely an oxymoron. An EIS is a process, a contract is not. Also, agency contract procurement specialists are used to dealing with construction contracts in which the number of two-by-fours are specified for a given project. Although contractual changes for construction projects are not uncommon, the participants have a fairly clear idea of the type of facility (its function, size, number of stories, etc.) they are dealing with. For an EIS, however, the "facility" or scope of an EIS is being "designed" as the EIS process proceeds. For this reason, many agencies will not start the EIS process until scoping is completed. Then, at least most of the issues to be addressed in an EIS are known before the participants have signed a contract stating what issues and alternatives will be addressed by the consultant.

> **An EIS is a process, a contract is not.**

Even if contracts must be inflexible, participants in the contract should recognize that they are dealing with a process rather than a product (most contracts focus on a product—in this case an EIS). If the scope of an EIS changes because of public input or an additional issue that was not known at the time the EIS started, a change in the contract scope of work (a contract "change order") may be necessary. Changes to a contract normally require additional budget, and they sometimes lengthen the schedule. They also generate paperwork. Therefore, the subject of change orders is not a happy one and the tendency is to assign blame, usually to the consultant, for the cost and delay. Change orders are not a sign of failure on the part of the consultant or any other participant. They reflect the nature of EISs, not contracts. The potential for change should be acknowledged at the start of the EIS process, and when changes to the contract are necessary, they should be expeditiously performed[120] (also see "Controlling Scope, Budget, and Schedule," below).

> **Change orders are not a sign of failure. . . .**
> **They reflect the nature of EISs, not contracts.**

---

[120]This assumes that the necessity for a change is not in dispute.

Role of the Consultant

If a consultant is retained to prepare an EIS, the consultant's basic role is to assist the lead agency in meeting its obligations under NEPA or a SEPA. This means that the consultant must be very familiar with the federal or state EIS process and the agency's EIS process. An EIS consulting firm generally offers its experience and advice as well as its technical capability to prepare an EIS. The relationship between the lead agency, consultant, and project proponent (paying client) will vary depending on the situation. As with other situations, relationships often depend on the personalities and management styles of the individuals who are involved in the process. If a lead agency's project manager does not have much experience in managing EISs, or prefers to be uninvolved in the process, a consultant may receive little specific direction and be expected to make decisions or perform tasks normally performed by a lead agency. A project manager for a lead agency who has substantial experience in managing EISs may not expect advice from a consultant, and he or she may provide strong direction to the consultant on how to accomplish tasks at hand.

The project proponent, on the other hand, may expect a great deal of advice from a consultant on, for example, how to incorporate the EIS process into the overall project planning process, what concerns to expect from land and natural resource agencies, and what mitigation measures might be most effective for a proposed project. An EIS consultant, however, is supposed to be unbiased and cannot participate as a member of a proponent's team that is actively seeking approval on the proponent's project. An EIS project manager is not the same as a proponent's project manager. Therefore, an EIS consultant must not cross the line of being an unbiased third party and, for example, sign permit applications on behalf of the proponent or act as the proponent's representative. For federal EISs, an EIS consultant must sign a disclosure statement that the consultant has no financial or other interest in the outcome of the proposal under evaluation in the EIS.

CONTROLLING SCOPE, BUDGET, AND SCHEDULE

Having said that an EIS process is dynamic and subject to change, it may sound as though there is no way to control the process or the EIS's scope, budget, and schedule. Actually, there are many ways, and most of them relate to good project management on the part of both the consultant's and lead agency's project managers. But even the best project managers can only reduce the number of contract change orders or their severity, not eliminate them entirely.

The scope, budget, and schedule are interrelated: Changes in one will usually cause changes in the others. For example, if a task or issue is added to the scope, the additional work could necessitate an increase in budget for staff and materials to accomplish the task, and time might have to be added to the schedule to accommodate the additional work. A change in the schedule may cause the budget to change, but not necessarily change tasks within the scope (time added to a schedule may result in additional project management and administrative costs, even if tasks have not changed). A contract change order normally results from the addition of a fairly substantial task. However, small additions, or requests that are small tasks, incrementally add up to a substantial amount of time and effort that is not being compensated by the terms of the existing contract. "Substantial" is a subjective word. The requesters of work (the lead agency and project proponent) tend to view their requests as not requiring much time or effort, while the requestee (the consultant) may view the actual amount of time required to accomplish the tasks as substantial.

The need to make changes to scope, budget, and schedule is reduced if they are prepared well to begin with. First, they must be realistic. A scope of work should consist of tasks that clearly identify responsibilities that can be accomplished and are measurable. The budget should relate directly to the tasks in the scope of work and allow staff sufficient time and resources to accomplish the scope of work. The schedule should allow time to accomplish the specific tasks within the scope of work, review internal drafts of the EIS, have meetings, prepare the EIS for printing, and conduct public reviews and hearings. Second, they should be as detailed as possible and specific regarding tasks and expectations. If the scope of work is detailed, the budget and schedule will be easier to prepare and the relationship between them is more evident. If one is modified, it is easy to see where the others must also be modified.

Controlling Scope

An EIS's scope includes topics or issues to be analyzed as well as the geographic scope of analysis. The level of effort to accomplish each task, or address each issue, is described in a contract scope of work.

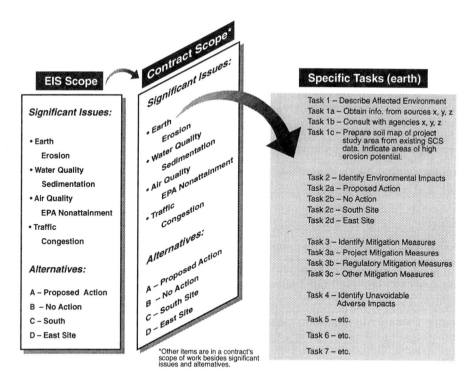

■ **FIGURE 9.2.** EIS scope and contract scope.

Figure 9.2 shows the relationship between an EIS's scope and contract scope of work in simplified form. All of the issues and alternatives identified in an EIS's scope should be included in the contract scope of work. However, contract scopes of work will normally contain more information than is shown in the figure, such as a description of the proposed action on which the analysis is based, any assumption and methodologies to be included in the analysis, any field work that will be conducted, number and length of meetings, number and types of graphics, and so forth. It is also important to state what will not be included in the analyses to avoid misunderstandings or incorrect assumptions about the level of effort for each task. Assumptions and premises for the analyses in an EIS should be made a part of the scope of work.

The scope of an EIS is determined by the lead agency, with input from the public who responds to scoping notices and provides comments on Draft EISs. Regulations emphasize the need for public involvement to help define the scope of EISs. New issues, alternatives, or mitigation measures may be added to EISs because of public concerns. Lead agencies and consultants who have substantial experience with EISs often anticipate public issues and concerns so that public input results in few surprises. If the lead agency and consultant prepare a thorough scope of work, and public comments do not change the scope of work, an EIS's scope may still change because of in-house agency reviewers' comments.

Of all the EIS contract participants, a lead agency's project manager has the largest role in controlling the scope of an EIS. This is because a lead agency makes the decisions regarding the scope of an EIS, and it is the lead agency's project manager that receives pressure from in-house reviewers, the public, or other agencies to add topics to, or expand, an EIS's scope. Lead agency project managers should evaluate requests for additional topics or increases in level of effort to determine if the requested work would substantially improve the analyses or result in a more informed decision by the agency decisionmaker. But in some cases, lead agency project managers do not take the time to evaluate the requests in light of the EIS's agreed-upon scope and elect to add whatever is requested, rationalizing that more information will improve an EIS.

The larger the number of individuals that are involved in reviewing an EIS under preparation, the greater the likelihood that tasks and issues will continue to be added to the EIS over time. The EIS's scope becomes a "creeping scope." A lead agency may have several specialists (e.g., biologist, archaeologist, hydrologist, traffic engineer, attorney) who review internal drafts of EISs. Each specialist reviews the EIS with his or her specialty's regulatory and agency program mandates in mind. Since each person's focus is different, he or she will make suggestions that add or expand the scope of the EIS in accordance with his or her specialty. This tendency is unconscious; that is, reviewers are not intentionally trying to change an EIS's scope. But the result is that incremental tasks, multiplied by many reviewers, will increase the level of effort beyond that specified in an EIS's contract. At some point, the EIS consultant must request a contract change order to receive compensation for the additional work. For large EISs with many internal reviewers, change orders may be necessary many times if the EIS's scope is continually creeping.

To control an EIS's scope and prevent it from creeping, the lead agency's project manager must first be able to recognize the potential for the scope to creep and, second, consciously evaluate suggestions from colleagues so that any out-of-scope tasks are identified as such. Third, the project manager should evaluate the out-of-scope tasks to determine if they are truly necessary. To determine if a task is truly necessary, the project manager should consult the EIS's scope of work. This assumes that the scope of the EIS is reflected in the EIS contract scope of work and that the EIS's scope was developed carefully to include NEPA's requirements, a SEPA's requirements, and the agency's requirements. In other words, the EIS scope of work should be the lead agency project manager's blueprint for the EIS. After all, the scope of work was signed by the EIS contract participants after much thought, discussion, and agreement about the details of the scope and methodology of the EIS. The scope of work should provide a focus for an EIS. If an EIS's focus is not maintained, its scope will be subject to the various focuses of the reviewers and will constantly change. Being focused does not mean being rigid, it means being aware and making conscious decisions to change or not change an EIS's scope.

Too frequently, the EIS contract parties (lead agency, project proponent, and consultant) upon completing the contract will file it away and not look at it for months. To maintain the focus of an EIS, the contract participants should regularly refer to the scope of work.

Controlling Budget

Budget is controlled by all participants in an EIS contract: by the consultant in how the consultant spends time and resources, by the lead agency in how the agency maintains the scope of the EIS and manages its work, and by the project proponent by how well the proposed project is defined and how involved the proponent is in the EIS process. An essential consideration in controlling budget is that actions of the lead agency and project proponent affect how the consultant spends time and resources.

Actions of the lead agency and project proponent affect how the consultant spends time and resources.

EIS budgets consist of staff time and materials used to accomplish tasks in a scope of work. The cost for staff's time is by far the largest part of an EIS's budget, although printing costs can be substantial if many hundreds of copies of a large EIS are being made. Since time is the biggest part of an EIS's budget, EIS team members must know how much time they have to accomplish their assigned tasks. To include sufficient time, it is a good idea to get team members' estimates of the amount of time needed to accomplish tasks when preparing the budget. Once the contract is approved, the project manager should discuss any adjustments to the budget with team members for a clear understanding of the level of effort and budget for each task. Then during preparation of the EIS, the project manager must monitor (e.g., weekly) each team member's use of the budget and periodically discuss task progress in relation to the available budget. Performance of tasks and maintenance of budget are not just the project manager's responsibilities, they are also EIS team members' responsibilities. One way to control the budget is to choose team members who perform tasks within budget.

Creeping scope, discussed above, adds staff time and budget to an EIS contract. Large changes to a scope of work will necessitate additional budget, sometimes doubling or tripling the original budget. Changes to a scope of work are not limited to additional tasks or expansions of tasks. Changes in an EIS's format and in basic assumptions regarding the analyses will also result in a need for more budget if the change takes place after the EIS's format or analyses were completed. The consultant must then revise a substantial amount of work that was already done. To control the budget by reducing changes in scope, the lead agency project manager should obtain in-house agreement on the EIS's foundation and framework (its assumptions and premises, scope, and methodology) before the team begins work on the EIS. During internal review of preliminary EISs, agency staff should consider the EIS's premises, scope, and methodology when making suggestions or review comments. The agency project manager should coordinate staff comments, to resolve conflicting directions, before giving them to the consultant.

Once work on an EIS has begun, any changes to the description of the proposed action or project will necessitate revisions, which in turn affect the budget. A proposed project that is well-defined and has complete information regarding its construction, design, and operation generally requires few changes during the course of EIS preparation. Also, if a project proponent submits a proposal that is unreasonable or pushes the envelope of a reviewing agency's authority, the proposal will most likely be revised during the course of preparing the EIS and will require changes to scope and budget. A project proponent or agency that makes frequent requests of the consultant for written reports, research of files, and other sundry tasks that are outside the contract scope of work is using the consultant services without compensation. If a task takes 10 minutes, most consultants will absorb the cost; however, repeated requests and tasks that take several hours will eventually result in a request for additional funds. Budget problems often result because more than one person representing the project proponent, as well as more than one lead agency employee, are requesting various consultant personnel to perform tasks which cumulatively consume substantial time and budget. These are the unseen out-of-scope tasks because no single person is controlling the requests being made to the consultant. A solution would be to designate a person with the lead agency (the project manager) and also designate a person for the project proponent who reviews requests to the consultant and submits those within the contract's scope to the consultant's project manager. The disadvantage of this solution is that communication bottlenecks at the designated person, particularly if he or she is unavailable and an immediate request of the consultant is desired. The consultant's project manager should keep records of out-of-scope tasks to submit to the project proponent (or paying client), with the monthly invoice, whether or not a request for additional budget is made at that time. For substantial out-of-scope requests, of course, the consultant's project manager should immediately discuss the request's affect on the budget with the client(s).

The budget should be tied to the EIS schedule as well as the scope of work. If the schedule is lengthened because of additional work or delays that were outside the consultant's responsibility, the budget may have to be increased for additional project management costs. Conversely, if an EIS is completed ahead of schedule, project management costs should be less than originally estimated. A long hiatus, during which work has stopped on the EIS for some reason, may also require additional budget to start up the project again. Team members will have been working on other projects and cannot simply pick up where a project left off if several months have gone by. Some of the momentum is lost when a project is temporarily shelved, some information is forgotten, and the EIS team members must get reacquainted with the project. Avoid delays and stay on schedule to control the budget.

Controlling Schedule

An EIS consultant is responsible for accomplishing tasks in a scope of work in accordance with the contract schedule. Within the schedule, however, may be tasks that are the responsibility of others (e.g., lead agency and proponent) which are to be completed within a specified period of time. As with the budget, discussed above, the lead agency and project proponent can assist or detract from maintaining the schedule, depending on whether they adhere to the schedule or make changes to the scope of work.

A consultant's project manager should prepare an internal schedule to accomplish all of the tasks in an EIS scope of work. The internal schedule is usually more detailed than the contract schedule, but the milestones or events in the contract schedule are part of the consultant's internal schedule. The consultant's internal schedule should have dates by which all EIS team members are to have their tasks accomplished, dates for EIS sections to go to word processing and to editing, dates for internal draft EISs to receive quality control review, and so forth. The schedule not only informs team members of due dates, it is also a planning tool so that the resources of the firm, such as word processing, can be available for the EIS when they are needed. Delays in consultants' schedules invariably occur from external causes, so that given all of the EIS schedules being managed in a firm, the allocation of resources can be a constant and highly stressful juggling act. Consequently, most consultants are highly motivated to maintain schedules.

Many people and agencies are involved in an EIS process. Therefore, to control the schedule—or, more accurately, to keep everyone on schedule—requires coordination and regular contact with those who are participating in the process. Generally, both the consultant and lead agency are actively coordinating with agencies and colleagues during an EIS process. Delays occur if the lead agency or an agency or source outside an EIS team does not provide information by an expected date. Those who are involved with EISs know how important it is to maintain strong ties and good working relationships with colleagues. Most agency personnel are professionals who are responsive and provide information by a requested date, but maintaining a friendly professional relationship can sometimes make a difference in responsiveness.

A schedule can be lengthened by a lead agency if the agency takes longer to review internal drafts of EISs than was originally scheduled, or if the agency decides to have more internal reviews of an EIS. The first case may affect only the schedule. But the second case affects the schedule and the budget because an additional version of the internal EIS must be prepared and reviewed. Changes to the scope of an EIS also may necessitate additional in-house versions of an EIS for review. Therefore, maintaining the scope of work will also maintain the budget and schedule.

> **Maintaining the scope of work will also maintain the budget and schedule.**

If a project proponent changes the description of a proposed action, the scope of work, budget, and schedule will most likely increase. Or, during preparation of an EIS, if the description of the proposed action is insufficient, the lead agency may request additional information from the proponent regarding the proposed project's construction, design, or operation. An EIS schedule may be delayed for the amount of time it takes the project proponent to prepare the information, and the consultant to incorporate and analyze it, in the EIS.

Schedule delays and additions to budget are a function of when, during the preparation of an EIS, a change in scope takes place as well as the magnitude of the change. If the change affects work that has already been done, so that the analyses must be revised or new analyses must be done, the affect on the schedule and budget will be greater than if the change took place before much work was completed.

QUALITY CONTROL

A universal quality control process does not exist. A consulting firm or agency may have its own quality control process, but many do not.

"Quality control" may be a one-time editorial review by an editor, or by a senior reviewer if a professional editor is not available.

The quality of an EIS is a function of the scope of work and is an important part of project management. An EIS is a complex document that consists of analyses prepared by many authors. Its format and content are prescribed by regulations, and it is subject to numerous internal reviews by many people. Each time an in-house version of an EIS is reviewed, the review comments cause portions of the EIS to be revised. During an EIS's preparation and during the many revisions of an EIS, the quality of the document is maintained through a quality control process that is primarily the responsibility of the project manager (with assistance from an editor). Quality control of an EIS is an ongoing process to maintain the integrity of the document, the accuracy of the analyses, the consistency of the methodology and conclusions throughout the EIS, the consistency of the format, and the focus and style in the EIS so it is written with one "voice." A technical editor provides much of the quality control, as well as grammatical and spelling corrections, when an EIS approaches the production phase or the point at which an EIS is becoming a document. A project manager, however, is responsible for the accuracy of analyses and consistency of methodology and conclusions and is responsible for an EIS's quality throughout the EIS process.

MAINTAINING THE ADMINISTRATIVE RECORD

EISs and decisions by a lead agency are subject to review by the courts. The courts will review the administrative record, which generally consists of the EIS and supporting studies. Courts will sometimes also consider other related documentation of events and information. Therefore, during preparation of EISs, team members should be keeping records in a project notebook, file, or equivalent method, of discussions, events, direction regarding the EIS by agencies—anything related to the EIS.

Even if an EIS is not reviewed by the courts, we should maintain the administrative record for accountability of our time and actions. We should know why we did certain things, but we may not remember the reasons unless we record them. Keep daily logs for each project. Document phone conversations (give copies to those who need to know).

In particular, keep records of instructions, directions from lead agencies or natural resource agencies regarding a preferred methodology, and what to include in an EIS's analyses.

EIS team members should maintain an organized project filing system. Correspondence should be filed in some methodical fashion—for example, chronologically and by agency. Contracts and contract-related information would be in a separate file. Information regarding the proposed project (the project description, drawings, and other related information) could be filed with older versions at the back and newer versions on top. It is very important to clearly mark old or obsolete information so that it doesn't inadvertently get used in the latest version of the EIS. An EIS can go through numerous versions before it is completed. Each version should be numbered and dated. Records and documents should be boxed and stored until it is clear that no court reviews will take place.

EFFECTIVE EIS TEAM MANAGEMENT

An interdisciplinary (ID) team, or EIS team, is comprised of technical specialists in the science and planning disciplines including wildlife, fish, and wetland biologists; geologists; soil scientists; hydrologists; urban planners; traffic engineers; archaeologists; and historians. Specialists who evaluate effects of noise, socioeconomic impacts, and impacts on visual quality also may be members of an ID team. An architectural or legal firm might be members of the team who provide specialized areas of expertise that are used by the ID team members in their analyses. Production staff such as graphic artists, word processors, and editors are members of the team that create the physical form of an EIS. Some firms and agencies have in-house capabilities to duplicate copies of the EIS for distribution, while others send a camera-ready copy of the EIS to a print shop for duplication. All of these specialists are members of an EIS team who become involved at various stages in an EIS process. The EIS project manager is involved throughout the entire process and is responsible for getting the appropriate people involved at the appropriate time. To effectively manage an EIS team requires (among others things) leadership, planning, organization, communication, and experience.

Leadership

A project manager is a team leader with the ability to strategize the most effective approach for an EIS, make decisions, and provide guidance with diplomacy within the group and with other groups and agencies. To lead a team, a project manager must be available to the team members to answer questions and help solve problems early, before problems start to affect an EIS's budget, schedule, or quality. As the leader of an EIS team, a project manager must know how team members are carrying out their responsibilities because the leader is the person responsible for all aspects of EIS preparation—"The Buck Stops Here."

Planning

A project manager is not required to have a background in planning or any other particular educational background. But he or she usually has a technical background in one of the disciplines involved in EISs. A project manager identifies the various events that take place during preparation of an EIS, and he or she plans ahead so that participants in the process will know what they will have to do and when they will be required to do it. Some actions depend on the actions of others, and those "critical paths" must be planned to avoid delaying the entire process. A project manager may need to negotiate project priorities with other staff members to ensure that an agency's or firm's resources are available when they are needed. Detailed scheduling helps plan an agency's or firm's resources.

Organization

Effective project managers are typically well-organized. They plan and schedule actions, and they waste little time because they know how tasks must be accomplished. Preparing an EIS with specialists who have widely varying backgrounds and abilities requires a project manager to orchestrate the team members' actions in an organized and systematic manner. By being organized, a project manager avoids crisis management, or situations where someone must stop what they are working on to take care of an issue that was overlooked. Team members will quickly tire of a "crisis du jour" type of management, and their confidence in a project manager will diminish.

Communication

Because EISs require team work, and team work requires a great deal of communication at all levels, communication is the key to effective project management. Project managers are frequently on the telephone and in meetings, and they frequently write memos to pass information to other team members. If an EIS team consists of a large number of individuals (a dozen or more is not uncommon), simply communicating with them can take a surprising amount of time. With today's "E-mail" and "voice mail," communication has become easier than it was 10 years ago, but the number of people with whom we must communicate has grown. As regulatory requirements increase, time spent in communication with regulatory agencies also increases. Analyses in EISs are becoming more sophisticated and specialized, and they require additional subconsultants with the experience to perform technical or specialized tasks. EIS subconsultants are team members too and must be kept apprised of schedule changes, changes to the project, budget status, and other important information.

Experience

No one should have the role of a project manager without first having the requisite experience. This seems obvious, but agencies and companies have given project management duties to people who may have had years of experience reviewing EISs, or writing sections in EISs, but did not have the communication, organization, planning, and leadership experience to be an effective project manager. Normally, a person obtains this experience, first, by preparing analyses in EISs for some years. Then, if the person is interested in project management, he or she usually has roles as an assistant project manager under the guidance and training of an experienced project manager. Training in and exposure to EISs do not always translate into requisite experience, however. There are those who do well in their technical area of specialty but do not plan, organize, communicate, or lead well and, therefore, should not be project managers.

EFFECTIVE EIS TEAM PARTICIPATION

The success of an EIS team depends on the team members as much as on the project manager or team leader. Assuming that a project manager provides the proper guidance and leadership discussed above, the team members must do their part for the EIS process to operate effectively. The criteria for effective EIS team participation are similar to the criteria for effective project management, but the magnitude is less because the team members' areas of responsibility are narrower than a project manager's. For example, since communication is not in one direction, a team member must communicate also, but not with as many people and not as frequently as the project manager.

Understanding the Proposed Action and Alternatives

To be effective in analyzing a proposed action, team members must thoroughly understand the who, what, when, where, and why of a proposal and its alternatives. Because a team member's role in an EIS process is generally a narrow one, such as the analyses of water quality impacts, there is a tendency to only look for project features that affect the area of analysis (water). But team members should have a thorough understanding of the proposed action and its alternatives, the EIS's assumptions and premises, its scope, and methodology before beginning the analyses. Understanding the details of a proposed action, its construction, and its operation helps to identify potential environmental impacts, such as indirect impacts, that may be overlooked if a team member does not understand the entire proposal. An EIS's assumptions and premises, its scope, and methodology affect the analyses of alternatives and mitigation measures. If team members do not understand the assumptions and premises of an EIS, they may have incorrect assumptions that affect their analyses and conclusions.

Organization

Being organized saves time for team members as well as for project managers. Conversely, being disorganized not only causes confusion and wasted time, it raises the frustration and stress level that is normally high in an EIS process. Most EIS team members are responsible for more than one project at any given time. Therefore, organization can be critical for keeping the requirements of one project separate from those of another. Even for one project, notes, data, budget and schedule information, and various tasks required to carry out the analyses can clutter desks or files in short order. The key is to be organized from the beginning of a project, not after becoming overwhelmed with paper.

Being organized is not just having a neat desk of course, it's a state of mind. To be organized in the way you go about doing your work, you must know what needs to be done and be methodical in the way you do them. Being methodical, in turn, is being systematic when preparing EISs as directed by NEPA: "Utilize a systematic, interdisciplinary approach. . . ."[121]

Communication

As with project managers, communication is critical for EIS team members to effectively carry out their responsibilities. Communication takes place between EIS team members, agencies, other consultants, and service companies that will be providing support during production of an EIS (e.g., laboratories, print shops). Team members routinely communicate with colleagues, but communication with those who are not familiar with the EIS process sometimes presents a problem. If you are asking for information, the way you phrase your question will affect the type of answer you will get. You may know what type of information you are looking for, but the person you are talking to may not. For example, if you want to know what types of activities are taking place on a proposed project site to prepare a land use analysis, you might ask the landowner "What is the land being used for?" The landowner's response would be "Nothing, it's vacant." The landowner is responding to the question in terms of a built or *developed* use. You need to explain that you want to know what *activities*, sanctioned or not, might be taking place on the site and give the landowner specific examples: Is the site used for pasture? Do neighbors ride bikes, horses, or hike on the site (i.e., trespass)? Are there any public easements on the site? Is active erosion taking place on the site (e.g., from a poorly designed outfall)? Since most people aren't familiar with an EIS process, communication will be effective only if you communicate so a person understands the type of information you need.

Team members must have a certain amount of initiative and tenacity when looking for information. Sometimes a person designated as a contact is not the most knowledgeable person on a subject. To make effective use of time and get the most complete or thorough in-

---

[121]NEPA Section 102(2)(A); and 40 CFR §1501.2(a).

formation, find out who has the most information about the topic you're analyzing. Make sure the source for the information is reputable, with the background and qualifications for the subject under discussion.

TIPS

On EISs With Short Fuses

If an unrealistic schedule for an EIS is driving the EIS process because, for example, the project proponent did not allow sufficient time for an EIS process in his or her overall project planning, or the scope of the EIS has increased but the schedule has not, the EIS team may be placed under considerable pressure to complete an EIS in less time than would normally be required. Shortcuts can be devised but at high cost to team members because all shortcuts are intensive uses of time and resources. The team works overtime, sends faxes and uses couriers to expedite the transmittal of information and approvals, and has weekend working sessions when, for example, information from the lead agency or proponent arrives late Friday afternoon. Quality control of the EIS is compromised if time is not available (one cannot perform "faster" quality control).

A "shortcut" that is often proposed by project proponents and some lead agencies is to provide sections of an EIS for review as they are completed. Don't do it, especially if there are many reviewers and many EIS sections. It sounds attractive in theory, but it does not save time; it actually takes more time, costs more, and creates confusion. Fallacies with this approach are as follows:

■ Each section of an EIS is not necessarily an independent analysis. An EIS is prepared using an interdisciplinary approach; therefore, some information related to a given topic may be in another section which should be referenced by the author. If reviewers don't have all of the information, their comments can be based on a partial or incorrect understanding of the analyses. Also, reviewers are more likely to forget the EIS's assumptions, scope, and methodology when reviewing separate sections.
■ Review comments are not simply "replace x with y." Changes in one section often affect the analyses in other sections. While a section is being reviewed, other sections are still being prepared. Therefore, new sections are being sent out for review while review comments

are being sent back to the EIS team and while the team is making revisions to sections that have been reviewed. Often review comments on new sections contradict review comments on earlier sections as the reviewers get more information. It quickly becomes confusing to reviewers and authors alike.

■ If one section or analysis is provided for review as soon as it is written, the basic approach to EISs, an interdisciplinary approach, is bypassed because the ID team has not had the opportunity to review and collaborate on their analyses.

On Managing EISs for Controversial Proposals

If a proposal has the potential for significant environmental impacts, the proposal will most likely be controversial. The degree of controversy will normally be related to the type of proposal, extent of potential environmental impacts, and interests (e.g., financial) of those who may be affected by the proposal. A proposal to construct a large maximum security detention facility (prison) will be controversial to local residents, while a proposal to allow off-shore drilling through federal leases will be controversial regionally and nationally as well as locally.

Since most EISs are controversial, all EISs should be managed with attention to the details discussed in this book starting with Chapter 4. For an EIS that is particularly controversial, however, the EIS team must be prepared to deal with the controversy. This requires the team to anticipate issues and plan for how to respond to them. A person who is designated as the lead agency's point-of-contact must be prepared for a variety of forums in which controversial issues may be raised, such as phone calls, letters, personal visits, or public meetings. Other team members who may have contact with the public in the field, or who may receive calls, should be aware of the proper responses. In some cases, public inquiries to consultants or other team members are redirected to the lead agency point-of-contact or public information specialist. In this way, the public receives consistent (and hopefully accurate) replies regarding controversial issues. News media normally are directed by team members to the lead agency for information.

On Time Wasters

If possible, the best course of action is to avoid people who waste time. But sometimes the person who wastes time is the lead agency's or consultant's project manager, or someone else with whom you frequently must communicate. When calling, have a list of topics ready and try to keep the conversation on those topics. Tell the person what topics you need to discuss and that your time is limited. Call 10 or 15 minutes before lunch or quitting time. Do anything, because time is precious. One can't dispense entirely with polite conversation, but try to keep it to a minimum. Be blunt: "We really have to make arrangements today for next month's public meeting," without being rude.

To Project Proponents

Avoid election years to announce your proposed project and start an EIS process unless you are certain of full public support of your project. Since an EIS is a public process with news media coverage, politicians will take advantage of the opportunity to grandstand against a project. More misinformation seems to get circulated that way.

CHAPTER SUMMARY OF TIPS

■ To ensure that an EIS meets expectations of the lead agency or project proponent, or both:

Identify the goals of the EIS process and major factors that affect the EIS: methodologies to be used, expertise that will be required, the schedule, and the cost.

Specify the responsibilities of the lead agency in management of the EIS process and any information or input to the EIS that will be provided by the agency and proponent.

■ To minimize contract change orders:

Wait until the scoping process is completed before signing a contract that specifies the range of issues, alternatives, and other requirements of the EIS.

Prepare realistic and detailed contract scopes of work, budgets, and schedules.

In the contract scope of work include assumptions, premises, and tasks that will not be performed.

■ To maintain an unbiased position, EIS consultants should not sign permit applications or act as a representative of the project proponent.

■ To determine if a requested task is within an EIS's scope, consult the contract scope of work.

■ To maintain focus on an EIS's scope, do not file away the contract scope of work—refer to it frequently.

■ To allow sufficient time and budget for tasks, get team members' estimates of how long they will need to accomplish tasks when preparing a contract budget.

■ To control the budget, choose team members who perform tasks within budget.

■ To control scope and budget:

The agency project manager should coordinate staff comments, to resolve conflicting directions, before giving them to the consultant.

Prepare well-defined descriptions of proposed actions regarding their design, construction, and operation.

Prepare rational projects that do not push the envelope of a reviewing agency's authority.

Avoid requests of a consultant that are not included in the scope of work (e.g., if monthly status reports are in the scope of work, don't ask for weekly reports).

Designate a person with the consultant, such as the ID team leader or project manager, and a person at the lead agency to coordinate requests for information or tasks.

Keep records of out-of-scope tasks to submit to the project proponent with monthly invoices, whether or not a request for additional budget is made.

■ To avoid inadvertently using outdated information in the latest version of an EIS, clearly mark old or obsolete information and graphics. Number and date each version of an EIS.

■ To avoid delays, keep all team members, including EIS subconsultants, apprised of schedule changes, changes to the project, budget status, and other important information.

■ To identify all potential environmental impacts, including indirect impacts, understand the details of a proposed action, not just the portions that affect your area of speciality or interest.

■ To effectively manage time, get organized and stay organized from the beginning of an EIS.

■ To communicate effectively with people who are not familiar with an EIS process, describe the type of information that is being sought.

■ To make effective use of time and get the most complete information, search for and avail yourself of the person who has the most information about the topic under analysis.

CHAPTER 10

EIS PREPARATION

The Ancient Mariner *would not have taken so well*
if it had been called The Old Sailor.
Samuel Butler

INTRODUCTION

Despite regulatory guidelines for the content and format of environmental impact statements (EISs), and guidelines for the EIS process, there is much flexibility in how an EIS can be prepared. Flexibility is necessary because of the almost infinite variety of (1) types of proposed actions, (2) environmental issues for a given area, (3) public concerns, and (4) political agendas that ultimately affect EIS preparation. There is no single correct way to prepare EISs. This chapter will address an approach, a methodical one, to preparing EISs that can be used for any type of EIS. The basic requirements remain the same, although the content and subject matter of EISs can be as unique as the projects, locations, and people involved in the process.

The reason for using a methodical approach when preparing EISs is that it organizes the process into logical steps and provides an element of control over a process that can readily become chaotic. A methodical approach simply makes the job easier in the long run, although it takes more preparation (to get organized) and planning in the beginning.

RESPONSIBILITY FOR PREPARATION OF EISs

For federal EISs, only an agency or a contractor selected by an agency may directly prepare an EIS.[122] Contracts to prepare EISs are discussed in Chapter 9.

Normally, a federal agency prepares an EIS under the National Environmental Policy Act (NEPA), and a state or local agency prepares an EIS under a state environmental policy act (SEPA). In 1975, Congress amended NEPA to provide an exception for federal EISs prepared by state agencies or officials with statewide jurisdiction when the federal action is funding. The exception applies mainly to Federal Highway Administration funding for state highways. A state agency or official with statewide jurisdiction may prepare a federal EIS when the responsible federal official provides guidance, participation, and independent evaluation of the EIS before it is approved and adopted by the federal agency.

A second exception is for applicants for community development block grants from the Department of Housing and Urban Development (HUD). If an applicant assumes the status of a responsible federal official, it may assume the agency's NEPA obligations regarding the block grant.[123] HUD must ensure that the applicant complies with the Council on Environmental Quality (CEQ) regulations and HUD regulations implementing Congress' delegation of NEPA obligations, but does not provide independent review of an EIS's substance. The applicant, not HUD, is responsible for the substantive content of the EIS.[124]

GETTING STARTED

Before beginning preparation of an EIS, an agency's or consultant's EIS project manager should have a scope of work which identifies the

---

[122]40 CFR §1506.5(c).

[123]42 U.S.C. 5304(g) (1988); Brandon v. Pierce (1984); National Center for Preservation Law v. Landrieu (1980).

[124]Brandon v. Pierce (1984); Atlantic Terminal Urban Renewal Area Coalition v. New York City Department of Environmental Protection (1989); Colony Federal Savings & Loan Association v. Harris (1980) (Fogleman, 1990).

tasks required to prepare the EIS, a schedule for the EIS process, and a budget, or a contract (if there is one) that consists of these items. In most cases, the agency staff or EIS team members for the project were identified during preparation of the scope of work. If not, the team members who have the qualifications necessary to accomplish the scope of work will need to be chosen and assigned their responsibilities (discussed later in this chapter).

To get organized, collect the materials and supplies you will need to begin work on the EIS. At a minimum, you will need the EIS scope of work, schedule and budget, the EIS regulations applicable to the proposed project, information describing the proposed project, new files, and notebooks. A computer, calculator, and calendar are also required.

Set up files for the project, such as contract information, description of the proposed action, public scoping comments, public comments on the Draft EIS, in-house review comments, correspondence, and regulations applicable to the proposal. Make a list of data or publications you will need and where they can be obtained. Keep track of any information that is borrowed. Note who loaned it to you and when, so you can return the information when you're done, and record the date of return. Keep a notebook in which you record the publications, personal communications, and other data sources which you will need to add to the bibliography or references section of the EIS. Set aside an area on your desk or credenza where project notebooks, contract scope of work, and items that you will frequently use are within easy reach—not in a file drawer.

For complex and large EISs, some EIS teams have a "war room"—a room dedicated for an EIS in which maps with overlays, charts, and other graphics are mounted on a wall, and a conference table and blackboard are available for the team to brainstorm, solve problems, and generally work together for their analyses. The schedule and milestones are also displayed to permit all team members to keep informed on the progress of the EIS and any revisions to the schedule or milestones.

WHAT TO TITLE AN EIS

An EIS's title gives the public a first impression of the proposed project and the EIS that is hard to change. The first thing most people will see is the title of an EIS—normally the name of the proposed project. The wrong title can cause the EIS team to battle a misimpression through the entire EIS process, because once the public gets the wrong idea or mental image of the project it is almost impossible to change it.

An EIS's title should convey, as succinctly as possible, what the proposed project or action is. It may be prudent, however, to make the title longer if it will make the nature (especially the magnitude) of the project clearer to the public. If only the name of a facility is on the cover of an EIS, it gives the impression that the proposed project is new construction of an entire facility. For example, "Acme General Hospital, Environmental Impact Statement" implies a new hospital. If the proposed action is to expand an existing facility, the title should be "Proposed Expansion of Acme General Hospital, Environmental Impact Statement."

The title may or may not be critical, depending on the public sensitivity regarding the type of action. As an example of a sensitive proposal, the U.S. Navy prepared an EIS for a facility on a barge that would be towed to sea and emit electromagnetic pulses to test the effects on naval ships' electronic equipment. Because the purpose of the test was to simulate the electromagnetic pulse from a high-altitude nuclear reaction, many public reviewers had various nuclear activity concerns, including radioactivity, even though the EIS clearly stated that the tests were simulations and did not involve any nuclear reactions. The title of the EIS was "Electromagnetic Pulse Radiation Environment Simulator for Ships (EMPRESS II)" (the name of the project) and the name of the project office on the cover of the EIS was "Theater Nuclear Warfare Program."

Whatever the title of the proposed project or action, some discussion with the project proponent may be required to develop an accurate project title for the EIS that will not cause confusion or misunderstanding of the proposal for public reviewers.

ANTICIPATING ISSUES

EIS consultants and agency personnel who successfully anticipate local impacts and concerns are usually familiar with the area by virtue of having lived and worked there for some time. Therefore, agencies often select local EIS consultants rather than one from across the country. Familiarity with an area is also useful in designing an approach to the EIS process that meets the needs or concerns of the public, and it makes the EIS team more effective in addressing the issues and concerns in the EIS.

Agencies that have responsibilities for EISs will sometimes agree on a scope of work for an EIS, and will select an EIS consultant, prior to completing a public scoping process. Although the scope of work may have to be modified after scoping to include additional subjects for analyses in an EIS, the majority of significant environmental issues that will be of concern to the public, and that should be addressed in an EIS, can be anticipated. To identify potential issues related to a proposed action, the EIS team evaluates the types of environmental impacts that are expected to result if the proposed action were implemented in a particular location. The proposed action may involve the same type of construction, design, and operation regardless of where it is located, but the impacts and issues related to a proposal in one location will not necessarily be the same impacts and issues for the proposal in another location.

> **The impacts and issues related to a proposal in one location will not necessarily be the same impacts and issues for the proposal in another location.**

To anticipate issues that should be addressed in an EIS, identify environmentally sensitive areas and land use restrictions, understand public sentiments for a particular area, and have a detailed description of the proposed action. The issues to be addressed in an EIS can be anticipated by considering how the proposed action's construction or operation might impact an area in terms of its

- Existing conditions (e.g., soil, water, air)
- Environmentally regulated areas (e.g., wetlands, endangered species habitats, historic resources)
- Land use regulations (e.g., ordinances, plans, and policies)
- Any public concerns about the type of proposal or location

OUTLINING AN EIS

Always prepare an outline early in the process for an EIS. Outlines are a must to visualize how the document will be organized and how it will appear when it is completed. EIS team members should refer to it to know how to organize their analyses, to be aware of other topics and issues that are being analyzed in the EIS, and to know what topics may have a bearing on other topics so that team members can confer about those topics. Give a copy of the outline to the editor and word processor.

The format of an EIS will vary depending on whether the EIS is a state or federal EIS (see Chapter 5) and on the guidelines of the lead agency. Therefore, an EIS's outline will depend on a lead agency's EIS regulations and on the issues and alternatives that are to be addressed in the EIS (see Chapter 12).

PREPARING AN INTERNAL SCHEDULE WITH MILESTONES

A contract scope of work and schedule should be detailed enough for the contractual parties to understand everyone's responsibilities or obligations. However, a more detailed in-house scope of work may be necessary to ensure that team members know when an event must take place or when a work product is due. To prepare a schedule with milestones for in-house use, review the contract scope of work and schedule (if there is a contract) to identify the various steps in the EIS process and in the preparation of the EIS. Highlight the major milestones, or events in the process, and prepare a separate list of the events with the dates on which they are scheduled to take place. Then identify the specific tasks that must take place for the ID team to be able to accomplish their assignments by the scheduled dates. For example, if all of the assigned analyses are to be submitted to word processing on March 15, note whether the authors should give their analyses to the word processor or to the project manager. Actually, the project manager should receive and compile the materials before giving them to the word processor. How long will it take to compile and review the materials? If you estimate that it will take two days, then the due date for the ID team to have their analyses to you, the project manager, is by 8:00 a.m. on March 13. Be sure to specify the time of day as well as the date, or you may receive something at the end of the date you set (e.g., 5:00 p.m., March 13), when you had intended to have that day to review the material. An internal schedule with milestones for production of an EIS could look like the following:

Team members submit analyses, including graphics, to PM	September 16, 5:00 p.m.
PM review and compile analyses	September 17–23
PM gives all sections to word processing	September 24, 8:00 a.m.
Word processing finished, return EIS to PM	September 30, 5:00 p.m.
PM backchecks word processing, submits EIS to editing	October 3, 8:00 a.m.
Editing submits EIS to word processing	October 7, 8:00 a.m.
Word processing returns EIS to editing for backcheck	October 9, 8:00 a.m.
Editing returns EIS to word processing for final pickups	October 10, 8:00 a.m.
Word processing returns EIS to editing/ PM for QA/QC and preparation for printing	October 11, 8:00 a.m.
Send EIS to print shop for copying	October 11, 4:00 p.m.
Print 15 copies for internal review	October 11–13, 5:00 p.m.
Print shop returns EIS, check copies	October 14, 9:00 a.m.
Deliver copies to lead agency and project proponent	October 14, 12:00 p.m.
EIS team reviews preliminary EIS, comments to PM	October 28, 5:00 p.m.
EIS team reviews comments/discusses revisions with PM	November 2, 8:00 a.m.
PM discusses questions, has meeting with lead agency	November 2–5
EIS team revises EIS	November 5–8
EIS team submits revised sections to PM	November 8, 5:00 p.m.
PM reviews analyses, compiles changes	November 9–10
PM gives revised EIS to word processing	November 11, 8:00 a.m.
Word processing returns EIS to PM	November 13, 5:00 p.m.

PM backchecks EIS, gives to editing	November 14, 5:00 p.m.
Editing/PM provides QA/QC	November 15–17
PM gives EIS to word processing for final pickups	November 18, 8:00 a.m.
Word processing returns EIS to PM for backcheck	November 18, 3:00 p.m.
Editing/PM prepare EIS for printing	November 18, 3–5 p.m.
Print 350 copies of EIS	November 18–23, 5:00 p.m.
Word processing prepares mailing labels	November 19–20
Lead agency provides cover letter or memo for EIS	November 23, 5:00 p.m.
Check printed copies of EIS and distribute	November 24

The above schedule shows work progressing on consecutive days without weekends or holidays. As detailed as the schedule may appear, completion of graphics and other tasks are not shown.

Dates for review periods have traditionally been identified by counting the number of days on a calendar. The start date for counting the number of public review days for a Draft EIS is the date the notice of the EIS's availability is published in the *Federal Register* or state equivalent. When a public review period is 45, 60, or 90 days or longer, and several in-house review periods of two to three weeks each are involved, the number of days for review must be counted on a calendar to determine the date for the end of each review period. Rather than count days on a calendar, some calculators will provide an end date when you enter the start date (day, month, and year) and the desired number of days. Also, if you want to know how many days are between two dates, enter the dates and the calculator will tell you the number of days between those dates. The calculator counts calendar days which is fine for public review periods. But you'll need to check the calendar to see if the date that is the end of the review period falls on a weekend or holiday, and add a day or two if you want the end date to be a work day. If you are only counting work days (i.e., excluding weekends and holidays) to count, for example, the number of days to allow for completion of various EIS team tasks, you will need a calendar.

The milestones, or tasks and events that were identified in the scope of work, and the schedule will probably change a number of times.

Any revisions to the schedule and milestones should be made and distributed immediately to EIS team members so team members are kept up to date.

IDENTIFYING NECESSARY INFORMATION

To find information to use in an EIS, one must first know what information is needed. A project manager or team leader should provide guidance and suggestions to those who are learning about EIS preparation, but basically the process is similar to research projects that might have been required in college. Identify the objectives of the study and the information necessary to fulfill those objectives. It becomes easier to identify information after preparing a number of analyses in EISs because the process of identifying information is the same for most EISs.

To prepare an analysis in an EIS the first step, or objective, is to identify and describe the existing environment of your subject or topic. If the subject is vegetation, you will need information that describes the existing vegetation, its types, condition, existing impacts or problem areas, presence of species of concern such as endangered species, and so forth, within the EIS's study area (its geographic scope). The next step is to identify and describe the expected environmental impacts or consequences to vegetation in the study area if the proposed project or action were to be implemented. Therefore, you will need information regarding the proposed project—where it would be sited in the study area, how it would be designed and operated, what construction would entail (e.g., excavation, filling), what new infrastructure would be required (e.g., water, power, sewer lines, schools, roads) and where they would be located. Any existing information regarding environmental impacts to vegetation from similar types of projects would also be helpful. Natural resource agencies sometimes prepare studies of the condition of a particular resource that has been affected by certain activities (e.g., cattle grazing impacts on range land). If the proposed project being analyzed in an EIS is a proposal to allow cattle grazing, an existing study on cattle grazing impacts would be relevant to the analysis of potential impacts to vegetation.[125]

---

[125] The type of environment and vegetation evaluated in the study should be similar to the EIS's study area. Impacts in mountain rangeland may not be the same as impacts in desert rangeland.

General background information that should be available to the EIS team includes: comprehensive plans, master plans, and other land use plans for the study area; natural resource management plans for the study area; EISs and EAs prepared for other projects in the area; and EISs and environmental assessments (EAs) prepared for projects similar to the one you are evaluating in the EIS.

Early in an EIS process, team members should meet with natural resource and land use agencies to discuss the proposed project, obtain the agencies' input and concerns, and obtain information that the agencies might have that is relevant to a proposed project and the study area. That is also the purpose of a scoping process; however, if scoping is limited to correspondence by mail, many letters from agencies during a scoping process are generic because a proposed project is rarely adequately described in a scoping notice. Also, agency letters, being somewhat formal, cause agencies to posture and present the agency's mandates, rather than provide information helpful to an EIS process. Most agency personnel, when consulted early and openly, are very helpful and can identify necessary information as well as provide additional contacts or sources of information.

Information Resources

Information that can be used to prepare EISs comes in many forms. There is no limit to types of information that might be usable as long as the information is relevant to the proposed action and to the study area. Information may be verbal or written; sources may be local residents or technical experts. Information may come from agency reports or publications, scientific journals, books, unpublished data, and other EISs. Most EIS team members develop an information network whereby the same cadre of sources are relied upon to provide information for EISs. The following are some general resources that normally have information used in the preparation of EISs.

Agency and Proponent Data A lead agency and project proponent often accumulate information about a proposed project or action and the study area. During the planning process for the project, the agency and proponent may have identified and collected many of the regulations, reports, maps, and background data that will be useful for an EIS. In addition, the proponent may have had studies prepared to help design or site the proposed project. These include feasibility studies, engineering, and geotechnical reports. In some cases, the proponent has had studies prepared for some of the natural resource concerns identified by resource agencies (e.g., wetland delineations).

Applicable Regulations Some agency publications for regulations, particularly those that govern land use and natural resources, provide information about resources as well as the rules or restrictions that control proposed activities. Sensitive or critical environmental areas (e.g., wetlands, landslide hazards, endangered species habitat) may be mapped and discussed in attachments to agency rules, ordinances, or other regulations. Comprehensive land use plans, and similar land and natural resource management plans, are prepared by agencies in response to various federal, state, and local regulatory requirements. One would need to identify the agencies with jurisdiction in the EIS's study area to find the applicable land or natural resource management plans.

Existing Information Most of the information in EISs comes from existing information, information that was collected or provided by other sources or from previous works. Reports, studies, and other publications prepared by agencies are usually available from the agency. If an agency has run out of copies, it will sometimes make copies or you may be able to borrow a report from someone who knows you. Libraries will have copies of some agency publications, as well as other technical reports or books that are relevant to the subject or area of study. Check the references or bibliography section of publications you already have to find additional information. As noted above, the lead agency or project proponent will frequently have collected much of the information that is available on a subject or study area.

Experts on a Subject A great deal of information is available from in-house colleagues or from experts with universities, professional organizations, and other agencies. Experts may be members of an EIS team or have no relation to the EIS. Depending on your relationship with an expert, and his or her willingness to take the time to discuss the proposed project and potential environmental impacts, you may obtain insights and current information that are not available in existing literature. Always evaluate sources of information for the relevance of their work to the analyses in the EIS, the relevance of their professional credentials or experience, and any indication of bias if the topic is a controversial one. If the topic is controversial, you should obtain and present the views of experts (written or verbal) on all sides of a controversy.

Local residents and persons who are not affiliated with an organization may also have information relevant to an EIS. On occasion, local residents may know more about ongoing activities in the study area and existing environmental problems than agencies. Some individuals have personal interests or hobbies that, after many years, have earned them reputations as experts in a particular subject or field. "Self-taught" experts often document their observations in accordance with accepted professional standards. Although documented information is preferred in EISs, relevant anecdotal information can also be used. An EIS should not rely too heavily on anecdotal information, however, and should note that the information is anecdotal when it is used.

Federal Government Data Sources   Numerous agencies of the federal government conduct environmental data programs, including interagency initiatives that address coordination needs as well as programs that address specific agency missions. The Environmental Protection Agency (EPA) identified 83 environmental data programs in 25 different federal agencies. At the international level, agencies and organizations compile data and environmental trends information. Some of these U.S. and international data sources are listed in Appendix D.

Incomplete or Unavailable Information
When evaluating the potential environmental effects of a proposal, the analysis includes reasonably foreseeable impacts and actions. However, because the analysis may be evaluating some period of time into the future (the time period depends on the scope of the EIS), or technological information may be lacking, a certain amount of "reasonable speculation" will be necessary. If information were lacking about possible future conditions, actions, or impacts, EISs used to include a worst-case analysis. This was a scenario that evaluated the environmental

impacts of a proposal that, once implemented, might result in an event of low probability but catastrophic consequences. For example, a supertanker might lose all of its oil, or terrorists might sabotage a nuclear power plant. Worst-case analyses, however, seemed to encourage endless hypothesis and speculation. Therefore, the CEQ revoked its requirement for a worst-case analysis and amended its regulations in 1986 to include the following:

- An agency shall always make clear that information is incomplete or unavailable.
- If the incomplete information is essential to a reasoned choice among alternatives and the overall costs of obtaining it are not exorbitant, the agency shall include the information in the EIS.
- If the information cannot be obtained because the costs of obtaining it are exorbitant or the means to obtain it are not known, the agency shall include within the EIS:

A statement that such information is incomplete or unavailable

A statement of the relevance of the incomplete or unavailable information to evaluating reasonably foreseeable significant adverse impacts

A summary of existing credible scientific evidence which is relevant to evaluating the reasonably foreseeable significant adverse impacts

The agency's evaluation of such impacts based upon theoretical approaches or research methods generally accepted in the scientific community

"Reasonably foreseeable" impacts includes impacts which have catastrophic consequences, even if their probability of occurrence is low, provided that the analysis of impacts is supported by credible scientific evidence, is not based on pure conjecture, and is within the rule of reason.[126]

THE INTERDISCIPLINARY (ID) TEAM

The CEQ regulations require that EISs "be prepared using an interdisciplinary approach which will insure the integrated use of the natu-

---

[126]40 CFR §1502.22.

ral and social sciences and the environmental design arts." The disciplines of the preparers shall be appropriate to the scope and issues identified in the scoping process."[127] Thus, many agencies refer to the team of professionals involved in the preparation of an EIS as the "ID team." An interdisciplinary approach is one in which the members of an ID team consult with each other during preparation of an EIS to "integrate" their analyses. Compared to a multidisciplinary approach, one in which the team members conduct their analyses without consulting with each other, interdisciplinary team members' analyses benefit from the teamwork. One of the advantages of an ID team is that the combined capability of the team is greater than the sum of the individual team members' experience.

> **The combined capability of the team is greater than the sum of the individual team members' experience.**

An ID team does not have to be in the same room, working together, throughout an EIS process. Team members should, however, get together several times during the process to discuss the proposed project and alternatives, the scope of the EIS and its assumptions and methodologies, the study area, available information, and contacts for information. Also, team members should frequently consult with each other to discuss mutual issues and find solutions to problems that affect one or more of the topics in an EIS's scope. If all of the ID team members are not in the same office, time and budget should be allocated for the team to convene as a group at certain stages in the process. Individual team members should meet as the need arises (this may be as simple as walking down the hall to a colleague's office). If it is not necessary to be in the same room to discuss an issue, telephone conversations may be preferable to meetings as long as the objectives of the discussion are met by the end of the conversation.

When working as an ID team, team members share information and experiences. If one team member learns of a change in an agency's requirements or obtains new information that is relevant to another team member's analysis, he or she should notify the other team member of the new information. If a team member's analysis has some bear-

---

[127]40 CFR §1502.6.

ing or overlaps with another team member's analysis, they should discuss their analyses and conclusions and review each other's work to ensure consistency. To do this, team members must know what subjects or topics are in the EIS's scope, and what team members have been assigned the responsibility for analyses of the topics. For example, to discuss the potential for a reduction in recharge of groundwater from soil compaction and creation of impermeable surfaces, the person who is analyzing hydrologic impacts may need to discuss the project's potential to cause soil compaction with the person who is analyzing impacts to soil.

Selecting the Right Professionals

Either a consultant or lead agency would select the members of an ID team. A lead agency that is also a proponent for an agency project has the option of developing an in-house ID team or of having an EIS prepared by a consultant's ID team. Land use and natural resource management agencies, such as the U.S. Forest Service and Bureau of Land Management (BLM), once had many, if not all, of the science and planning disciplines that were required for an ID team. When the Forest Service evaluated a proposed action, it would select in-house staff to prepare an EIS. After numerous budget cuts and reductions in staff over the years, agencies now have a fraction of the in-house expertise they once had. Therefore, EISs for many agency proposals and private proposals are prepared by consultants.

An EIS's scope, or the subjects to be analyzed, must be known to select the members of an ID team. If the subjects or issues in the scope of an EIS are traffic, air quality, wildlife, and cultural resources, for example, the team members might be a transportation engineer, an air quality specialist, a wildlife biologist, and an archaeologist. A project manager or ID team leader is also required, unless it happens that the ID team leader is also one of the specialists on the team. If so, that person would have a dual role. The larger EIS consulting firms have most of the expertise normally required in EISs. Smaller firms will "team" with other firms to create an ID team that consists of all of the disciplines required for a particular EIS's scope. Large firms will also team with a small firm if the firm has staff whose credentials or experience are particularly suited for an EIS's scope, or if the lead agency is requiring small firm participation as a contract requirement.

When selecting a consultant's EIS or ID team, lead agencies consider the reputation of the firm or firms on the team, the firms' and staffs' qualifications, and experience with EISs and similar types of proposals. Agencies have additional evaluation criteria such as the location of a firm, and ability of a firm to perform within budget and schedule, depending on the needs and concerns of the agency. Agency evaluation criteria are primarily related to the requirements of the contract. But a consultant's team will ultimately be an agency's ID team, and the agency is, of course, a team member. Therefore, the individual members of an ID team, their experience and qualifications, and ability to work effectively as a team with the lead agency may be more important to a lead agency than the firm's background (a firm is nothing more than its people). In fact, because several EIS consulting firms competing for the same contract may show staff with similar credentials and experience, agencies often choose a firm based on the "chemistry" of the team members, and whether the agency's staff feel they can work comfortably with the consultant's staff.

While it may seem obvious that an ID team should consist of people who have the experience and professional credentials appropriate for an EIS's scope, people come with varying degrees of experience and credentials. There is no prescribed period of time for experience or number of credentials that make a person suddenly "qualified." Everyone was once a member of an ID team for the first time, but that didn't mean the person wasn't qualified. The role of a person required for an ID team should be considered as well as the subject or topic for which that person will be responsible. A senior-level scientist with a Ph.D., although eminently qualified, may not be the right person for a small role that does not require in-depth analysis, or a role as an assistant to another senior-level person. It would not be an efficient use of staff or budget. A senior-level scientist may not be appropriate for the role of a team member at all if that person prefers to be a research scientist— that is, prefers to work alone and without the give-and-take that is required in team work. A "qualified" person is one whose education, experience, and credentials fulfill technical requirements (the EIS's topic or subject) as well as the role required of a person as a team member.

Communication

If communication seems to be mentioned frequently in this book, it's because communication is so vital to the EIS process. There is no such thing as too much communication within an ID team (if the communication is related to the EIS). The primary flow of communication is between a team leader, or project manager, and the rest of the ID team. The team leader receives direction from the lead agency and obtains information from the project proponent. The team leader, in turn, gives direction and information to other team members. The rest of the ID team members, if they are using an integrated approach to their analyses, are communicating with each other on a daily basis. All ID team members must communicate externally with state, local, and federal agencies, numerous information sources, and contacts to obtain information, as well as with vendors or service companies during production of an EIS.

Communication takes place by way of correspondence and conversations in meetings or in telephone conversations. Correspondence, especially between agencies, tends to be formal, while one-on-one conversations are less formal and usually get more accomplished. Generally, communication at the working levels of organizations is more effective, and accomplishes more, than that at the higher levels. Therefore, if you must discuss preferences of methodologies for analyses in an EIS, professional standards to use, applicability of regulations, and other issues with agencies, you should start discussions with the working-level staff. After an understanding has been reached at that level, you can request a letter to document an agreement regarding methodology or other requirements related to the analyses in an EIS. An agency's management level may have to be involved in your communication, however, if the discussions are regarding a matter that affects agency policy or is unusual or precedent-setting.

Always document verbal communication in the form of minutes of a meeting, a record (e.g., fill out a form) of a telephone conversation, a memorandum, or a letter. An EIS process takes months or years to complete, and conversations can be forgotten. Also, agency personnel may change during an EIS process, and it may be helpful to new agency staff to see any documentation regarding the agency's position, concerns, or areas of emphasis for an EIS's analyses. If communications with the agency had been progressing well, the documentation of past communications will, hopefully, allow new personnel to continue where the last person left off.

Once you understand the EIS process, deciding what to communicate, who to communicate with, and how to communicate it generally becomes obvious. But many people don't give this much thought— it comes naturally after a while, or it doesn't happen at all. To communicate effectively, ID team members must be aware of what *other* team members need to know as well as communicate what *they* need to know. Communication should be a habit, a part of the business of being an ID team member who is integrating "the natural and social sciences and the environmental design arts."

The following are 10 ID team communication commandments:

1. Share knowledge. Don't assume that anyone knows something or has some information; ask that person if he or she requires the information.
2. Don't assume that someone else is doing something that needs to be done; ask the team leader or ask the person who would logically be concerned about the subject.
3. Don't hesitate to ask questions. Everyone does something for the first time.
4. Make sure you understand the roles of the ID team members, the lines of authority, and the lines of communication. If you don't, ask the team leader.
5. If you are unsure about a proper course of action, or the proper protocol to use, ask the team leader.
6. If you ask questions and don't get a satisfactory answer, ask someone else who is knowledgeable on the subject, but start with your team leader. Document your sources.
7. If you ask questions and get conflicting answers, get those people together on the phone or in a meeting to discuss the issue and get consensus. The team leader normally would have the final say on a matter related to an EIS; but broader policy decisions, or technical issues outside your team's expertise, may have to go to a higher level or to outside your group or agency.
8. If you identify a problem, suggest solutions to the problem.
9. Don't leave the team leader, or lead agency, out of the communication loop.

10. Be persistent. Sometimes communication requires repeated telephone calls and enduring the "bureaucratic runaround."[128]

Flexibility

Team members of an ID team may have to be added or deleted if the scope of an EIS changes during the EIS process. Changes in an ID team are more likely to take place if the team was developed before public scoping was completed. An EIS's scope may also change following public review of the Draft EIS. Issues may be added to the EIS's scope or a different level of technical analyses than was originally envisioned may become necessary. Team members may need to be added who have the expertise to address the additional issue or to conduct the technical analyses. If an issue is being deemphasized, the team member who was responsible for that issue might be dropped from the team.

ID team members need to be flexible and understand that the EIS process may change their level of involvement in an EIS, or whether they are involved at all. Flexibility is necessary, but being dropped from a team, or suddenly added, can cause frustration and problems in maintaining consistent workloads. The consultant's and lead agency's management staff should recognize the need for flexibility in ID team assignments and help reallocate workload in the company or agency when team assignments change. In most cases, topics are added to the EIS's scope rather than deleted.

Organizations that have "backups," or more than one person with similar capabilities can adjust more readily to a revision of the ID team. However, in the current climate of agency reductions and heavy competition among environmental consultants, few organizations can afford redundancy in their staff composition.

Maintaining Consistency of Analysis

Consistency in an EIS from one analysis to another is maintained when team members use an interdisciplinary approach, communicating and consulting with each other, during their analyses. Consistency encompasses many things, depending on an EIS's scope. The basic considerations for consistency include the EIS's study area, assump-

---

[128]A sense of humor helps.

tions and premises, and scope and methodologies that are the foundation and framework of an EIS (discussed in Chapter 8). All of the various analyses (e.g., air, water, soil, wildlife, vegetation, cultural resources, public services) should be within the same study area unless there is a specific reason to have one subject's study area different from the rest. If a proposed project includes an erosion control and water quality treatment system, both the soil and water analyses should include the assumption that the system would be in place if the proposal were implemented. If the timeframe for the analyses in an EIS is identified as 5 years into the future, all of the analyses should use the 5-year timeframe, not 20 years for the wildlife analyses and 10 years for the public services analyses. This doesn't mean that the individual analyses cannot have differences in their assumptions or methodologies, but any differences should be for a good reason and differences should be explained in the analyses.

The team should agree upon and use the same units of measurement.[129] If the wildlife habitat analysis is in metric units (meters) and the vegetation analysis is in English units (feet), the readers must mentally convert the units to visualize similar dimensions from one analysis to another. Even if consistency were not a professional issue, this would be a discourtesy to the readers. If water quality is discussed in both the water section and the fish section, the regulatory standards for water quality used in each analysis should be the same. Standards of a state, for example, may be different from federal standards. If both standards are used in the analyses, the reason for using both standards should be explained.

Consistency is not in and of itself "good," or an inviolate requirement. The purpose for maintaining consistency is to reduce the readers' confusion and enhance their understanding of the subject that is being discussed in an EIS. The only requirements are that an author make a conscious decision to be inconsistent with the rest of the EIS and that the inconsistency is technically necessary or improves the readers' understanding of the subject. Lastly, an author should explain the reason for the difference in his or her analysis in, for example, describing the methodology used for the analysis.

---

[129]The ID team should know whether a lead agency requires analyses to be in metric units.

THE KICKOFF MEETING

A kickoff meeting is a meeting of all of the EIS team members that usually takes place after a contract has been signed, or an authorization to start work on an EIS has been received. Some team leaders, or project managers, will have only the technical ID team members at a kickoff meeting. However, it is worthwhile to have everyone who will be involved attend the meeting, including editors, word processors, graphic artists, and subconsultants. Everyone performs better with a common understanding of a project's requirements such as its schedule, scope of work, and any unique aspects of the project (and less mistakes are made). The lead agency project manager should also be at this meeting; however, if an EIS is being prepared by a consultant, the consultant's project manager often meets with the lead agency's staff first, then has a kickoff meeting with the consultant's ID team. The advantage of having the lead agency's project manager at a kickoff meeting is that it provides an opportunity for the rest of the ID team to ask him or her questions directly, such as who appropriate contacts at the agency might be for specific subjects.

The purpose of a kickoff meeting is for the ID team to convene as a group and introduce members of the team, as well as for team members to gain an understanding of their roles, the proposed action and alternatives, assumptions and premises, and scope and methodology of the EIS. The contract scope of work, schedule, and budget should be explained and understood by the team members. The team leader, who is responsible for arranging and managing the kickoff meeting, must provide information that is needed by the team to begin work on the EIS. Since most clients want work to begin "immediately," once the decision to proceed has been made, an ID team leader is under pressure to find information and prepare for the kickoff meeting in a short amount of time. Ideally, a team leader should have at least five working days to adequately prepare for a kickoff meeting, but frequently is given only a day or two. If information was made available some time prior to the award of a contract or authorization to proceed, and approval to start was anticipated, a team leader may be at least partially prepared before being given the notice to start work. Items to distribute and discuss at a kickoff meeting include the following:

- Proposed action description
  Construction methods (including excavation, fill)
  Phasing/timing
  Features of design
  Features of operation
  Any unique features of the project
  Mitigation measures
- Alternatives to the proposed action (e.g., alternative sites, designs)
- EIS foundation
  Assumptions and premises
- EIS framework
  Scope and methodology
- Contract information[130]
  Scope of work
  Schedule and milestones
  Budget
- Glossary of terms to be used (including terms unique to the project)
- Public issues, sensitivities
  Copies of scoping comments (if scoping has been completed)
- Outline/organization of EIS (including outline of summary section)
- Project (and any alternative site) location map(s)
- Base map(s) of site(s) to be used by team members
- Style manual (company or agency)
- Any protocol concerns (e.g., classified or proprietary information)
- Proper treatment of news media or public inquiries
- Known data sources, previous studies (and who has them)
- Any field visits or field work (and their schedules)
- Questions and answers

Depending on how long an EIS process is going to take, at least two or more progress meetings (part-way through the Draft EIS, and at the start of the Final EIS) with the ID team should be held to discuss the team members' progress. A team leader's role is to monitor

---

[130]If the contract does not provide sufficient detail, an in-house schedule, scope of work, and budget may be necessary that specifies the tasks, number of hours or budget, and due dates for each team member.

the progress of team members on a continual basis, but progress meetings keep team members apprised of the "big picture" of how the overall EIS is changing or progressing. A team leader should have discussions regularly with individual team members, especially those who will have the most work or the most difficult issues. The team leader should also have progress meetings with the lead agency and project proponent if they do not attend the ID team's progress meetings. For any meeting (make it productive) have an agenda with clear objectives, and adhere to it to accomplish those objectives.

CHAPTER SUMMARY OF TIPS

■ To maximize ID team interaction on complex and large EISs, dedicate a room in which maps with overlays, charts, and other graphics are mounted on the wall and in which a conference table and blackboard are available for the team to brainstorm and work together for their analyses.

■ To prevent misimpressions regarding the nature of a proposal, create an accurate title for an EIS.

■ To design an effective approach to the EIS process, and have an EIS team that effectively addresses issues, select an EIS team that is familiar with the study area and the public's concerns.

■ To anticipate issues for an EIS, identify environmentally sensitive areas and land use restrictions, understand public sentiments, and have a detailed description of the proposed action.

■ To minimize reformatting and rewriting internal versions of an EIS, always prepare an outline and provide it to the ID team early in the process so they may refer to it to know how to organize their analyses, be aware of other topics and issues in the EIS, and know what topics may have a bearing on other topics so that team members can confer about those topics.

■ To save time identifying the end date of public and internal review periods, and the number of days between two dates, use a calculator that computes the dates and number of days in a matter of seconds.

■ To keep EIS team members on schedule, revise schedules and milestones promptly as changes occur, and distribute the revisions to team members.

■ To obtain information from agency personnel or other organizations, consult with them informally and early in the EIS process (avoid last-minute requests).

■ To reduce time spent looking for data, ask the lead agency and project proponent for information they may have already collected regarding the study area and impacts of similar types of proposals.

■ To prepare an EIS using a systematic interdisciplinary approach, EIS team members should frequently consult with each other to discuss and review each other's work on mutual issues.

■ To select qualified ID team members, consider the appropriateness of the individual's role on the team as well as the person's education and experience.

■ To accomplish the most with regard to identifying or negotiating an EIS's methodologies, professional standards, applicability of regulations, and other issues with agencies, start discussions with the working-level staff.

■ To maintain the administrative record, use an organized system for filing and keeping track of information and documents. Always document verbal communication (an EIS process can take over a year to complete and conversations can be forgotten).

■ To communicate effectively as ID team members, be aware of what *other* team members need to know as well as communicating what you need to know.

■ To maintain consistency in an EIS, the various analyses (e.g., air, water, soil, wildlife, vegetation, cultural resources, public services) should be consistent with elements of an EIS's foundation and framework (unless there is a specific reason for differences).

■ To ensure that all EIS team members have a common understanding of an EIS, include everyone who will be involved in the EIS at the kickoff meeting, including editors, word processors, graphic artists, and subconsultants.

CHAPTER 11

PUBLIC INVOLVEMENT

People's minds are changed through observation and not through argument.
Will Rogers

INTRODUCTION

One purpose of environmental impact statements (EISs) is to include the public in an agency's decisionmaking process. Prior to the enactment of regulations requiring EISs, agencies made decisions that affected the environment with little or no input from the public, and with no disclosure to the public that such actions were being contemplated. Currently, when preparing environmental documentation under the National Environmental Policy Act (NEPA) or a state environmental policy act (SEPA), all federal agencies and many state and local agencies must involve the public to varying degrees depending on the type of proposed action and the agency's regulations.

There are several ways by which the public is included in an agency's EIS process—through public meetings and hearings, scoping notices, review of Draft and Final EISs, and agency decision notices. Public involvement or participation is the actual interaction of the public with the lead agency, and public notification is the agency's notice to the public that certain actions are taking place as part of the EIS process. Public notification may or may not result in public involvement or participation depending on whether anyone responds to an agency's notice.

The amount of public participation and the methods used to include the public are decided by the lead agency. Although the Council on Environmental Quality (CEQ) (federal) regulations strongly encourage public participation, the amount is not always specified. For example, the CEQ regulations require a public scoping process for EISs but do not state how long the scoping process should last. Many agencies allow at least 3 weeks for public scoping, but the actual amount of time will be specified in a scoping notice or notice of intent to prepare an EIS. Scoping meetings, or meetings to which the public is invited at the start of an EIS process, are optional. The regulations do, however, identify specific ways the public must be notified of agency actions, as well as additional and optional methods of public notification.

The public is also involved in the review of a Draft EIS. The CEQ and state EIS regulations specify the minimum amount of time that Draft EISs must be available for public review. A public hearing on a Draft EIS is optional, or only required if a certain number of people or certain types of agencies request the hearing, but most agencies will hold a hearing to take verbal comments on a Draft EIS without being requested to do so.

A Final EIS is provided to those who request it and who commented on the Draft EIS. Although federal and some state agency regulations prohibit lead agencies from making a decision on a proposed action for a specified time after issuance of a Final EIS, there is no public comment period for a federal Final EIS.[131] The public, however, is not precluded from commenting on a Final EIS. Public comments on a Final EIS are provided to the decisionmaker along with the EIS and other information related to the proposed action.

Public scoping is encouraged, but not required, for a federal environmental assessment (EA). A finding of no significant impact (FONSI), prepared after an EA concludes that a proposal will not result in sig-

---

[131]Agencies may elect to provide public review periods for Final EISs [40 CFR §1503.1(b)].

nificant environmental impacts, is made available to the public. Also, SEPAs do not require scoping on state equivalents of an EA (e.g., environmental checklist).

Agencies may decide to have more public participation than is required by regulations. For proposed actions that affect local land uses, such as comprehensive land use plans, agencies may invite local citizens to participate in advisory or working groups. The groups may evaluate various land use scenarios or alternatives to ensure that the neighborhoods' wishes are considered in plans that guide future growth or development.

WHEN TO INVOLVE THE PUBLIC

Lead agencies must provide public involvement and public notification as part of an EIS process. Public involvement requires time, preparation, and budget. EIS regulations encourage agencies to maximize public involvement; however, it is up to the agencies to determine how much public involvement is reasonable or can be accomplished within an agency's or proposed project's limitations. The ideal situation is one in which an agency or project proponent has all the time and money it needs for an EIS. And in some cases, the schedule, scope of work, and budget for an EIS have included "maximized" public involvement from the beginning of an EIS process. But in most cases, agencies feel it is reasonable to have "adequate" public involvement. There are no guidelines on how much public involvement is adequate or how much is considered maximum, but there are some guidelines on when, in the EIS process, public involvement is required. These times for public participation are at the start of the EIS process (scoping) and during review of a Draft EIS. Generally, if an agency provides the required public involvement, it is considered "adequate."

> **Public involvement requires time, preparation, and budget.**

Public notification takes place concurrently with, but more frequently than, public involvement. Scoping notifies the public that a proposed action is being considered by a lead agency; a threshold determination, such as a negative declaration or a finding of no significant impact notifies the public that the lead agency has determined that a proposed action will not result in significant environmental impacts; a notice of intent notifies the public that an EIS will be prepared; a notice of availability of a Draft EIS notifies the public that a Draft EIS is available and the public may comment on the Draft EIS; a notice of availability of a Final EIS notifies the public that a Final EIS has been prepared and is available; and a record of decision notifies the public that the lead agency has made a decision on a proposed action. SEPA processes have similar public notices.

The public may respond to an agency's public notice and, for example, disagree with an agency's threshold determination (e.g., a determination of no significant environmental impact). If the public's disagreement with an agency's notice is supported by facts or reasonable argument, an agency may change its decision and issue a new public notice to prepare an EIS. Thus, in effect, public notifications are a form of public involvement if the public responds to the notices.

SCOPING

A scoping process invites the public to tell a lead agency what the public thinks should be addressed in an EA or EIS. The public's concerns about a proposed project or action—its potential environmental impacts related to the project's location, construction, design, and operation—help a lead agency focus its analyses on the topics of most concern and on potentially significant environmental impacts. Some members of the public copy the possible environmental topics or issues listed in EIS regulations, and they submit them to a lead agency to include in an EIS's scope. However, the purpose of scoping is not to find as many topics as possible to address in an EIS, it is to narrow the scope of analyses in an EIS so that the truly significant issues receive the majority of the agency's attention and the insignificant issues are eliminated from detailed study. Therefore, comments from the public during scoping should not be mere laundry lists of topics, and they should include the reasons that the commentors feel an issue is significant and warrants analysis in an EA or EIS.

> **The purpose of scoping is not to find as many topics as possible to address in an EIS, it is to narrow the scope of analyses.**

THRESHOLD DETERMINATION

Scoping will also help an agency make its threshold determination of whether a proposed action will result in significant environmental impacts and, therefore, whether an EIS will be required. If an agency has already decided that an EIS is necessary, the threshold determination has been made and there is no need for the public to help make the determination. Under some agencies' procedures, however, the public (usually the project sponsor) may disagree with an agency's determination that an EIS is necessary. The agency may change its threshold determination and decide to not prepare an EIS, although the reverse is more common. An agency that issued a threshold determination to not prepare an EIS is more likely to receive public comments that an EIS should be prepared, and the agency may change its threshold determination from one that did not require an EIS to one that requires an EIS.

Because public participation is an integral part of the EIS process, and the most common complaint in lawsuits is that an agency should have prepared an EIS, agencies should invite public comments prior to making a threshold determination to not prepare an EIS. A critical factor in judicial review is the degree to which the public participated in the threshold determination.

Scoping assists an agency in making its threshold determination, but public comments alone do not determine whether a proposal will have significant impacts, nor does lack of public comment indicate that a proposal will not have significant environmental impacts. A threshold determination is a judgment call by the lead agency which must be based on facts and strong rationale to not be considered arbitrary and capricious by the courts. An agency must demonstrate that it identified and evaluated the environmental issues sufficiently to support its conclusion that the environmental impacts of a proposed action would not be significant.

REVIEW OF DRAFT EISs

A Draft EIS is "published," "circulated," "distributed," or "issued" for public review (the terms are synonymous). The lead agency will prepare a mailing list consisting of (a) those who have requested a copy of the Draft EIS, (b) individuals, groups, and organizations who are known to be interested in or potentially affected by a proposed action, and (c) agencies who have jurisdiction by law or expertise. Copies of the Draft EIS are usually sent to local libraries and other public offices as well. If a Draft EIS is large, summaries may be sent to some of the address-ees in lieu of the entire document. Technical appendices are normally only sent to those who have requested them or who are providing tech-nical review of the Draft EIS (e.g., to agencies with jurisdiction or ex-pertise).

EIS regulations prescribe minimum public review periods for Draft EISs that range from 30 days (some states) to 45 days (federal). For large EISs, lead agencies frequently allow longer public review peri-ods, or they will extend the review period (e.g., 15 days or longer) if requested. Longer review periods and extensions of review periods are at the lead agency's discretion; although for federal EISs, the EPA can exercise discretion over EIS review periods for a compelling reason.

The purpose for public review of a Draft EIS is similar to the purpose for public scoping—to provide comments to the lead agency regarding the Draft EIS. An agency sometimes requests comments without being specific as to the types of comments that would be help-ful to the agency. As a result, the public may provide vague or general-ized comments rather than specific comments for which an agency can provide specific responses. Comments should relate to the analyses in the Draft EIS such as

■ The appropriateness of the study area—whether the potentially af-fected areas were included in the analyses
■ The significant issues addressed in the Draft EIS—whether they were identified and adequately analyzed
■ The range of alternatives—whether an appropriate array of alter-natives to the proposed action were presented and evaluated
■ The range of mitigation measures—whether mitigation for poten-tial environmental impacts were adequately identified and evalu-ated
■ The data used in the analyses—whether the conclusions in the analy-ses were supported by appropriate and accurate data
■ The methodology—whether the methodology and approach in the analyses were appropriate for the technical issues addressed in the Draft EIS

Agencies should request the public to support its comments with rationale; for example, why is the use of particular data inappropriate, or why is a particular analysis inadequate? Public comments may be provided to a lead agency in writing by mailing letters to an agency contact person, by providing written comments at a public hearing on the Draft EIS, or by providing verbal comments by speaking at the public hearing.

A lead agency responds to written and verbal public comments on a Draft EIS by preparing a Final EIS. A Final EIS may contain corrections to the Draft EIS, additional information, additional alternatives and mitigation measures, clarifications, and explanations of why further analysis is not warranted. In some cases a Final EIS is a republication of the Draft EIS, with changes in response to public comments; and in other cases the Final EIS consists only of an errata sheet, responses to comments, and copies of the public comments. A Final EIS is publicly distributed to those who requested it and those who commented on the Draft EIS. Any comments on a Final EIS should be given to the decisionmaker, regardless of whether public comments were solicited.

HOW MUCH PUBLIC INVOLVEMENT IS ENOUGH?

A formula does not exist for how much public involvement to provide in an EIS process. Agencies that frequently interact with the public, particularly land managing agencies, seem to have a good sense of how much public involvement to provide for its proposed actions—generally, more than the minimum amount required by EIS regulations. For a land management proposal, agencies often work closely with the public, informing them of the various problems and options for solutions, requesting participation in workshops or meetings, and providing frequent updates in the form of newsletters or other correspondence. For a private proposal, depending on the type of proposal, the public involvement may not be quite as intense because an agency may not have as many options for a private proposal as it has for its own.

When deciding how much public involvement is enough, one must ask, "enough for what?" What is the objective of public involvement for the EIS? To meet regulatory requirements, to satisfy the lead agency's requirements, or to obtain the public's assistance in decisionmaking?

An EIS project manager for a lead agency should be familiar with the agency's public involvement policies. Many agencies have detailed community involvement or citizen participation procedures or requirements. It may be possible to combine EIS and planning public involvement processes if the EIS is evaluating a planning or land management proposal. Also, what would be the benefit of public involvement to the EIS? Could the public assist in identifying alternatives, additional sources of data, information about the existing environment, or potential environmental mitigation measures? If the intent of the lead agency and project proponent is to provide the minimum or "required" public involvement, the involvement identified in federal or state EIS regulations will generally be "enough" from a *procedural* standpoint. However, from the standpoint of the *intent* of regulations—for example, to provide public involvement to the "maximum extent practicable"—more than the required public involvement may be advised. This is particularly true if a proposed action is highly controversial, if it has safety or health risks, or if the environmental effects are uncertain or unknown. Since judicial review may include compliance with the intent of the law as well as the procedural requirements of the law, and controversial or risky proposals are likely to be challenged in court, the lead agency and project proponent should not limit public involvement to the minimum level.

METHODS FOR INVOLVING THE PUBLIC

The most common methods for involving the public in an EIS process are through meetings, hearings, and correspondence. Many options are available within those methods for the types of meetings (e.g., formal, informal, working groups, advisory groups) and techniques of public participation that best accomplish the objectives of the agency with regard to the EIS process. Many agencies have procedures established for public involvement, including the preparation of a public involvement plan. Public involvement methods in one area, or for one type of proposed action, may not be effective in another area or for a different type of proposal. The subject of public involvement techniques is a large one that is described in detail in numerous publications and agency training materials. Some agencies have specialists trained in developing and conducting public involvement processes. A public involvement specialist would be an appropriate member of an EIS or interdisciplinary (ID) team. For the purposes of this discussion, however, the various methods or techniques will not be analyzed because they are numerous, and their appropriateness will depend on factors such as

■ The agency's objectives for public involvement
■ The type of proposed action
■ The affected area
■ Known public concerns or issues related to the area or proposed action
■ The amount of time and budget available for public involvement

> **Public involvement methods in one area,
> or for one type of proposed action, may not
> be effective in another area or
> for a different type of proposal.**

Whatever public involvement methods or techniques may be chosen, do not lose sight of the fact that the process is for an EIS and not for a traditional planning process. The main difference is one of focus. An EIS process is focused on the environmental issues related to a proposed action. The differences between (1) a planning (e.g., land use) proposal that may be the subject of an EIS and the overall planning process and (2) the relationship between the EIS process and planning process should be clarified so participants understand their roles in either or both processes. For an EIS process, public participants may provide advice or suggestions to the decisionmaker, but do not make any decisions. An agency decisionmaker does not delegate any of his or her authority to the public (compared with a planning board comprised of citizens that do have decisionmaking authority).

HOW TO USE PUBLIC INPUT

To determine how to use public input, one must first understand that public input is often based on personal values and perceptions. It does not matter to most citizens that their values and perceptions are subjective and do not necessarily reflect the values or perceptions of other people. Agency personnel also have values and perceptions that are a result of their background, training, and the mission of the agency. An EIS process itself inherently assumes certain values—for example, concern for the environment.[132] The dilemma of an EIS team is how to respond to public input that places a proposal or its environmental impacts in the context of personal values and perceptions that may not reflect "reality" as the team understands it. An EIS team deals with facts and must be unbiased in its analyses of a proposal. The team is neither in support of or against a proposal in an EIS, and if they have personal opinions (as opposed to professional opinions), they strive to keep those opinions from biasing the analyses. Agency decision-makers make any value judgments, whereas the EIS team simply presents the facts.

Yet public input does not always deal with the facts. Even agencies who are reviewing an EIS may have motives or agency objectives that are not related to the facts under evaluation. An EIS process is sometimes viewed as an opportunity for citizens, groups, and agencies to be heard by the lead agency and to provide input such as

■ Past grievances against the lead agency or project proponent
■ Disagreements regarding agency policies unrelated to the proposal
■ Objections to a proposal on moral grounds (e.g., anti-gambling, anti-military)

The majority of public comments on a proposal or EIS are from citizens, agencies, or groups who are opposed to a proposal under evaluation in an EIS. Opposition can be based on a wide variety of perceived threats to

■ Jobs, business, property value
■ Security of home or family
■ Health or safety
■ Quality of life
■ Environmental values

Sometimes just the potential for change is unwanted or perceived to be a threat.

---

[132]EIS laws were written by legislators who had certain perceptions of the EIS process. But the lead agencies and consultants who must implement the laws have very different perceptions of EISs and the process.

An overwhelming majority of public comments on any EIS are negative. Most people who are supportive of a proposal, or who have no feelings one way or the other, will not take the time to attend meetings, write letters, or review EISs. They usually observe silently from the sidelines. Those who do participate because they are supportive of a proposal will often do so because they have something to gain, such as financial profit. This is not to say that economic incentives are good or bad. Many who express opposition to a proposal do so because of economic considerations as well. The role of an ID team is to recognize and address public concerns, without assigning values (of good or bad) to those concerns. In some cases the only appropriate response to a public concern may be that the concern is outside the scope of the EIS. For example, an EIS's scope normally does not include moral issues because an EIS's purpose is to evaluate environmental impacts. Psychological health or stress is also not required to be addressed in an EIS because the relationship between the effect (psychological stress) and the change in the physical environment caused by the action (the "causal chain") is "too attenuated."[133]

Agencies may not maximize public participation in their EIS processes because, in many cases, the negative input is perceived as counterproductive. Nothing useful is obtained from public comments that simply state opposition. The regulations, which encourage public participation, are written as though citizens, agencies, and interest groups understand the EIS process and are eager to review the facts and provide constructive comments to improve an EIS and an agency's decisions. This is seldom the case.

The public generally does not understand the purpose for their involvement in an EIS process. Because they are asked to "provide input" or "provide comments," they assume that their comments should be whether they are in support of or against a proposed project. If a lead agency is to receive useful comments, the agency must explain a number of things to the public, such as

---

[133]Damage to psychological health was alledged from the risk of a nuclear accident. But the courts held that *risk* of an accident is not an effect on the physical world [Metropolitan Edison Company v. People Against Nuclear Energy (1983)] (Fogleman, 1990).

- How the public involvement process works (e.g., scoping, draft EIS review) and where the public is in the process at that particular time
- What the public's role is in the process, and what members of the public have been invited to participate
- How the public's comments will assist the lead agency in making its decision
- What types of input the lead agency is seeking (e.g., suggestions for alternative sites or designs for the proposal, suggestions for possible mitigation measures, specific comments on the facts in the EIS such as errors or omissions)
- Differences between the public's involvement in the EIS process and any other public involvement process the agency commonly conducts

An agency's explanation of public involvement can be provided in a cover letter transmitting a scoping notice or EIS, as well as in a separate section of an EIS. Although EIS regulations do not require a discussion of the public involvement process that was provided for an EIS, such a discussion would disclose the agency's effort to include the public and would inform members of the public of the total public involvement process when they may have been involved in only certain parts of the process.

The use of public input is partly a function of the type of proposed action and the attitudes of a lead agency. In some cases, agencies do not solicit public input to improve an EIS or the agency's decision. Public input is viewed as a requirement, and the responses to public comments or input are a formality provided in a Final EIS. In this case, there is no real interaction between the agency and public, nor is the public's input used. On the other hand, some proposals are simply more conducive to public participation. Land use policies and guidelines for growth and development are proposals for which agencies tend to include the public on a regular basis. One reason is that the public is more accepting of invitations to participate in land management proposals. They are comfortable as participants in an EIS process that deals with a subject to which they can relate. In contrast, highly technical or threatening proposals (e.g., anything nuclear) seem difficult for the public to participate in because (1) many members of the public may have negative feelings about the type of proposal and (2) they may be uncomfortable as "participants" because of a reluctance to be involved in, or suspicion of, such a proposal. If a lead agency is seeking more than the minimum public participation, it must convey to the public that being participants in an EIS process does not necessarily mean they must be supportive of the proposal under evaluation. The environmental impacts of the proposal are being evaluated, not whether the proposal should be approved or disapproved.

> **The environmental impacts of the proposal are being evaluated, not whether the proposal should be approved.**

HOW TO CONDUCT PUBLIC HEARINGS OR MEETINGS

The purpose of public hearings or meetings is to allow the public to make verbal comments regarding an EIS and the proposal under evaluation in an EIS. The focus of a hearing or meeting should be on the EIS process rather than a proposal's approval or disapproval. For EISs prepared at the local level (e.g., city, county), a local agency may be conducting other hearings or meetings, separate from the EIS process, to address the merits of the proposal.

Public hearings or meetings are held at various points within an EIS process, most commonly during public scoping and during public review of a Draft EIS. Additional public meetings may be held by a lead agency depending on the agency's public involvement policies or public involvement plan for a particular proposal.

Public hearings are normally more formal than public meetings. A hearing may have (1) a public official who presides over the hearing, (2) ground rules on how the hearing will be conducted, (3) a speaker's microphone and podium, and (4) a court reporter or tape recording of the proceedings, or both. Ground rules, to provide everyone with an opportunity to speak, include how speakers will be called to speak, how speakers are to identify themselves, and how much time speakers have to make their comments. Speakers are invited to comment, but not ask questions unless the question is for clarification of the hearing's procedures or the EIS process (e.g., when does the public comment period close?). Public meetings are less formal, but their format will vary widely depending on the objectives of the meetings and the lead agency's goals. All meetings should have some ground rules to prevent chaos and to prevent the most vociferous individuals from being the only ones heard. But informal meetings frequently permit dialogue in the form of questions and answers rather than limiting comments to statements.

Some meetings are working sessions or meetings between members of the public and the EIS team. Working meetings are generally not open to participation by all "the public," because there would be too many people to get anything accomplished. A public meeting participant may be someone who is designated by his or her peers or neighbors to reflect their concerns in the working meetings. In some cases, an elected representative is involved in working meetings; however, it may be a good idea to have other (nonpolitical) members of the public involved in the working meetings as well.

"Meeting" will be used generically to mean "hearing or meeting" for the following discussions. First, the lead agency must evaluate the need for a public meeting, identify its objectives, and, if necessary, prepare a plan for accomplishing those objectives. A plan should be prepared if there will be multiple public meetings, or if the meeting will be large. The project manager or ID team leader usually makes arrangements for the meeting location, time, equipment and supplies, and people needed to run the meetings (e.g., speakers, assistants, audio-visual specialists). The following are considerations when planning for a public meeting:

- Where should meetings be held—what cities or towns? If a proposed project has 10 alternative sites in 10 different towns, it may not be possible to have a public meeting in every town. Select locations that are central for the most people.
- How many people are expected to attend? Will you need a small room, a community center, or a school auditorium? It is better to have too large a room than one that is too small. The room should be in a neutral location if possible, rather than in the proponent's or opponent's facility.
- When should meetings be held? The time should be when the most people are likely to attend, or when they are less likely to have other competing obligations or interests. Meetings are usually in the

evening to allow participants to attend after work. What nights are best? Avoid evenings for popular sports events (e.g., Monday night football), and avoid Friday nights. Most public meetings fall between Tuesday and Thursday.

■ How will you record the name and addresses of those who attend the meeting? Some agencies use sign-up sheets, while others distribute sign-up cards. If a large number of participants are expected, cards save time because people are not lined up waiting to fill out sign-up sheets, and the cards can be organized to be used in various ways during and after the meeting. Sign-up cards are given to everyone who attends the meeting; they request a participant to write his or her name and address, to indicate whether the person wants to receive a copy of the EIS, and whether the person wishes to speak at the meeting. The cards need to be collected before the meeting begins so the meeting facilitator will have the cards of people who want to speak.

■ Who from the EIS team will attend the meetings? Will those people be available on the dates and times of the meetings? How will the EIS team members get to the meetings? Car pooling is a good idea because parking may be at a premium if a large number of people are expected to attend the meetings. If a large number of people are expected to attend, the local police may appreciate being notified of the event or may need to provide traffic control.

■ Will visual aids be used at the meetings? If so, what types of visual aids will be used and who will prepare them? If projectors are necessary, will the meeting room have them or will the EIS team bring all visual aid equipment? Does the meeting room have an audio system and electrical outlets in the necessary locations for visual or audio equipment? Someone should be in charge of audio and visual equipment to ensure that the meeting rooms have everything necessary and that they are all in working order.

■ Is the press expected to attend? In some cases an area is reserved for the press and their equipment to minimize interference with other attendees and the proceedings of the meeting.

After meeting arrangements have been made, the public must be notified of the meeting. The public should receive at least 2 weeks' notice prior to a scheduled meeting. If the purpose of a meeting is to take comments on a federal Draft EIS, an agency must make the EIS available to the public at least 15 days in advance of the meeting. Notification can be by mail (e.g., to adjacent property owners, and to those who have requested notification), notices in the public media (newspapers, radio, television), and posting of notices (on proposed project sites or in areas where the notices will be seen by a large number of interested people). For proposed projects that affect a large area or several areas, meetings should be held at several locations to reduce public participants' travel time and encourage participation.

Meeting notifications should include, at a minimum, the purpose of the meeting, the meeting's time and location, and a phone number and contact person. Include information explaining the public's role in the EIS process, as discussed above under "How to Use Public Input." It will help the public understand the process and prepare for the meeting, and it will increase the likelihood of obtaining comments useful for the EIS process.

Importance of Meeting Facilitators

A meeting requires a facilitator, someone who sets the tone for the meeting, introduces agency representatives or consultants and explains their roles in the meeting, explains the meeting procedures, and manages the meeting. Since the main purpose of a public meeting is to allow participants to make comments, the facilitator spends most of his time guiding the speakers in the meeting. For example, the facilitator will tell participants

■ How much time they will have to make their comments (usually 3 to 5 minutes depending on the number of people who wish to speak and the number of hours available for the meeting)

■ How they will be called to speak (usually in the order in which sign-up cards or sheets are received by the facilitator)

■ Whether they will come to the front of the room to a microphone (some meeting facilities have remote or portable microphones that do not require speakers to leave their seats)

■ To whom they are to direct their comments (usually to the decisionmaker who is sitting at the front of the room)

■ How the speaker will be notified that his or her time is about to elapse (the facilitator may have a stopwatch and usually provides a 1-minute or 30-second warning)

Some agencies designate an agency official as the facilitator, others have the decisionmaker for the EIS or the consultant for the EIS as the facilitator, and still others prefer someone who is not associated with the EIS or proposed project but may be an agency employee (e.g., a public affairs specialist). An EIS consultant, the facilitator, or the agency's project manager may explain the proposed action and the EIS process. The workload for the facilitator is lightened, however, if someone else presents project- and EIS-related information.

The facilitator has an important role in a public meeting—he or she is in a position of some authority, and should maintain control of the meeting. A facilitator's role is similar to that of a judge who ensures that the ground rules are followed and enlists cooperation from the audience. To do this, a facilitator should have certain qualities and capabilities, such as:

Credibility A total stranger to the audience may or may not have credibility at the beginning of a meeting. A high-ranking agency official's credibility (by virtue of his or her position) will depend on the background and values of the meeting participants. If someone has a poor reputation with the public, that person should not be the meeting facilitator, regardless of how important that person may be within the agency. If the agency as a whole does not enjoy credibility with the public, an outside or neutral facilitator may be advised. The most credible facilitator is often someone who is local, known to the meeting participants, and respected.

Empathy A facilitator who can understand other people's positions, motives, and concerns will be able to set the proper tone for the meeting. The facilitator represents an agency by virtue of his or her role; therefore, his or her tone may be assumed to reflect the agency's tone. The way remarks are made, or information conveyed, tells meeting participants if a person is defensive or if a person genuinely cares about their concerns. A facilitator must understand that meeting participants may not be familiar with the EIS process or meeting procedures, and he or she should provide careful explanations and provide fair and even-handed treatment of all participants. Whether a participant rambles beyond the time limit, or attempts to disrupt the meeting, the facilitator should treat everyone with respect and courtesy even while asking the speaker to summarize or conclude his statements or to cease the disruptive activity.

Resolution Although a facilitator should be empathetic, he or she also must guide the meeting participants with some degree of firmness or resolution. The amount of control that must be exercised will depend on the situation and the cooperativeness of the participants. Most people understand the need for ground rules if they are fair and explained ahead of time. However, it is not unusual for a few people to be carried away by emotions (such as anger) and to ignore the ground rules. Some facilitators who anticipate "troublemakers" at a public meeting will arrange to have law enforcement officers present to help maintain order. Usually, the presence of the officers will maintain order, without requiring any action on their part. Most facilitators, however, will simply attempt to reason with an unruly meeting participant and appeal to the person's sense of fairness, or request courtesy toward fellow meeting participants. Participants who arrive late (there are always stragglers), after the ground rules have been discussed, may need to have the ground rules repeated. For those who want to speak longer than the allotted time, facilitators can offer to let them speak again after everyone else has had a turn to speak.

Public Speaking Skills A public meeting facilitator must be comfortable in the spotlight and in the role of a facilitator. A person who is uncomfortable speaking to and interacting with a large group of people will not appear credible, empathetic, or resolute. Signs of discomfort include mumbling, not looking at the audience (talking to the podium), fumbling with papers, and other actions that indicate the person would rather be doing something else. Some people have a natural affinity for speaking to groups; but effective facilitators are usually well-prepared, so the public speaking part of their duties is unencumbered by distractions such as deciding what to do next. Facilitators may be formal or informal—no single style is best. A facilitator should, however, be flexible and able to think on his or her feet if something unexpected happens.

A Decisionmaker's Role in Public Meetings

A lead agency's decisionmaker should attend EIS scoping meetings and meetings to take comments on a Draft EIS because public comments are one of a decisionmaker's considerations in making his or her decision. A decisionmaker could read the public comments in a transcript or an EIS, but his or her attendance at a meeting indicates to the meeting participants that the decisionmaker cares enough to be there personally. Also, the participants should be speaking to the decisionmaker rather than speaking to the audience. This may require a participant to have his or her back to the audience in order to face the decisionmaker, but the purpose of the public meeting is not for the meeting participants to confer with each other—it is to provide comments to the lead agency's decisionmaker. The meeting facilitator should make clear to participants that the decisionmaker is there to take comments, not to answer questions. If this point is not clear, participants who are upset with a proposed action will ask questions that could place the decisionmaker in a defensive position. If participants ask questions anyway, the facilitator should remind them to change their questions to statements.

The Meeting Place

Agencies will go to varying lengths to encourage attendance at meetings. Some will provide the bare basics, while others will provide cookies and coffee for more of a "town meeting" atmosphere. As stated previously, a meeting place should be large enough for the anticipated number of people. The number of people who might attend may be hard to predict; however, the more controversial a proposal, the greater the likelihood for a large attendance. A meeting place that allows for the room to be expanded into an adjacent room and that has a supply of folding chairs will provide backup for an overflow situation. Water fountains and restrooms should be available for the duration of the meeting (e.g., not locked up after 7:00 p.m.). Proper ventilation and heating and cooling must be provided for large numbers of people in a confined space. For these reasons, auditoriums are frequently used for public meetings because the facilities are already arranged for large numbers of people (although the seating is usually fixed). School auditoriums may provide adequate seating; however, parking may not be adequate at an elementary school auditorium, for example, because the students don't drive.

Meeting places may require a fee which may or may not be refundable. Legal paperwork such as waivers for liability may have to be filled out in advance before the room can be reserved. These are some of the details a project manager must anticipate and accommodate in an EIS schedule and budget.

UNDERSTANDING AND WORKING WITH IRATE CITIZENS

If you are the person designated as the contact for public inquiries regarding an EIS, you may have to deal with irate citizens during the EIS process. What are their concerns? You may only know part of them, or you may be making incorrect assumptions, so even if you have some preliminary ideas, the first step is to *listen*.

If you are talking to an irate citizen, let him or her do the talking. Don't be defensive and attempt to argue. Make notes of the person's issues or points, and when the person is finished speaking, address his or her points individually. An irate citizen most likely does not know or care what your responsibilities or authority are and may demand action that you may not be able to provide. Provide explanations, not justifications. If possible, provide solutions; but if you don't know answers to problems or questions, tell the person you don't know. Perhaps you can find suggestions from someone else in your organization or another agency that will address, if not solve, the person's problem. Tell the person what you will do and also whether you will call him or her, or provide the name and phone number of someone who may be able to help the person. But if you're not sure that someone else is the right person, don't send the person on a bureaucratic runaround.

Give citizens information they need to understand the process, not just the project. Explain the lead agency's decisionmaking process, how decisions are made, and who the decisionmaker will be, as well as where in the process the project is and what comes next after the hearing, after the EIS, after the decision, and so forth. The more citizens understand about the process, the better they can articulate their concerns.

Citizens may not know how to express a concern if it is a value or perception rather than a fact. Values and perceptions tend to be vaguely expressed. Explain that in order to respond, you must try to specify the source or cause of their concerns. Allow them to express their concerns in their own words, and then ask questions to clarify points so that you understand any underlying problem areas. Don't attempt to change their positions because values and perceptions are not subject to logic, but do correct any factual errors or misunderstandings.

A person who is very upset may say things that are insulting. No one should have to accept abuse, but understanding that the insults are not meant personally—you are the recipient because you represent an agency or a cause of frustration (whether you deserve it or not)—may raise your tolerance level.

Most of the time, discussing people's concerns in a nonpolarized (no one is right or wrong) atmosphere is beneficial. The citizens feel they are being heard, and their concerns are being addressed if an effort is made to understand their positions. You are obtaining a clearer understanding of citizen concerns which allows you to respond more accurately in the EIS. This does not mean that everyone will be satisfied. Not all problems have solutions that can be implemented (if it can't be implemented, it's not really a solution), nor is an EIS process the correct vehicle for solving all problems.

> **The EIS process is not the correct vehicle for solving all problems.**

CHAPTER SUMMARY OF TIPS

- ■ To support a threshold decision to not prepare an EIS, agencies should invite public comments regarding a proposal's potential environmental impacts.

- ■ To provide useful public comments regarding a Draft EIS, make comments as specific as possible regarding the appropriateness of the study area, significant issues, alternatives, mitigation measures, data used in the analyses, the methodology, and approach. Provide rationale for opinions and suggestions.

- ■ To determine the amount and type of public involvement for an EIS, identify the agency's objectives for public involvement (e.g., to meet minimum regulatory requirements or to obtain the public's assistance in decisionmaking).

- ■ To ensure effective public involvement, include a public involvement specialist as a member of an EIS or ID team.

- ■ To reduce potential confusion, explain differences between the public's role in an agency's planning process and in an EIS process when requesting public involvement.

- ■ To address public concerns that are outside the purpose or scope of an EIS, provide the reasons the concerns are outside an EIS's scope.

- To receive useful public comments, explain in public notices how the public involvement process works, what the public's role is in the process, what members of the public have been invited to participate, how the public's comments will assist the lead agency, and what types of input the lead agency is seeking.
- To encourage public participation, select meeting locations that are central for the most people.
- To provide adequate meeting facilities and encourage public participation:

    Select meeting locations that are neutral and accessible.
    Provide a meeting place of adequate size.
    Select a time and date when most people are likely to attend.

- To reduce time required to obtain names and addresses of meeting attendees, and expedite collection of names of those who wish to speak, distribute sign-up cards rather than attendance sheets.
- To reduce required parking at meeting locations, agency participants should carpool to the meeting.
- To reduce traffic-related problems for large attendance at public meetings, notify local police and request assistance.
- To reduce occurrences of technical difficulties at a meeting, obtain the services of an audio-visual specialist to check the meeting place before the meeting, and be in charge of audio and visual equipment during the meeting (e.g., bring extra bulbs for projectors, extension cords).
- To minimize interference with other attendees and the proceedings of the meeting, set aside an area for press and their equipment.
- To lighten the workload for a facilitator, have someone else describe project- and EIS-related information to meeting attendees.
- To provide the proper tone for a meeting, select a facilitator with qualities such as credibility, empathy, and resolution and with capabilities such as public speaking skills.
- To make clear to meeting attendees that their comments are directed to the agency's decisionmaker, the decisionmaker (or a representative) should attend public meetings.
- To avoid delays and surprises, inquire about the need for legal paperwork or payment of fees when making arrangements to use a meeting place.

CHAPTER 12

WRITING EISs

God told Moses he had good news and bad news.
"The good news first," said Moses.
"I'm planning to part the Red Sea to allow you and your people to
walk through and escape from Egypt," said God, adding,
"And when the Egyptian soldiers pursue, I'll send the water back on
top of them." "Wonderful," Moses responded,
"but what's the bad news?"
"You write the environmental impact statement."
Anonymous

INTRODUCTION

In this chapter we will address how to write environmental impact statements (EISs): the major sections of an EIS, how they relate to each other, and how to focus on significant environmental impacts, identify mitigation measures, recognize peculiar ways words are used in EISs, and analyze cumulative impacts. This chapter does not address how to analyze specific environmental impacts, although examples from actual analyses in EISs are discussed. Other publications are available that provide scientific protocols and detailed methodologies, including field work and computer modeling, for studying environmental impacts on, for example, air, water, and the biological environment.

PURPOSE AND CONTENT OF DRAFT AND FINAL EISs

The primary purpose of an EIS is to serve as an action-forcing device to ensure that the National Environmental Policy Act's (NEPA's) and state environmental policy acts' (SEPAs') policies and goals are included in agency programs and actions. Lead agencies (or their consultants) prepare Draft and Final EISs to provide the agency's decisionmaker information on the potential environmental consequences of his or her decision. Normally, a Draft and Final EIS together are called the "EIS." An EIS identifies and evaluates the potential environmental impacts of a proposed action and its alternatives, including the no-action alternative. After an EIS has been completed, the lead agency's decisionmaker chooses the proposed action or one of its alternatives and provides the rationale for the choice in a federal record of decision (or equivalent document for state EISs). Another major purpose of an EIS·is to include the public in, and obtain public input for, the EIS process so that a decisionmaker will take the public's environmental concerns into consideration when making his or her decision. Several public notices are issued during an EIS process, as required by regulations, to disclose to the public that an agency is considering an action that may affect the environment; that the agency intends to prepare, or does not intend to prepare, an EIS; that a Draft EIS is available; that a Final EIS is available; and that the agency has made a decision. An EIS, or an action's environmental consequences, is only one of many factors that a decisionmaker takes into consideration when making a decision. Other factors include economic considerations, technical feasibility, agency mission requirements, statutory requirements, and national interest. An agency decisionmaker is not required to give more weight to environmental considerations than other considerations.

A Draft EIS consists of the analyses of environmental issues that were identified during scoping and that are related to a proposed action and its alternatives. A Draft EIS describes the proposed action, alternatives, and the purpose and need for the action, explains the EIS's framework and methodology, identifies and evaluates significant environmental impacts and mitigation measures, and identifies unavoidable adverse environmental impacts and cumulative impacts.

Most people expect that a "final" document is a revised version of a "draft" document. For federal and some state EISs, however, the Final EIS does not have to include a revised Draft EIS. A Draft EIS and Final EIS will contain very different information unless the Final EIS includes the Draft EIS. For this reason an EIS is normally comprised of a Draft EIS and Final EIS. One must review both documents to obtain all information provided by the EIS process. If revisions to a Draft EIS are minor, a Final EIS may consist only of an errata sheet (for corrections or revisions to the Draft EIS), public comments on the Draft EIS, and the agency's responses to public comments.

Some lead agencies prefer to rewrite and republish a Draft EIS as part of a Final EIS, even if changes are minor, because it is easier to refer to one document after an EIS process is completed. The California Environmental Quality Act (CEQA) requires a Draft environmental impact report (EIR) or revised Draft EIR to be included with the Final EIR, although the Draft EIR does not have to be sent with the Final EIR to those who already received the Draft document. Revising and republishing a Draft EIS uses large amounts of money, time, energy, and paper. Why would large amounts of resources be necessary to make "minor" revisions? Because the revisions, corrections, or changes themselves may be minor, but there may be many of them. More importantly, a lead agency that is revising a Draft EIS while preparing a Final EIS tends to not limit revisions to those responding to public comments. The agency staff continue to revise the EIS even when public comments were not received on the points the agency is revising. If there are many internal (agency) reviewers, this can lead to rewriting a great deal of the EIS. What appears to be a simple change, late in the process, can have repercussions throughout an EIS because the sections in the document are interrelated and somewhat redundant. An interdisciplinary (ID) team leader or project manager is normally the only individual who reviews the entire EIS. Resource specialists focus on their area of specialty and are usually unaware of changes that will result in other sections because of changes they make. Therefore, if a change is not absolutely necessary, it is better left alone. The goal of the process is to foster excellent action, not continually refine a point. Considering that EISs are often hundreds of pages long and the mailing lists consist of hundreds of addresses, revising and reissuing a Draft EIS is a large waste of resources.[134]

Final EISs may include additional information as well as revisions to the Draft EIS. Including new material in a Final EIS does not necessarily require an agency to prepare a supplemental Draft EIS. The value of the new information, whether the information was discussed in the Draft EIS, and whether a decision is still pending, determine the need for a supplemental EIS (also discussed in Chapter 7).

MAJOR ENVIRONMENTAL SECTIONS OF AN EIS

Agency EIS implementing regulations indicate the preferred content or format for EISs. The major environmental categories and content of an EIS are basically the same for federal and state EISs, although the names of the categories or headings and their organization may vary (see Figure 5.4). Federal EIS regulations discuss the *Affected* Environment and Environmental *Consequences* as separate sections, while some state agencies discuss the *Existing* Environment and Environmental *Impacts* within the same section or environmental element. EISs have a Summary that addresses the major environmental issues, areas of controversy, mitigation measures, and conclusions. The Purpose and Need section describes the purpose and the need for the proposed action, and the Proposed Action and Alternatives section describes the features of the proposed action and its alternatives (described in Chapter 8). The majority of the environmental analysis in an EIS is in the description of the Affected or Existing Environment and in the Environmental Consequences or Impacts sections.

Natural and Built Environments

The topics or environmental elements to be analyzed in an EIS are commonly separated into two main categories, the Natural and the Built Environments. The elements of the Natural Environment include the following:

---

[134]If a Draft EIS was so poorly prepared that it requires major revision, it should be issued as a supplemental or revised Draft EIS rather than a Final EIS.

■ *Earth*
  Geology
  Soils
  Topography
  Unique physical features
  Erosion
■ *Air*
  Air quality
  Odor
■ *Climate*
■ *Water*
  Surface water movement/quantity/quality
  Runoff/absorption
  Floods
  Groundwater movement/quantity/quality
  Public water sources
■ *Plants*
  Unique or sensitive (threatened or endangered) species
  Numbers or diversity of species
■ *Animals (including fish)*
  Unique or sensitive (threatened or endangered) species
  Habitat for
  Numbers or diversity of species
  Fish or wildlife migration routes
■ *Energy and natural resources*
  Amount required, rate of use, efficiency
  Source/availability
  Nonrenewable resources
  Conservation and renewable resources
■ *Scenic resources*

The Elements of the Built Environment include the following:

■ *Environmental health*
  Noise
  Risk of explosion
  Releases or potential releases to the environment affecting public
    health such as toxic or hazardous materials

- *Land and shoreline use*
  Relationship to existing land use plans
  Housing
  Light and glare
  Aesthetics
  Recreation
  Historic and cultural preservation
  Agricultural land
- *Transportation*
  Transportation systems
  Vehicular traffic
  Waterborne, rail, and air traffic
  Parking
  Movement/circulation of people or goods
  Traffic hazards
- *Public services and utilities*
  Fire
  Police
  Schools
  Parks or other recreational facilities
  Maintenance
  Communications
  Water supply
  Storm water
  Sewer
  Solid waste[135]

All of the above elements do not have to be addressed in an EIS. The elements that should be included in an EIS are identified during scoping, and a lead agency may combine the elements or organize them in any manner to reduce duplication, improve readability, and focus on significant environmental impacts.

AFFECTED OR EXISTING ENVIRONMENT

Environmental Baseline

The existing environment that would be affected by the proposed action and its alternatives is described in an EIS to provide an envi-

---

[135]Washington SEPA Rules (WAC 197-11-444), undated.

ronmental baseline. The impacts of the proposed action and alternatives will be identified by using the existing environment as a baseline. A change, or the degree of change, in the environment is the difference in the environment between the baseline (the existing conditions) and the new environment (the altered conditions) caused by a project. The baseline, however, may not be as static as the name implies. If a study area is undergoing change—for example, the site may be rapidly changing from a pasture to a grassland habitat for wildlife—the existing environment may not stay in one condition throughout the course of EIS preparation. The EIS or ID team must decide whether to take a "snapshot in time" approach for defining the existing environment to avoid attempting to analyze a "moving target," or revisit the site several times during the course of the EIS and revise the analyses as the site changes. The environmental baseline, or the existing environment, is also the environment that would exist under the no-action alternative. Thus, the no-action alternative is also called the baseline against which the impacts of other alternatives are compared.

ID team members who have responsibility for analyzing the environmental elements or resources that are in the scope of an EIS should visit the proposed project site and any alternative sites to observe the condition of the resources they are evaluating, existing or ongoing environmental impacts such as erosion or pollution, "tolerated trespass" activities such as hiking and biking, and the presence of any sensitive species or critical areas. Also, existing environmental impacts on a site may be coming from an off-site source. For example, farming activities upwind and upstream from a proposed project site may be affecting air quality and water quality. Some of this type of information may be in existing literature, but there is no substitute for first-hand observation.

Information for the Existing Environment

One must know what the geographic or physical study area for the EIS will be before starting to look for information on the existing environment. At the kickoff meeting for the EIS, the project manager or ID team leader should have provided a detailed description of the study area including a legal description, a map showing the location of the proposed action and alternatives, and a map of the proposed action's site and any alternative sites. The site maps are usually topographic maps at a large enough scale to be used for illustrating the locations of significant features. Maps might be used to show stands and types of vegetation, habitat types, wetlands, soil types, steep or unstable slopes, floodplains, existing land uses (housing, agriculture, recreation), and other features. The study area for any field work should be staked or flagged by the project proponent if the boundaries are not clearly defined by existing landmarks. If a proposed action includes construction, the "footprint" of the facilities (buildings, roads, utility lines) should be flagged or staked for the ID team members to visualize, on the ground, where the facilities will be located, and what and how environmental elements will be affected.

The legal descriptions of the proposed project and alternative sites, or map of the study area, or both, can be sent to various agencies when requesting information for the existing environment. The agencies to contact depends on the element of the environment and the agencies with jurisdiction of the study area. For example, if the land is federally owned, and the subject of analyses is potential impacts to wildlife, contact federal land and natural resource managing agencies with jurisdiction over the study area such as the U.S. Forest Service, U.S. Fish and Wildlife Service, and state offices having similar responsibilities (such as a fish and game department). Those agencies in turn can suggest other data sources or information regarding types of vegetation, threatened or endangered species, and regulations that might be applicable to the proposed action. Some agencies now charge fees to search their databases in response to requests for information regarding any records of the presence of threatened and endangered species on a proposed project site. Agencies, libraries, and other sources for information are discussed in Chapter 10.

The description of the existing environment needs to be thorough, but EIS authors can get carried away with collecting information, particularly if there is a great deal of information available. Stay focused on the significant issues, and provide only information on the existing environment that is relevant to the analyses of the significant issues. If soils or diversity of plant and animal species are not significant issues, don't provide detailed lists of soil types, or lists of every plant and animal species within the study area. EIS regulations advise against collecting extraneous background data and producing encyclopedic EISs. Public reviewers have come to expect lists and mountains of background data because they have been routinely included in

EISs[136]; however, if they are not necessary to the evaluation and do not serve an EIS's purpose, lead agencies should discourage the practice.

ENVIRONMENTAL CONSEQUENCES OR IMPACTS

The discussion of environmental consequences of a proposed action and its alternatives is the most important part of the analyses; many reviewers go straight to this section and forego the rest. However, prior discussions in the EIS provide much of the rationale for the analyses. The EIS describes the proposal and alternatives; the purpose and need for the proposal; the framework, assumptions, and methodology of the EIS; and the existing environment. These discussions are integral parts of the environmental analyses. The consequences or impacts section states what the impacts of a proposed action and its alternatives are expected to be. This is the "bottom line," which tells the decisionmaker and the public the environmental consequences or impacts of the decision—the decision to choose the proposed action or one of the alternatives. But to fully understand how the analyses and conclusions were derived, the foundation and framework of an EIS must be read and understood as well.

Identifying types of impacts and discussing them may not be difficult. Stating what the environmental impacts or consequences will be, however, is inherently difficult because you are attempting to predict what *would* happen, *if* certain actions took place, *if* regulatory requirements were met, and *if* mitigation measures were implemented. Thus the sections in the EIS prior to the consequences section that provide the EIS's framework and foundation support the conclusions regarding the anticipated environmental impacts. Impacts are "anticipated," are "expected," or have "potential" for occurring. The impacts may have a "high" or "low" probability, they may be "likely" or "unlikely," and the impacts of one alternative relative to another alternative may be "greater" or "less," but impacts can rarely be predicted with crystal ball certainty (also see the use of "would," "could," and "will" later in this chapter).

---

[136]It is doubtful that the public enjoys reading extraneous background data, but the public frequently comments that a species list or some other background data was forgotten and needs to be added.

ALTERNATIVES

In the words of the CEQ regulations, the alternatives analysis is the "heart" of an EIS.[137] Its purpose is to compare the environmental impacts of the alternatives, including the proposed action, to each other so the decisionmaker can choose an alternative and know what the environmental consequences of that choice would be. If all of the alternatives take place in the same location as the proposed action, the description of the affected or existing environment will be the same and need not be repeated for each alternative. However, the environmental consequences or impacts of each alternative will be different, and the lead agency must decide whether the proposed action and alternatives will be shown as subheadings under each environmental element, or whether the environmental elements will be subheadings under the proposed action and each alternative. The two basic options for the format of discussing alternatives are shown below.

■ *Option 1*
    Environmental Consequences
      Natural Environment
        Air
          Alternative 1
          Alternative 2
          Alternative 3
        Water
          Alternative 1
          Alternative 2
          Alternative 3
      Built Environment
        Traffic
          Alternative 1
          Alternative 2
          Alternative 3
        Cultural Resources
          Alternative 1
          Alternative 2
          Alternative 3

---

[137]40 CFR §1502.14.

■ *Option 2*
　Environmental Consequences
　　Natural Environment
　　　Alternative 1
　　　　Air
　　　　Water
　　　Alternative 2
　　　　Air
　　　　Water
　　　Alternative 3
　　　　Air
　　　　Water
　　Built Environment
　　　Alternative 1
　　　　Traffic
　　　　Cultural Resources
　　　Alternative 2
　　　　Traffic
　　　　Cultural Resources
　　　Alternative 3
　　　　Traffic
　　　　Cultural Resources

The purpose of alternatives is to provide a range of methods to meet a proposal's objectives and potential environmental impacts for the decisionmaker to consider. Also, the impacts of the alternatives must be compared to each other in the analysis of environmental consequences or impacts section of an EIS. All ID team members should compare alternatives in the manner identified by the team leader or lead agency. For example, if a proposed action has two alternatives and the proposed action, the proposed action can be "Alternative 1" and the other alternatives can be numbered 2 through 3 (or lettered A through C) so that the title of each alternative does not have to be repeated throughout the analysis. The no-action alternative (let's call it Alternative 3) is the baseline or the existing environment alternative against which all of the other alternatives can be compared. The proposed action and each alternative would end with a statement such as "Alternative 1 would result in more impacts (quantify or describe) than Alternative 3, but fewer impacts (quantify or describe) than Alternative 2." In this case we have included the proposed action as an alternative, for Alternatives 1 through 3. One can also discuss the proposed action and its alternatives 1 through 2. The number of alternatives is the same, but they are simply presented differently—for example;

- Proposed action (Alternative 1)
- Thirty percent reduced project size (Alternative 2)
- No action (Alternative 3)

or

- Proposed action
- Thirty percent reduced project size (Alternative 1)
- No action (Alternative 2)

The CEQ criticizes EISs for not emphasizing the alternatives analysis of an EIS in terms of the lead agency's preferences and consideration of whether the alternatives meet the intent of NEPA. In a record of decision a lead agency may discuss any preferences among alternatives, discuss essential considerations balanced by the agency in its decision, and discuss how these considerations entered into its decision. But there are no guidelines for where or how an agency would provide similar information within the text of an EIS. Also, many EISs are prepared by consultants. Some roles of EIS preparation may not be appropriate to delegate to a consultant, because a consultant may not know an agency's authority or preference with regard to alternatives (any preferences are sometimes not identified until the Final EIS). In order to emphasize the alternatives section of an EIS, the lead agency must evaluate certain parts of the discussion that are within the agency's purview, such as the agency's preference with regard to alternatives, identification of any alternatives that would be "reasonable" yet outside the jurisdiction of the lead agency, and "essential considerations of national policy."[138]

The Environmentally Preferred Alternative

The environmentally preferred alternative is one that would promote NEPA's policies. It would result in the least environmental dam-

---

[138]40 CFR §1505.2b.

age and would best preserve and enhance natural, historic, or cultural resources. The environmentally preferred alternative is identified so that a decisionmaker is clearly faced with a choice between that alternative and the others. A record of decision is to identify the environmentally preferred alternative, and the decisionmaker must discuss how he or she balanced the benefits of the environmentally preferred alternative against the benefits of other alternatives.

Alternative Uses of Available Resources

NEPA and the CEQ regulations require agencies to "study, develop, and describe appropriate alternatives to recommended courses of action in any proposal which involves unresolved conflicts concerning alternative uses of available resources."[139] This statement receives little attention by most agencies, perhaps because of its vagueness. One is not sure when or how this analysis would be required, or what "unresolved conflicts" and "available resources" mean. If the conflicts and resources were significant issues, an EIS would be prepared and alternatives would be identified and evaluated under an EIS process. The requirement is generally taken to mean that if such unresolved conflicts regarding a proposal were known, an agency is obliged to develop proposal alternatives that address alternative uses of the resources regardless of whether an EIS is prepared for the proposal. The regulations do not elaborate on the method for an alternatives analysis (for insignificant-conflicts-on-uses-of-resources issues) outside an EIS process.

FOCUS ON SIGNIFICANT IMPACTS

Whether an action or impact is "significant" will always be debatable among agencies, the public, and the courts because the word is subjective. Measurable thresholds beyond which an action's impacts become significant do not exist. Significance is not necessarily related to the dollar value of a proposal or its physical size. Generally, the courts have rejected specific size or monetary factors as a guide to determining the significance of an action. Therefore, significance is largely a function of an individual's perception and values. From the public's

---

[139]40 CFR §1501.2(c).

perspective, whether an issue is significant depends on an individual's perception and potential for being affected by the issue. A homeowner will perceive significant personal impact if his home would be demolished for a new highway. The removal of his or her home will be compensated, but personal or emotional impact may still remain. An impact may not be considered significant nationally, but may be perceived as significant locally. As another example, a judge ruled that a U.S. Navy proposal to capture and train dolphins was a "major federal action with an effect on the environment."[140] The action undoubtedly would have had a major effect from the dolphins' perspectives, but it seems a stretch to say the effect on the environment would be major, even if dolphins are a part of the environment[141] (a "major" action and its "significance" are usually viewed together).

The lead agency (or consultant) ID team and staff within the lead agency that reviews internal drafts of the EIS, and staff with other agencies that review EISs, will most likely have different perspectives on significance. The ID team and lead agency staff should identify, discuss, and resolve any differences in perspective early in the preparation of the EIS's foundation and framework (described in Chapter 8). An EIS team's ability to focus on significant issues, and to keep an EIS focused throughout its development, depends on how well the EIS's scope was established and whether the EIS's assumptions and scope received internal agreement.[142]

The EIS regulations, the CEQ, and those who provide EIS training emphasize the importance of focusing on the significant impacts or issues in an EIS, and eliminating from detailed study issues which are not significant. If lead agencies followed this advice, EISs would not be as long and would not require as much effort by reviewers to ferret out the truly important or significant issues. Lead agencies, however, routinely include detail regarding insignificant issues. Some of the reasons are as follows:

---

[140]Progressive Animal Welfare Society, et al. v. Department of the Navy (1989).

[141]Regulations authorize the Secretary of Defense to take up to 25 marine mammals each year with concurrence by the Department of Commerce (DOC). DOC had concurred with the Navy's proposal.

[142]Focusing on significant impacts may be agreeable in concept, but during an EIS process the conviction is frequently lost.

■ Agency reviewers who are responsible for certain resources naturally feel that their resource is "significant," and therefore should be included or expanded in an EIS. Lead agency staff and other reviewing agency staff who have statutory or agency mandates to further the goals of a particular resource review EISs to ensure that their resource is not overlooked or underemphasized. Many of these reviewers, who may be experts in their fields, are not familiar with the EIS process and view brief discussions of issues as inadequate, rather than an attempt to focus on only significant issues. Also, the reviewer may have his or her own agenda (e.g., to increase the importance of a program) for the analysis in an EIS. Lead agencies tend to add detail for insignificant issues for a variety of reasons including internal politics, aversion to conflict, and a belief that "more" is "better."

■ Agencies tend to err on the conservative side, and they add more than may be required to avoid criticism that the EIS is inadequate and avoid litigation. Authors on NEPA law observe that the legislators who wrote and passed NEPA probably did not envision the law as one that would create the huge number of lawsuits, or size of EISs, that it has. One legislator said he had thought an EIS would be around 15 pages.

■ Agencies seem more willing to add detail for insignificant issues to EISs that are funded by private proponents—not necessarily because the budget is not a concern, but to avoid the appearance of being biased for the proposed project or being on the proponent's side. Agencies are sensitive to public criticism of being biased and seem reluctant to take a position that may be perceived as in the project proponent's favor.

During scoping or review of an EIS, reviewers may request the addition of environmental elements to the EIS. First, one must determine if the issue is an effect of the proposal. If a proposed action will significantly affect the environment, all of the action's specific effects on the environment, whether or not significant, must be addressed. But if someone asks for information regarding an issue that is not an effect of the proposal, that issue need not be analyzed (e.g., if the proposal does not affect soil, potential impacts to soil do not have to be evaluated). To prepare an EIS that focuses on significant impacts, the entire EIS team, especially the lead agency, must first stay focused on the effects of the proposal throughout preparation of the EIS. When a proposal and its characteristics are well defined at the beginning of an EIS process, the significant effects and issues are more apparent. It is one thing to know that a topic is not a significant issue, but how do you know how much analysis is enough? Nonsignificant issues should be briefly discussed, or briefly analyzed in an EIS, enough to identify the expected impacts and mitigation measures, where it is feasible to do so.[143] The evaluation process for a nonsignificant issue is basically the same as for a significant issue, however the level of detail would be less.

Identifying Significant Impacts

If significance is subjective and a function of perception, how can an ID team decide what makes an impact "significant" for the purpose of an EIS? There is no clear answer because significance depends on the circumstances of the proposal and the affected environment. Significance will depend in large degree on the environmental baseline or the condition of the existing environment. An impact in a relatively pristine environment may be considered significant (in terms of the degree of change from the baseline environment), while the same impact may not be significant in a more degraded environment. An action can, however, have significant impacts even if the environment is already degraded. If a resource has been depleted so that little of the original remains, even a small impact on that resource may be significant.

Determining whether proposed actions significantly affect the environment, according to the CEQ regulations, requires consideration of context and intensity. The same guidelines can be used in evaluating the significance of impacts.

Context   The significance of an action must be analyzed in contexts such as the affected region, the affected interests, and the locality. Significance varies with the setting of the proposed action. For example, in the case of a site-specific action, significance would usually depend upon the effects in the locale rather than in the world. Both short- and long-term effects are relevant.

---

[143]46 FR 18031.

Intensity Intensity refers to the severity of impact. The following should be considered in evaluating intensity:

- Impacts that may be both beneficial and adverse
- The degree to which the proposed action affects public health or safety
- Unique characteristics of the geographic area such as proximity to historic or cultural resources, parks, prime farmland, wetlands, wild and scenic rivers, or ecologically critical areas
- The degree to which effects on the quality of the environment are likely to be highly controversial
- The degree to which the possible effects on the environment are highly uncertain or involve unique or unknown risks
- The degree to which the action may establish a precedent for future actions with significant effects or represents a decision in principle about a future consideration
- Whether the action is related to other actions with individually insignificant but cumulatively significant impacts
- The degree to which the action may adversely affect districts, sites, highways, structures, or objects listed in or eligible for listing in the National Register of Historic Places or may cause loss or destruction of significant scientific, cultural, or historical resources
- The degree to which the action may adversely affect an endangered or threatened species or its habitat that has been determined to be critical under the Endangered Species Act
- Whether the action threatens a violation of federal, state, or local law or requirements imposed for the protection of the environment[144]

How thoroughly must one evaluate significant impacts? According to the courts, the discussion should be a "reasonably thorough" discussion of "probable environmental consequences" and does not have to be a complete evaluation of them.[145] A "reasonably thorough" discussion is still subjective, and what is reasonable in one case may not be reasonable in another. But the evaluation does not have to be exhaustive.

---

[144]40 CFR §1508.27.

[145]Trout Unlimited v. Morton (1974); Citizens for a Better Henderson v. Hodel (1985); Stop H-3 Association v. Dole (1984) (Fogleman, 1990).

ENVIRONMENTAL IMPACT ANALYSES

The existing environment, including existing environmental impacts, must be carefully evaluated and described to identify the types and amount of impact that a proposed action may cause. Environmental impact is the change in the environment caused by an action. To identify change or impact to the environment, evaluate the features of a proposed action—its proposed construction, design, and operation—to determine

■ The elements of the environment that would be affected by the action (scoping)
■ The condition of the existing environment for an environmental element or resource (defining the affected or existing environment)
■ The changes the action will cause if the action were implemented (identifying environmental consequences or impacts)

If a proposed action will not change an existing environmental element—its condition, number, and diversity of resources remain the same—there is no impact to that element. Environmental changes or impacts may be beneficial as well as adverse. If an environmental element is in poor condition or is having adverse impacts in the existing environment, a proposed action could cause changes or impacts that are beneficial if the changes improve the element's condition or stops the adverse environmental impacts. A beneficial effect of a proposed action may also be that it stops a potential or future adverse environmental impact from taking place. For example, a prescribed burn in a forest, which reduces the amount of fuel (dense undergrowth or brush), reduces the potential for a devastating forest fire in the future.

As a simplified example of how to evaluate a proposed project's change or effect on the existing environment, let's assume we are evaluating the increase in impervious surface area that would result from the proposal. If the existing impervious area of the proposed project site is 50 acres and the proposed project includes 150 acres of impervious area, but the 150 acres would cover the existing 50 acres, the net change would be 100 additional acres of impervious surface. For this example one needed to know the existing condition with regard to impervious surface area, the location or footprint of the impervious area, and the amount and location of impervious area that would be created by the proposed project. This example is only a mathematical exercise to identify change to the environment. It lacks an explanation of the relevance or significance of the change—the increase in 100 acres of impervious surface area in the study area.

Change Versus Impact

A change to an environmental element that degrades or pollutes the quality of that element is generally considered an adverse impact, such as degradation of air quality or water quality. But what of a change that increases an environmental element, such as noise? If a proposed action will result in an increase of noise, is that change an "impact," and, if so, is it an adverse impact? Most EIS authors automatically say "yes" because noise is unwanted sound. But what if there is no one to hear the sound, and therefore no one to not want the sound? A change in the environment would take place, but the change is not necessarily an "adverse impact." For an impact to take place, there must be a receptor. In some types of analyses, one must go a step further after describing a change (e.g., an increase of 3 dBA),[146] to evaluate what the impact of that change will be. Who or what will be affected by the change, and if no one will be affected (including animals), is there an impact? In this case, "change" of an element and "impact" are not synonymous because the change is a measurement, not a change (impact) to a resource. Water and air are resources, noise is not. The method of identifying impacts depends on the type of environmental element. One should not automatically label changes to the environment "adverse impacts," nor should one assume that because an action increases something, it is adverse. Whether a proposed action results in environmental impacts depends on the particular circumstances of the existing environment. Whether an impact is adverse or significant also depends on the particular circumstances, especially a consideration of who or what would be affected by the change.

> One should not automatically label changes to the environment "adverse impacts," nor should one assume that because an action increases something, it is adverse.

---

[146]Three decibels on the "A"-weighted scale, a scale for sound within the range of human hearing.

The further removed an impact is from the cause, that is, the more steps there are between the cause of an impact and the impact, the more difficult it is to predict what the impact will be and how significant it will be. For the example above of the impervious surface area, a potential impact might be reduced recharge to an aquifer that flows to a stream and that supports an important fishery. The question of how the impervious surface area might affect the fishery would require a fairly detailed analysis to answer and would require available information on the watershed as a whole, not just the proposed project site. Whether the reduction in recharge to the aquifer from the proposed project site would cause a significant impact to fish depends on a number of factors including the amount of groundwater contributed to the stream from the 100 acres in relation to the amount being contributed from the overall watershed. The actual amount and flow rate of groundwater may not be known without installing and observing monitoring wells; however, this may not be necessary if it can be shown that the 100 acres of impervious surface area is a small percentage of overall surface area that is providing recharge to the aquifer, and therefore the impacts are expected to be insignificant. This is a very simple example for illustration, and the actual analyses would require considerably more discussion.

> **The further removed an impact is from the cause, that is, the more steps there are between the cause of an impact and the impact, the more difficult it is to predict what the impact will be.**

Accuracy Versus Assumptions

ID team members are responsible for the accuracy of their analyses and conclusions. First, team members must be qualified to analyze a particular environmental element, and, second, the analyses must be based on knowledge, not assumptions. The following is an example.

A contractor who was providing NEPA training to U.S. Navy personnel was explaining the concept of cumulative impacts. The contractor, who was a biologist by training, used a military firing range as an example, and she said that if the number of firearms at the range was doubled, the noise level would double. Noise analysis was not her area of expertise, and she made an incorrect assumption. Noise is not additive, it increases logarithmically. If the existing noise level produced by 20 firearms was 70 dBA, and you doubled the number of firearms (for a total of 40), the new noise level might be only 73 dBA, not 140 dBA. Before using data or making conclusions regarding a discipline outside your area of expertise, discuss the subject and verify your understanding with a qualified person.

Assumptions regarding the potential environmental impacts of a proposed action are easy to make. This is lazy impact analysis, although lack of time and budget are familiar excuses. Team members must know the details related to each feature of the action—where it would be located, how it would be sited or situated, how it will be designed, how it will be operated, and other details. If it is a type of facility with which you are not familiar, visit a similar facility if possible. Beware of "boilerplate" impact analyses that provide the types of environmental impacts expected from certain types of development. They are based on numerous assumptions which may not apply to the specific project and specific locations under analyses.

> **Beware of "boilerplate" impact analyses that provide the types of environmental impacts expected from certain types of development.**

As an example, some of the ID team members for an EIS on a proposed thoroughbred racetrack had never seen a horse racing facility. A team member who was evaluating potential impacts to wildlife described the existing environment as if it were pristine wildlife habitat. The site was cattle pasture, with various farm buildings, surrounded by heavily trafficked roads and two major railroad lines. The team member stated that the only wildlife that was expected to use the site if the racetrack were built are those animals that could adapt to an urban environment such as mice, rats, moles, raccoons, coyotes, and skunks. However, the site was already in a largely urbanized environment, and the "urban wildlife" species were on the site in its existing condition. A study of a racetrack that the proposed racetrack was replacing identified a much larger diversity of wildlife species using the track than was assumed by the team member. Ponds, open fields, limited hours of operation, and the fact that the racetrack only operated in the summer made the facility more attractive to wildlife than would be expected for most commercial or industrial facilities. The team member had used a "boilerplate" approach for construction of a generic facility on a generic site, and the analysis did not reflect the specific circumstances of the existing environment or the changed environment.

Using a federal district court case as another example, the Governor of Puerto Rico and several private citizens sued the U.S. Navy for naval training around the island of Vieques, Puerto Rico, alleging that the training violated the Endangered Species Act and Marine Mammal Protection Act.[147] The endangered species included the brown pelican, the manatee, and several species of sea turtles. Considering that the training activity included Navy and Marine Corps use of Vieques for amphibious landings as well as for air-to-ground, ship-to-shore and artillery training, one could easily assume adverse environmental impacts to marine animals. The following are portions of District Judge Torruella's written opinions.

With regard to the brown pelican:

If pelicans are as susceptible to military activity as is alleged one wonders why they established a nesting colony in such close vicinity to Vieques' most active military zone. . . . In fact, the major disturbance to this nesting colony is brought about by visits of fishermen who go into the cay to collect snails. By restricting the presence of humans in this area, Defendant Navy has de facto provided a refuge for the pelicans (and other wildlife).

With regard to sea turtles:

The record shows that the greatest threat to these species in Vieques, as throughout the Caribbean, has been the unrestricted fishing that has taken place. There is evidence that this fishing, although presently illegal under the Endangered Species Act, is still taking place around Vieques, together with the poaching of sea turtle eggs and nesting adults. Defendant Navy's presence on Vieques, together with the restrictive nature of its activities, has had some measure of benefit to the turtle population by precluding some of the illegal fishing and egg poaching.

With regard to manatees:

The evidence presented demonstrates that the manatee is found in larger numbers and concentrations in Vieques than any other area of Puerto

---

[147]Barcelo v. Brown (1979) (Wellington, 1988).

Rico except the Naval Reservation at Roosevelt Roads, across Vieques Sound. We do not deem it coincidental that both these areas are under the control of Defendant Navy.

The U.S. Fish and Wildlife Service provided a biological opinion regarding the Navy's activities on Vieques, which stated in part:

> Based on the Team's on-site inspection, information in the December 1979 Draft Environmental Impact Statement, reports of contractors employed by the Navy for environmental studies, and other pertinent reports, it is our Biological Opinion that naval activities associated with training at Vieques Island are not likely to jeopardize the continued existence of the manatee, brown pelican, loggerhead turtle, green turtle, leatherback turtle or hawksbill turtle, or adversely modify habitat essential to these species existence. Cumulative effects were considered in reaching this opinion but we felt they did not apply in this case.

The National Marine Fisheries Service's biological opinion also stated that the Navy's activities were not likely to jeopardize turtles or whales.[148]

Methodology

An EIS does not have to use the best scientific methodology available, but the methodology should have a reasonable basis that takes into account relevant considerations. An EIS may have an overall methodology (discussed in Chapter 8), and the analyses for each resource or subject area may also have a methodology. EIS authors should be aware of the EIS's overall methodology to make the individual analyses consistent with the methodology of the EIS. Provide an explanation of the methodology at the beginning of an analysis of an environmental element or subject. If the methodology is long or complex, it may be placed in the appendix, and a note in the text should refer the reader to the appendix for additional information.

To analyze the impacts of the proposed action and the impacts of the alternatives, the methodology of analysis must be consistent; that is, the same methodology should be used for all of the alternatives. First, describe the affected or existing environment. The affected envi-

---

[148]Joseph A. Wellington, *Naval Readiness, Operational Training, and Environmental Protection: Achieving an Appropriate Balance Between Competing National Interests,* unpublished, 1988.

ronment is the description of the environment within the study area, and the description is by type of environmental element (e.g., air, water, traffic, cultural resources). If you are responsible for analyzing the impacts of a proposal on your assigned resource or topic—water, for example—begin by describing all of the existing water resources in the study area. Commonly, the description of water resources is divided into surface water and groundwater, and each of those categories are further divided into water quantity and water quality. Therefore, describe the existing water quantity and quality for both surface water and groundwater.

Team members should be given a description of existing land uses within the study area early in the process. If this information is not available for the kickoff meeting, the person who is writing the land use section of the EIS should give a copy of his or her existing environment section to other team members as soon as possible. The "existing environment" discussion for land use should describe existing human activity on the site, including unauthorized uses or trespassing, because those activities may have environmental impacts associated with them. To do a credible job of environmental impact analysis, the team should visit the project site to evaluate its condition prior to starting the EIS. The team should also be familiar with the features of the proposed project and its alternatives, the proposed "footprints" of the project and alternatives on the site, how construction would take place, what the project would look like, and how it would operate after completion.

Describing Impacts

Be very specific in describing impacts, and quantify impacts if possible. Describe impacts with the knowledge that for each impact you identify you must address mitigation measures. Mitigation for environmental impacts may or may not be practical, or may only be partially effective. However, to be enforceable and measurable as to their success, any mitigation measures should be specific; therefore the impacts must be specific. For example, if you say there would be "adverse water quality impacts" to a stream, how will you describe mitigation with any kind of specificity? What is the impact? Sedimentation? Nutrients? Heavy metals? What will cause the impact? Construction? Animals? Traffic? Once specific impacts are identified, specific mitigation measures such as the construction of siltation screens, catch basins, bioswales, or a buffer zone to keep horses and cattle some distance from a stream can be identified as well.

Also, impacts must be specific in order to compare and contrast the impacts of the proposed action and alternatives. If the environmental consequences or impacts of an action or its alternatives are too general or vague, the decisionmaker would not be able to distinguish differences between the impacts of one alternative relative to another.

Lastly, describing specific impacts is useful to reviewers who are being requested to provide input to the EIS. If you wrote only that a proposed project would result in construction noise, the usefulness of the information to provide meaningful response is limited. However, if you wrote that "Construction noise of up to 100 dBA at 50 feet would be heard during daylight hours from June to September," a reviewer could express a specific concern that the noise level would disturb migratory waterfowl that rest nearby starting in late August. A suggested mitigation measure for the potential disturbance to waterfowl might be that construction end by mid-August.

Quantify impacts as much as possible. Develop the habit of thinking and writing spatially, in three dimensions, and in timeframes. To say that a project would result in an increase of impacts without quantifying the impacts is vague and doesn't provide a strong basis for the analysis' conclusions. Accurate quantification is not always possible; however, a range of impacts (low end to high end) might be possible, or a statement that sets an upper limit such as "a maximum of approximately 2000 square feet." Professionals are expected to use their judgment, based on the facts and standards of their profession, and make conclusions about whether impacts would be significant. How long would impacts last? Would they occur only during certain times of the day, only during construction, or for the life of the project? Where are the closest receptors of the impact, how many are there, and how sensitive are they to the impact? Will an impact's effects decrease or increase over time or over space? Quantify the time or distance.

Place environmental impacts in a context that the average person can understand, so that a reader can tell if a noise increase of 3 dBA is of any concern; for example "a change of less than 3 dBA is normally not detectable by the human ear." Otherwise, a reader would say "So what?" So what if there is an increase or decrease in something, what does it mean? Don't make the reader guess whether there is any significance to an impact or why it was identified. The significance of an impact must be explained.

> **So what if there is an increase or decrease in something, what does it mean?**

If you provide data in your analysis, you also are obliged to explain the significance of the data in a way that is understandable to a lay person. EISs are prepared by engineers, scientists, and planners in areas that are so specialized that a scientist wouldn't understand what a traffic engineer had written, or what another scientist had written, if the subject were another area of specialty. Write in plain language so that any adult can understand the material. Some technical words or concepts may be necessary in order to conduct the analysis, but always explain what a word means, or what a technical concept is, the first time it is used.

If a proposed project has built-in features (features that are part of the project) that would mitigate impacts, identify the impacts that would result with mitigation measures in place. Some authors approach the impact analysis as though the proposed project were a generic project, with no special features, and then add mitigation as an afterthought. This is similar to the "boilerplate" approach discussed above. The author first identifies impacts that would be expected from any construction project, for example, and then discusses mitigation measures that would be required by regulations or that are built-in features of the proposed project. The analysis should, however, identify the impacts of the proposed project with the mitigation measures that would apply to that project and skip the boilerplate impacts.

To identify anticipated impacts after the implementation of mitigation measures may require an evaluation of how effective those mitigation measures are expected to be. If a mitigation measure is known to be only partially effective, is difficult to implement or maintain, or is untested, it may not be relied upon with confidence to mitigate the environmental impact. The effectiveness of mitigation measures is also a function of how well the measure is designed, implemented, and maintained. For the purposes of analysis in an EIS, one would normally assume that a mitigation measure would be properly designed, implemented, and maintained in the absence of known technical difficulties. If difficulties are known, monitoring plans and follow-up actions may be necessary to ensure the effectiveness of mitigation measures, particularly if the regulatory procedures of agencies with jurisdiction do not include monitoring or enforcement.

ANALYSES OF CULTURAL RESOURCES AND ENDANGERED SPECIES

An EA or EIS must evaluate the degree to which a proposed action will comply with other laws, including the laws on historic preservation and endangered species. Whether these issues will be in the scope of an EIS depends on the type of proposal and its location. For most proposals that include clearing, demolition, renovation, or construction, lead agencies must evaluate whether cultural or historic resources would be affected. Cultural and historic resources can be found in any type of environment, built or natural, urban or rural, on land or under water. Proposals that are normally categorically excluded from the requirement to prepare an EA or EIS are exempted from the exclusion if there is a possibility of affecting cultural or historic resources. In other words, a proposal that has the potential to affect historic or cultural resources cannot be categorically excluded. Thus, the degree to which a proposal complies with environmental laws must be evaluated whether or not an EA or EIS is prepared.

> **The degree to which a proposal complies with environmental laws must be evaluated whether or not an EA or EIS is prepared.**

The same exemption from categorical exclusion applies if a proposal has the potential to affect threatened or endangered species or their habitat. Threatened and endangered species are not as common in the built or urban environment because of lack of habitat, sensitivity to human activities, and pollution. Both plant and animal species (including fish and marine mammals) can be listed as threatened or endangered. Many lead agencies no longer show on maps in EISs the specific location of any cultural resources or endangered species in order to protect those resources. The following is a general approach for analyses of cultural and historic resources and endangered species in EISs. The details of the process and time required to consult with agencies will vary depending on the proposed action and the resource of concern.

Cultural or Historic Resources

The analysis of impacts to cultural or historic resources normally requires an archaeologist or historian, or both. If the area that would be affected by a proposed project is an undeveloped area, an archaeologist is normally retained to conduct a literature and field evaluation to determine the likelihood for the project to affect any cultural resources. In this case the resources may not be visible, because anything on the surface may have been collected or destroyed, leaving resources that are buried. If existing literature and the field analyses conclude that the potential for the presence of cultural resources is low, further analysis is normally not required. However, because the field work may have been over a widely spaced grid (and resources might have been missed), a standard requirement is for construction to halt if any resources are encountered and for the agency with jurisdiction to be notified. If there is a moderate or high likelihood for the presence of cultural resources, additional field work to look for, identify, and excavate any resources may be necessary, or an archaeologist may be retained to observe project-related excavation. An area that has been farmed is not necessarily free of cultural resources because the resources may be at depths greater than the reach of farm equipment.

A developed area may require evaluation by a historian or architectural historian to determine whether any structures are potentially eligible for listing in the National or State Registers of Historic Places, or both. A structure that is potentially eligible does not necessarily have to be preserved. Other options are available, such as photographic documentation, which are negotiated between agencies such as the state Historic Preservation Office, federal Advisory Council on Historic Preservation, the land managing agency, any affected American Indian tribes, and permitting agency. For the purpose of EISs, where better examples of a building's features exist in other retained historic buildings, or where features integral to the building's historic, cultural, or architectural significance would not be affected, the environmental effects may be determined to be insignificant.[149]

---

[149]Preservation Coalition, Inc. v. Pierce (1982) (Fogleman, 1990).

If a lead agency does not have an archaeologist or historian on staff, the analysis for an EIS is normally prepared by a consultant. If a consultant is not familiar with the EIS process, the process, including its purpose, assumptions, scope, methodology, format, and language of EISs should be explained. The objectives of the consultant's task, as well as his or her role as part of the EIS or ID team, should be clear. In particular, consultants should understand that an EIS does not necessarily provide all of the information required to comply with historic preservation laws (unless the lead agency wants it to). Compliance with the historic preservation laws, however, is a separate process in accordance with the requirements of the laws and the agencies with jurisdiction. Therefore, the consultant's scope of work for an EIS may be different from other types of contracts or studies regarding cultural resources.

Threatened or Endangered Species

The CEQ regulations require agencies to evaluate the degree to which actions may adversely affect endangered or threatened species or critical habitat, as required by the Endangered Species Act. Biologists with expertise in plants or animals normally evaluate literature and contact federal and state agencies that administer the provisions of the Endangered Species Act for information regarding the presence of endangered or threatened species and their habitat in the study area.

EIS team members should visit the study area to observe the condition of the existing environment. But if a sensitive species is known or suspected to be in the study area, field work to verify the species' presence is necessary. If a study area is in a marine environment, field studies may include dive surveys to, for example, evaluate fish spawning habitat. Evaluation of literature and field work provide information used to describe the existing environment and identify potential impacts. The existence of endangered species in an area or the possibility that an endangered species may be present does not necessarily mean that the environmental effects of an action will be significant. Factors such as the type of species, its location and sensitivities, the existing environment, the type of proposed project, and its impacts all determine whether a proposed action may affect a species.

Consultation with the U.S. Fish and Wildlife Service (USFWS) is normally conducted for federally listed species during the EIS process if it has not already been done. Consultation may be informal or formal, and a biological assessment may be prepared and included in an appendix to the EIS. The consultation process, however, is not part of the EIS process, although it may be taking place at the same time as the EIS process. The biological assessment concludes with an opinion on whether a species or its habitat would be jeopardized by the proposed action. The USFWS and U.S. National Marine Fisheries Service (NMFS) (for certain marine species) review the biological assessment and issue biological opinions stating the USFWS and NMFS opinions regarding whether the species or its habitat would be jeopardized by a proposed action. An EIS discusses the issues related to endangered species, including consultation with other agencies. Compliance with the Endangered Species Act, however, is a process conducted in accordance with the requirements of the act and the agencies with jurisdiction. Most EIS lead agencies prefer to have completed consultation prior to completing an EIS so that the agencies' opinions and any mitigation measures are identified in the EIS.

MITIGATION MEASURES

Mitigation measures are actions that reduce the adverse environmental impacts of a proposed action. Adverse impacts are the environmental "problems" and mitigation measures are the "solutions" in an EIS. Not all environmental problems have solutions, however, and mitigation is discussed in the CEQ regulations and by the courts as required where "practicable" and "where it is feasible." Impacts do not have to be mitigated completely, to a level of no impact. Mitigation measures must be "reasonable" and capable of being implemented. Also, mitigation should relate to the specific impacts of a proposal. If a proposal is contributing to an environmental impact that is caused by many other sources, it may not be reasonable to require a proposed project to mitigate for more than its share of the impact.

The CEQ regulations provide the following examples of mitigation:

■ Avoiding the impact by not taking a certain action or parts of an action
■ Minimizing impacts by limiting the degree or magnitude of the action and its implementation

■ Rectifying the impact by repairing, rehabilitating, or restoring the affected environment
■ Reducing or eliminating the impact over time by preservation and maintenance operations during the life of the action
■ Compensating for the impact by replacing or providing substitute resources or environments[150]

In addition, agencies with jurisdiction over a proposed action may identify mitigation measures in accordance with their regulations, policies, and procedures. As noted earlier, the courts have determined that lead agencies do not have to adopt the mitigation measures in an EIS. However, if agencies with jurisdiction have permitting authority over a proposed action, their mitigation measures will most likely be conditions of their permits. A lead agency commits to mitigation measures that are included in its record of decision or that are made conditions of its permit or other approval.

Some reviewers of EISs feel that regulatory requirements that control environmental impacts should not be considered mitigation measures. In other words, only an action that is in addition to regulatory requirements is considered a mitigation measure. This, however, is an unreasonably restrictive definition of mitigation measures. Any action that reduces, avoids, or compensates environmental impacts is a mitigation measure regardless of whether the action is required by regulations.

In some cases, a proposed action's environmental impacts may be mitigated by actions unrelated to the proposal. For example, if a state highway department were making improvements to roads in the study area of a proposed action, traffic impacts caused by the proposal may be mitigated by the highway department's actions. Existing conditions may also mitigate a proposal's environmental impacts if those conditions serve to offset or reduce the proposal's impacts. On the other hand, one court disregarded mitigation measures that were not project-related.[151]

The discussion of mitigation measures and environmental impacts are interrelated. First, impacts cannot be accurately identified

---

[150]40 CFR §1508.20.
[151]Preservation Coalition, Inc. v. Pierce (Mandelker, 1995).

without considering the mitigation measures that would be implemented as part of a proposed action or as known requirements by law or by agency policies and procedures. Some agencies object to discussing mitigation at the same time that impacts are discussed because the discussion of mitigation seems to be an attempt to excuse the environmental impact. However, to limit discussions of environmental impacts only to the impacts section, and mitigation only to the mitigation section, is an artificial restriction that compartmentalizes an EIS. Just because EIS regulations identify distinct topics that must be in an EIS does not mean that authors have to slavishly limit their discussion to the topic of the section. If, by including an explanation of how a feature of a proposed action will mitigate the impact, a reader will better understand the impact of a proposed action, the mitigation should be discussed.[152] The focus of an impact section should be impacts, and the focus of a mitigation section should be mitigation, but some overlap is inevitable. Redundancy is not forbidden as long as it serves a purpose and improves understanding.

Second, mitigation measures and impacts are parts of the alternatives analysis. Alternatives are developed to provide a range of environmental impacts, and mitigation measures affect the range of impacts. If an alternative to a proposed action is a project of reduced size or scope, that alternative is also a mitigation measure that results in less environmental impact than the proposed action. Thus, the interrelatedness of mitigation measures and impacts is evident in identifying and analyzing alternatives.

Third, mitigation measures should be discussed for impacts that were not addressed by, or in addition to, features of the proposed action, regulations, and the alternatives. At this point in an EIS, most of the mitigation measures for a proposed action's impacts will have been identified. EIS authors may have difficulty identifying additional mitigation measures. The regulations that control proposed actions are so numerous and detailed that few potential environmental impacts are not addressed by existing requirements that mitigate impacts. Also,

---

[152]40 CFR §1502.14(f) states (in discussing the impacts of alternatives) "Include appropriate mitigation measures not already included in the proposed action or alternatives."

some difficulty lies in defining what "reasonable" or "practicable" mitigation measures might be for a given proposed action. EIS authors look to professional guidelines such as "best management practices," "best available control technology," and professional manuals to identify any additional mitigation measures.

The following should be considered when identifying and evaluating mitigation measures.

Appropriateness

Is the mitigation measure appropriate for the type, magnitude, and time of impact? Does the mitigation measure reflect the specific conditions of the existing environment and the impacts of the proposed action? Don't apply generic mitigation measures based on assumed environmental impacts. For example, not all animal species will react or be affected by a proposed action in the same way. Some species are more tolerant of human activity than others, and the same species may react differently to impacts during migration or breeding seasons. Some impacts might be more significant at night than during the day. Make the mitigation appropriate for all of the factors related to an impact.

How much mitigation is enough, and by whose standards (lead agency, commenting agency, or a private sector expert)? The lead agency has the final say regarding standards to use in the preparation of an EIS, although the courts may have the final say in the long run if the EIS is taken to task for inadequately considering mitigation measures. Provide a discussion of mitigation measures that is reasonable for the type and level of environmental impact. Significant environmental impacts should receive fairly detailed discussions of mitigation measures, while insignificant issues require the identification of mitigation measures where it is feasible to do so. Impacts which cannot be mitigated are unavoidable adverse impacts—both significant and insignificant.

Effectiveness

A mitigation measure's potential effectiveness should be evaluated in terms of how effective the method was in the past for similar conditions and types of impacts. If a mitigation measure is untested or its effectiveness is uncertain, additional types of mitigation or additional quantity of mitigation may be necessary for possible partial or total failure. For example, wetland mitigation in the form of creation of new wetlands to replace filled wetlands is normally greater than one acre for one acre (e.g, two acres of new wetland for every acre filled). Evaluate "standard" mitigation measures to ensure that they are effective or necessary. There is no point in requiring that the first six inches of topsoil be stockpiled for later use in mitigating vegetation impacts if (a) the project site is in a desert, (b) it has pumice or sand, not topsoil, and (c) the requirement will not mitigate vegetation impacts. However, if a mitigation measure is a regulatory requirement, it may be necessary regardless of its effectiveness for a particular project or locale. If possible, discuss the effectiveness of a mitigation measure with the agency that has jurisdiction for the regulatory requirement to see if a waiver from the requirement or a more effective mitigation measure might be identified. Disclose the outcome of such discussions in the EIS's discussion of mitigation measures.

Implementation

For mitigation measures to be effective, it must be possible to implement them. A mitigation measure must be feasible and an agency must have the authority to implement or require mitigation. Mitigation measures, even for the same impact, frequently involve more than one agency with jurisdiction. If mitigation measures are developed in consultation with all the affected agencies, responsibilities can be identified and assigned as part of a mitigation plan. The mitigation measure must be technically feasible to implement and cost effective. The expected environmental benefit should be commensurate with the cost of implementing the mitigation. Also, the timing to implement a mitigation measure must be feasible in terms of a proposed action's requirements. Some public reviewers insist that a mitigation measure be in place and be proven effective, prior to agencies giving a project approval. The rationale is that once approval has been given and the project has created an environmental impact, it is too late to discover that a mitigation measure is not effective. However, to require a project proponent to implement mitigation before the proponent knows whether he or she will obtain approval is not reasonable, especially for mitigation that is costly and could take years. Other forms of commitments from the project proponent such as monetary bonds, legally binding agreements, and a mitigation plan that identifies the mitigation method and availability of resources to implement the mitigation are more reasonable requirements.

> **The expected environmental benefit of a mitigation measure should be commensurate with the cost of implementing the mitigation.**

Monitoring

Monitoring may be necessary to ensure that mitigation measures are properly implemented, operating, and maintained. Mitigation measures that are difficult to implement, that are untested, or that take place over a period of time often require monitoring. A monitoring plan, which may be part of a mitigation plan, identifies how monitoring will take place, who will conduct the monitoring, and how frequently. Contingencies are identified of the corrective actions that will be implemented if the mitigation measure starts to develop problems. The goal is to correct any problems early enough to ensure the mitigation measure's success. Of course, a mitigation measure must be properly designed and constructed or implemented to begin with.

Enforcement

Depending on the type of mitigation measures, and the agency with jurisdiction, mitigation measures may or may not be enforced. An agency that has jurisdiction and that required conditions in its approval or permit, ostensibly would have provisions to enforce the conditions of approval. Enforcement actions vary with agencies and include work stop orders and fines. However, some agencies do not follow through to monitor the actions of a project proponent and to enforce mitigation measures. Whether the problem is one of insufficient staff and budget, or lack of coordination between agencies, program managers, or enforcement staff, lack of enforcement may result in ineffective mitigation measures. Project proponents are not necessarily trying to avoid meeting the conditions of their approval. Many private project proponents do their best to meet conditions of their approval. Agencies that are proponents of their projects must frequently monitor and enforce their own actions. Monitoring and enforcing mitigation measures ensure that the details of a mitigation plan, identification of problems, and corrective actions are not overlooked.

Impacts of Mitigation Measures

Environmental mitigation measures are designed to improve existing, or reduce potential, environmental impacts; therefore, it is usually not necessary to analyze the environmental impacts of implementing a mitigation measure. In most cases the environmental impacts that result from a mitigation measure are only beneficial. In other cases, some adverse environmental impacts might occur, for example, during construction of a mitigation measure. Beneficial and adverse environmental impacts should be evaluated in an EIS if the impacts have not already been addressed—for example, if impacts from mitigation would be to resources that are not affected by the proposed action, or to an area other than the proposed action or alternative sites.

Mitigation Measures for Threshold Determinations

If mitigation measures would reduce impacts below the threshold of significance, and are therefore an agency's rationale for not requiring an EIS, approval of the proposed action must be conditioned on the mitigation measures. The mitigation measures must be made contractual obligations (or some other enforceable obligation) in order to mitigate significant environmental impacts below the threshold of significance.

SHORT-TERM USES VERSUS LONG-TERM PRODUCTIVITY

NEPA identifies two additional subject areas that are to be included in EISs; these are

■ The relationship between local short-term uses of man's environment and the maintenance and enhancement of long-term productivity
■ Any irreversible and irretrievable commitments of resources which would be involved in the proposed action should it be implemented

The CEQ regulations do not discuss these requirements in any further detail, and most preparers of EISs consider these requirements annoyances. The subjects make very long headings under which many EISs have very brief discussions. The intent of including these subjects in an EIS is to clearly show the tradeoffs involved in the decisionmaking process. To do this, an EIS team member would basically distill the conclusions regarding the various elements analyzed in an EIS to show how a proposal would in some ways use certain environmental resources, while in other ways benefit the environment. The wording with regard to "the maintenance and enhancement of long-term productivity" is ambiguous, however ("productivity" and "long-term" are subjective). One could assume that the phrase is referring to how the proposal will benefit environmental resources (if resources are enhanced or maintained, one would also assume that it would be for the foreseeable "long-term").

The meaning of "irreversible and irretrievable commitments of resources" has also led to much conjecture. Some EIS authors define "irreversible" as a change in an environmental resource or condition that could not be returned to its original condition, while "irretrievable" means use of a resource that can never be replaced. Minerals once mined are irretrievable; however, timber once harvested can be replanted. Timber, therefore, is not irretrievably committed; however, the forest habitat that was originally in place may be irreversibly changed.

Other EIS authors lump "irreversible" and "irretrievable" together and do not attempt to define their separate meanings. Common resources that are committed when a proposal is implemented include energy and natural resources (e.g., construction materials, land, wildlife habitat). If a proposal includes construction of a facility, some EISs have provided estimates of energy and materials that would be used for the proposed project (for construction and operation).

Legislation amending the California Environmental Quality Act (CEQA) deletes the requirement for a discussion on short-term use of the environment versus long-term productivity in an EIR.[153] The discussion of "any significant irreversible environmental changes which would be involved in the proposed action should it be implemented" is only required for EIRs prepared in connection with (a) the adoption, amendment, or enactment of a plan, policy, or ordinance of a public agency, (b) the adoption by a Local Agency Formation Commission of a

---

[153]The amendment was effective in 1994, but the 1995 CEQA guidelines have not been revised to reflect the change.

resolution making determinations, or (c) a project which will be subject to the requirement for preparing an EIS pursuant to the requirements of NEPA. The CEQA guidelines provide some insight into this section's contemplated discussion:

> Uses of nonrenewable resources during the initial and continued phases of the project may be irreversible since a large commitment of such resources makes removal or nonuse thereafter unlikely. Primary impacts and, particularly, secondary impacts (such as highway improvement which provides access to a previously inaccessible area) generally commit future generations to similar uses. Also, irreversible damage can result from environmental accidents associated with the project. Irretrievable commitments of resources should be evaluated to assure that such current consumption is justified.[154]

THE LANGUAGE OF EISs AND OTHER WRITING TIPS

EISs are written by or for agencies, so there is a tendency to use a "bureaucrateze" style of writing. This usually means writing in the third person, passive voice, and past tense. This style of writing produces unnecessary words and is often difficult to understand. Although some agencies are attempting to improve their employees' writing by providing writing courses, many bureaucrats simply object to a first person, active style of writing because they feel it is inappropriate for agencies or those in official capacities.

In addition to writing style, an EIS's format will affect the readability of an EIS. The format and content of EISs are provided by regulations[155] and frequently make a document's organization awkward by breaking up a subject into different sections and chapters. Readers must flip back and forth between sections if they are trying to understand the information for a particular subject.[156] Add agency acronyms, legalese terms, technical jargon, and environmental buzzwords, and an EIS can become hard to read.

---

[154]California Office of Planning and Research, *California Environmental Quality Act Guidelines*, 1995.

[155]There is some flexibility as long as the basic requirements are met; however, agencies are usually reluctant to stray very far from the regulations.

[156]Comments from reviewers frequently state "you forgot to include xyz"; and our response is, "it was included but it's in a different section."

Writing to Communicate

Writing is a form of communication. Educational institutions teach students to read and write, but not necessarily to communicate. Communication is usually learned through trial and error through the course of our lives, and nowhere does a poor choice of words in our writing come back to haunt us as from something we wrote in an EIS. Authors are amazed when they wrote something that is technically sound and innocent, yet somehow conveyed an incorrect or unintentional meaning to the reader. Effective communication requires knowing the audience, what their concerns may be, what misconceptions they may have, and the type of information they are looking for. Since an EIS is read by a wide variety of people with different educational, professional, and personal backgrounds, communicating what one really means to say is inherently difficult. Although 100 percent effective communication may not be possible, you can improve on conveying what you mean to convey by being precise in the choice of words, using plain and unbiased language, and showing (with visual aids such as maps, graphs, or other drawings) how something works or how you came to your conclusions. A rule of thumb is, if the editor doesn't understand something, the chances are that neither will most of the readers.

Precise Meaning

Many words have several meanings, or different meanings depending on their usage. You may know which meaning you intended, but the reader may not, and the context of the sentence may not make it clear. Slowly reread what you write, a day or more after writing it, and look for ambiguities or sentences that could be misunderstood. Place yourself in the role of a lay person who is trying to understand what you wrote, or an opponent to the proposed action who is looking for flaws in your rationale. Editors are helpful in identifying ambiguous words, but if they don't have the appropriate technical background, they may not be able to find all of the problems. Peer review, if time and budget allow, is another way of finding problems with word usage, but peers may make the same mistakes you make because they understand the subject. To choose the (write) (right) correct word, you must think about the meaning you are trying to convey as you choose the words.

Statements can be ambiguous if their relationship to other statements and their meanings are unclear. As an example, a common ambiguity in EISs occurs in the comparison of impacts of the various alternatives. An author will write "The air quality impacts of Alternative 2 would be less" or "would be reduced." Less than what or reduced from what? Less than the proposed action, the existing conditions, the no-action alternative, or another alternative? Why are the impacts less, and how much less are the impacts? The author may have described the impacts of all alternatives in previous sections, and therefore assumed that the statement is self-explanatory. However, authors must consider that readers are attempting to read and understand a large amount of information, not just the portion the author wrote. Authors should explain their rationale, and they should pull information from other sections into their discussion if it will assist the readers.

Plain and Unbiased Language

Plain language consists of commonly used words. This doesn't mean that one should use only simple or monosyllabic words, or that the analyses has to be short or uncomplicated, but it does require a conscious effort to exclude technical jargon or buzzwords as much as possible. Technical subjects will require some use of technical terminology in order to conduct the analyses. Define terms that are not commonly known, and explain the significance of facts or observations if, without the explanation, the significance would only be clear to someone with your specialty. A technical analysis written for specialists should be placed in an appendix, and the important points, rationale, and conclusions should be written in language that most adults can understand in the text of an EIS.

Unbiased language consists of words and statements that are factual and neutral with respect to an author's personal values. The choice of words can convey personal bias regarding a proposed action or other topic. Most authors do not deliberately choose biased words or deliberately make biased statements. The choice is unconscious and, because the author is unaware that he or she is biasing the analyses, someone else (usually a public reviewer) will point out the bias (to the surprise of the author). In most other forms of writing, bias is not an issue. If you write letters, articles for magazines, or books, your opinions or bias reflect your values and are accepted as such because the product is entirely yours. An EIS or the analysis you authored, however, is not "yours"; it does not belong to any one author but, instead, belongs to the lead agency. The language of an EIS should reflect its purpose, to present the facts—including both sides of any controversy—to the decisionmaker.

Neutral Words

Use of neutral words in EISs is similar to the use of unbiased language. Neutral words don't convey an emotion or show a bias in your attitude toward the subject. As an example, someone once wrote in an EA, "One acre of vegetation would be obliterated by the proposed action." The word "obliterated" carries more than just a description of the impact; it connotes violence and makes clear that the author does not like the proposed project. However, likes and dislikes are personal and should not enter the job of environmental analysis. "Removed" can be used in place of "obliterated," and the impact is still clear to the reader. Or for a more active voice, the sentence could be rewritten to say "The proposed action would remove one acre of vegetation from the site."

Another example is "Douglas squirrels were observed, and a garter snake was seen slithering down a mountain beaver hole at the wetland edge." "Slithering" has a negative connotation and could lead a reader to assume that the writer does not like snakes. It may not bias a reader against snakes, but the word is not one that professionals normally use, and the reader may wonder how professional the author was in his or her analysis.

Show, Don't Tell

Don't call an impact an "impact," but instead describe what the impact will be. A reader understands the impact better if you write that a wetland would be filled by a project, rather than a wetland would be "impacted." First of all "impacted," which means forced or wedged between something, isn't particularly accurate in describing environmental impacts; but it's fine if you're describing impacted teeth. Also, consider what "impact," which is pervasive in environmental work, means to most people who are not involved in environmental work. Consider what readers will visualize if you write "light and glare would impact passersby and traffic." Images of people blinded by light and glare come to mind. The reason is the word "impact" itself means a forceful blow or collision between objects. In this case, the author simply meant that light and glare would be visible to passersby. The "impact" is visibility, so use the word that describes the impact rather than using "impact" as a verb or noun. EIS regulations, as well as this book, are full of the word "impact" because they address impacts in general, but an EA or EIS should be full of words that describe specific impacts. Or, as editors are fond of saying to writers of stories, "Show, don't tell (SDT)." The context may be different, but the same principle applies.

> **Don't call an impact an "impact," but instead describe what the impact will be.**

Be specific when describing impacts: who, what, when, where, why, and how much. If there are existing impacts, what is causing it? How is it occurring in the existing environment? How much (quantify) is the existing impact? Then, describe the expected *difference* in impacts by comparing the impacts that would result from the proposed project to the existing impacts. Will there be an increase in impact? A decrease? How significant, or how important, is this difference when placed in the context of the locality or in the context of other impacts? If the proposed action were implemented, would the impact under analysis pose a health or safety risk? Most authors who use the word "impact" to describe impacts are not being analytical and are taking the easy route. An example is the following:

"The proposed subdivision would have an impact on fire and police services, schools, and water and electric utilities serving the area." This sentence doesn't describe what the impacts would be. Some authors use this type of sentence as a lead-in to a paragraph, as though it were necessary to alert the reader to the fact that impacts are being discussed there. In any case, it is only necessary to say what the impacts would be. Would the proposed project increase demands on these services? If so, would those demands be within the services' existing capacity, or would additional services be required to accommodate the project?

Other words commonly used to label impacts are "conflicts" and "problems"; for example, "the proposed project would cause *conflicts*" or "cause *problems*." Describe the potential conflicts or problems.

The following are more examples of nonspecific and specific descriptions of impacts:

Nonspecific	Specific
The project would have adverse impacts on wetlands.	The project would require 2.5 acres of the 10 wetland acres on the site to be filled.
Construction noise would adversely impact nearby residents.	Construction noise would be audible to nearby residents. See Table 12.1 to place project noise in context with other noise levels, and see Figure 12.1 for the range of expected noise levels.

■ **TABLE 12.1** Sound Pressure Levels of Representative Sounds and Noises

Source	Decibels	Description
Large rocket engine (nearby)	180	Pain threshold
Jet takeoff (nearby)	150	
Pneumatic riveter	130	Constant exposure endangers hearing
Jet takeoff (60 meters)	120	
Subway train	100	
Heavy truck (15 meters), and Niagara Falls	90	
Average factory	80	
Busy traffic	70	
Normal conversation (1 meter)	60	
Quiet office	50	Quiet
Library	40	
Soft whisper (5 meters)	30	Very quiet
Rustling leaves	20	
Normal breathing	10	Barely audible
Hearing threshold	0	

Source: Tipler (1976).

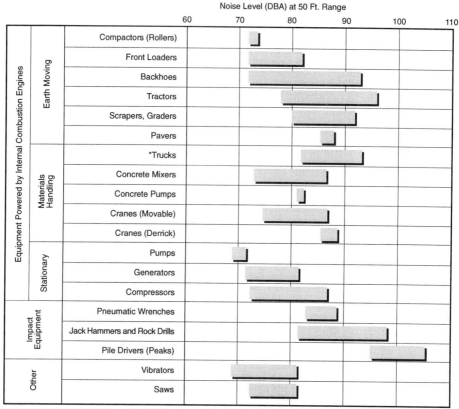

■ FIGURE 12.1. Construction equipment noise range. *Source:* U.S. Environmental Protection Agency, *Noise from Construction Equipment and Operations, Building Equipment, and Home Appliances*, NTID 300.1, December 31, 1971, Revised Washington State Department of Transportation District 1, February 1991.

Nonspecific

Wildlife habitat would be adversely impacted by the proposed project.

Specific

The proposed project would require the removal of approximately 6 acres of vegetation, which would result in the loss of 6 of the existing 48 acres of wildlife habitat. (Check the vegetation section of the EIS and make sure that the acreages are the same.)

The above examples are simplified; in the actual environmental document the discussion of wetlands, for example, would be more detailed (e.g., describe types of wetlands such as palustrine, emergent, scrub-shrub).

Do not describe a "potential" as an impact; for example, do not say that "a potential for sedimentation would occur." State what the potential is for a specific impact, not that "potential" itself is an impact.

Analyses: Quality and Time

The quality of environmental analyses is dependent on the available amount of time, data, and resources (including human resources). Generally, an experienced EIS team member will take less time to prepare a quality analysis than one who is inexperienced. As discussed earlier, a team member's experience is in two areas: in the subject of analysis and in the role of an EIS team member. A person may be an expert for a given subject, but without EIS experience, he or she may not understand the EIS process, his or her role in the process, the format of EISs, and approach to writing EISs. The quality of available data is also a factor in the quality of analyses. If data required for the analysis are available and appropriate for the analysis, the analysis may take less time to prepare than if data are unavailable (no one has studied the subject), difficult to obtain (it's a large study, there is only one copy, and it's classified), or unusable (too old, doesn't fit the study area or purposes of the study). Some time will be spent searching for relevant data, evaluating data, and discarding data. In the absence of data needed for an analysis, a lead agency must decide whether to spend the time and effort to create data by conducting original research (research not based on the work of others), field work, or computer modeling.

Staying Focused Within EIS Sections

Focus, as discussed here, refers to maintaining focus when analyzing a particular subject so that the relationships between cause and effect of environmental impacts are clear to the reader. Maintaining this focus is difficult in some cases because EISs are organized by subject areas; however, environmental impacts are interrelated to a great degree. Therefore, an author could easily wander into subject areas other than his or her own. For example, an area of vegetation also may be habitat for wildlife. Yet most EISs divide discussions into a chapter on vegetation and a chapter on wildlife. If you are analyzing impacts to vegetation, you might make a statement that a proposed project would remove a certain amount of vegetation. However, you should not wander into the realm of wildlife by stating that wildlife habitat would also be removed. The author of the wildlife chapter would discuss impacts on wildlife habitat from the removal of vegetation (unless you are the author for both the vegetation and wildlife analyses, and both topics are being discussed in one chapter).

In some cases, subjects are so interrelated that some overlap between sections is necessary. For example, an EIS that evaluates water quality issues and fishery issues will probably require some discussion of water quality in the fish chapter. The authors of the water quality and fish chapters should work closely together (in interdisciplinary fashion) to share data, discuss methodology, and decide who will discuss aspects of the analyses that overlap. Each author can then refer readers to the other's chapter for additional information on water quality or on fish.

Analyzing Impacts of the No-Action Alternative

Consider the following statement that is common in EISs: "Under the no-action alternative, the site would remain as it is currently." "Remain" for how long? What does "as it is currently" mean? How *is* it currently? This statement is most likely incorrect as written because some changes, natural or manmade, are inevitable on any site. In addition, few sites are completely without existing environmental impacts. Therefore, a statement that "no impacts would occur under the no-action alternative" is equally incorrect. The statement should say that under the no-action alternative, the *project's impacts* (specify which impacts you're talking about for the appropriate section of the EIS such as air, water, or vegetation) would not occur; for example, "Under the no-action alternative, the proposed project would not proceed, and 10 acres of vegetation would not be removed."

The no-action alternative should identify ongoing (existing) impacts and reasonably foreseeable future impacts that might occur without the project's implementation. Ongoing impacts include existing environmental impacts from activities or environmental problems (e.g., erosion, pollution) taking place within the study area. Also, include the consideration of environmental impacts that may result from an action that could reasonably take place if the proposed action does not take place. In one example, the impacts of the construction of over 2 million square feet of industrial park were evaluated under the no-action alternative (for each element in the EIS) for a thoroughbred horse racing facility. The project proponent for the horse racing facility had preliminary plans for an industrial park that could be constructed without requiring a wetland fill permit from the U.S. Army Corps of Engineers (COE), the lead agency for the EIS. Because the land was zoned for industrial use, and the proponent had such plans, the COE determined that a reasonably foreseeable action, under the no-action alternative for the horse racing facility, was an industrial park. In other words, if the horse racing facility were not approved (no action), the industrial park was a reasonably foreseeable action.

Impacts to Nonsentient[157] Objects

As discussed above, the significance of some impacts depends on the receptors of the impacts. Therefore, be specific in describing who or what will perceive the impacts you are analyzing. If the receptor is not human, it most likely is another animal because inanimate objects (nonsentient objects) cannot perceive impacts. The following are some examples:

"There would be no impacts on Witter Lake or associated wetlands, except noise." A lake or a wetland cannot hear noise. The author probably meant the animals that inhabit the lake or wetlands would hear noise, but we're not sure. Does this include the fish?

"Construction impacts would occur to nearby residences." This sentence is nonspecific and refers to nonsentient receptors (residences). The statement also does not specify the type of impact caused by construction. In this case the impact is noise. Even if this sentence were

---

[157]Incapable of sensing or feeling.

under the noise chapter, the sentence should specify that the individuals in three residences would hear noise caused by construction activities. Residences don't hear noise, the occupants of the residences (or the residents) hear noise. As written, the statement could be interpreted to mean that vibration from construction could cause the residences to crack or crumble.

"Houses would be buffered by Main Street." How and what does the street buffer? A buffer implies that an impact would be mitigated. How would a street mitigate an impact on houses? In this example, however, the author was only describing a physical separation between two sites, not a buffer in the sense of mitigating an impact.

Apologizing for Impacts

Don't overexplain. Some authors explain the same thing three times in one paragraph or make extraneous comments that add nothing to the point of the discussion. One gets the impression that the author is apologizing or insecure. State what the impacts will be, or how effective mitigation will be, then stop.

Omit Needless Words[158]

In the interest of reducing the length of EISs and increasing their clarity, find and eliminate words that are common in EISs, but totally unnecessary. For example, labeling impacts (e.g., direct, indirect, short-term, long-term) not only results in vague descriptions, it also adds unnecessary words. The usefulness of categorizing impacts into types is as guidance in regulations so that authors conducting environmental impact analysis will consider all of the possible types of impacts associated with a project. This doesn't mean that the impacts have to be labeled according to type. It is unnecessary to label most impacts, as in the following example: "*Direct impacts* would consist of noise and traffic impacts experienced by adjacent properties. . . ." If you're going to label impacts, these same impacts are also long- and short-term, as well as cumulative. The focus should not be on labeling types of impacts, but on evaluating and demonstrating whether the impacts are significant.

---

[158]Strunk and White, *The Elements of Style*, 1972.

Sometimes authors add words that are, if not incorrect, at least pointless or confusing. For example:

"Noise would disturb wildlife populations." Since the discussion had nothing to do with wildlife population or statistics, and only meant to say that wildlife would be disturbed by noise, "populations" should be deleted.

"The project would have impacts on water use patterns." Actually, the discussion was regarding water volumes and had nothing to do with "water use patterns."

The parlance of EISs includes unnecessary words shown in italics in the following examples:

Original	Change to
"The proposed action *would result in a reduction of* . . ."	"The proposed action would reduce . . ."
"The proposed action *would result in conversion of* . . ."	"The proposed action would convert . . ."
"Landscaping *would be employed to* minimize visual impacts."	"Landscaping would minimize visual impacts . . ."
"*Water would be used to* wash down trucks . . ."	"Trucks would be washed . . ."
"The manure bins *into which the waste would be deposited* would be constructed with a concrete floor . . ."	The manure bins would be constructed . . ."
"Separate removal procedures would follow for conventional solid waste such as paper, metal, and other debris *which can litter the stable area* . . ."	"Separate removal procedures would be provided for conventional solid waste . . ."
"Utilities would be buried *underground* . . ."	"Utilities would be buried . . ."

Individually, the above examples may seem picky or minor. But cumulatively, if all authors were to eliminate unnecessary words and describe impacts with specificity, the clarity of EISs would be greatly improved, and the length of EISs would be reduced.

Use of Auxiliary Verbs "Would," "Could," and "Will"

In the above examples you may be noticing the use of the word "would" as opposed to simply saying the proposal removes, or will remove, vegetation. There is a reason for this, even though the sentences could be grammatically strengthened by not using "would." The reason goes back to the discussion of the planning process and the fact that EISs are prepared early in the planning stage when detailed design information and conditions of permits, which may affect project design or operation, are not known.

"Would," "could," and "will" are used in the context of describing potential impacts, as well as in describing potential mitigation measures to reduce or avoid those impacts. In describing potential impacts, "will" requires a level of certainty that does not exist with regard to a proposal at the EIS stage. Whether the proposal "will" receive all required approvals and permits, obtain required funding, and meet any other requirements is unknown. Therefore, when describing the potential impacts that might result from a proposal, the word "would" is used, as in "the proposed project *would* result in an increase in. . . ." "Would" implies that an impact would occur (1) if the proposal received all required permits and approvals and (2) if the proposal were actually built. If you were to use the word "will," you would need to add "if the proposal received all required permits," and so on, every time you describe an impact because "will" provides strong assurance of something taking place.[159]

Similarly, when describing mitigation measures, "would" is used if the mitigation measure is part of the proposed action, as in "sedimentation control measures would be implemented" (if all approvals were received and the project were built). If a mitigation measure is not part of the proposal but an agency with permitting authority has the option of making a mitigation measure a condition of its permit, "could" is used because whether the mitigation measure will actually be required by an agency is unknown at the time an EIS is being pre-

---

[159]"Will" and "shall" implies commitments in a legal context.

pared. Critics of EISs have called the use of "would" or "could" wishy-washy and noncommittal. However, as stated earlier, agencies may use mitigation measures identified in an EIS as conditions of approval; however, the listing of mitigation measures in an EIS does not require an agency or project proponent to implement them. Said another way, an EIS does not make any commitments—it is wishy-washy by design. Commitments to implement mitigation are made in enforceable documents such as conditions of permits.

> **An EIS does not make any commitments.**

Lastly, reviewers of EISs have criticized agencies for using "will" because of the appearance that an agency has already made up its mind to proceed on a proposed action. Therefore, agencies are criticized whether their EISs say "would" or "will."

This does not mean one can never use "will" in an EIS. The guidance to use "would" applies when discussing possible impacts and mitigation measures associated with a proposed project. For other subjects or actions that have already been decided, there is no reason to not use "will," as in "a new freeway that *would* provide access to the proposed project has been approved and construction *will* begin this fall." Therefore, if your computer's word processing software has the "global search and replace" feature, you should not use it to change all "wills" to "woulds" after the document has been written. Some "wills" should remain.

Defensive Writing

Authors of EISs sometimes have a subtle, defensive tone in their writing. Perhaps this is because EISs are publicly reviewed, often by those vehemently opposed to a project; and after suffering repeated beatings by reviewers (words can be brutal too), authors unconsciously start to write as though defending themselves. A common example is in frequently using "not," as in writing "the proposed project would not be incompatible with local zoning ordinances." It is as though the author is anticipating accusations from the reader. Why not just say "the proposed project would be compatible with . . ."? It sounds more positive, as in saying "I am an honest person," instead of "I am not a crook."

Professional Editors

Too many people think that editors only check for grammar and spelling, and during budget cuts, editors are considered "nonessential." The rationale is that a supervisor or a senior-level person can take over the editing duties since they have to review the documents anyway. However, the best editors are trained in their profession and do more for "unifying" a document written by multiple authors than any other team member. The editor's job is not to rewrite poorly written documents, but to point out problem areas, missing information, contradictions or inconsistencies, and format problems, as well as grammar and spelling errors. Few non-editors can do this well, and, besides, using senior-level staff as editors is not an effective use of senior-level salaries. Although "excellent paperwork" is not the goal of the EIS process, the goal is to produce a document that effectively communicates to readers so that "excellent action" or informed decisionmaking can take place. Information that is presented clearly and without confusing errors and inconsistencies is more likely to be an effective decisionmaking tool, and the public also will appreciate it.

CONSULTATION WITH AGENCIES

Some environmental regulations, such as the Endangered Species Act, require consultation with agencies if a proposed action has the potential to affect the regulated resource. As described earlier, the consultation takes place in accordance with the requirements of the regulations and the agency with jurisdiction. However, consultation with agencies also can be a very useful method for obtaining information for an EIS, for discussing environmental issues, and for developing mitigation measures. Federal, state, and local agencies have employees with various technical expertise depending on the mission of the agency. Many professionals who are members of an EIS team have established networks of agency colleagues. Consultation with those colleagues may be as informal as a phone call to discuss an issue or ask a question, or involve a series of meetings with numerous agency representatives.

The level of consultation, as well as the agencies you contact, will depend on the issues related to a proposed project. If an issue crosses several jurisdictional boundaries, or is regulated by numerous agencies, you should make interagency consultation part of the EIS's scope of work. Consultation with numerous agencies normally requires time and effort that should be planned early in the EIS process.

Another time to consult with agencies is when making a threshold determination of whether to prepare an EIS. Agencies with jurisdiction by law or expertise should be consulted to determine whether a proposal has the potential to violate an environmental law or to affect protected resources such as wetlands, endangered species, or cultural resources.

WHEN TO DO RESEARCH OR CONDUCT FIELD STUDIES

Research is a basic requirement for the analyses in EISs. One must research available information on the existing environment, other studies on the environmental impacts of similar types of actions, and appropriate methods for conducting the analyses. Most of the research is to obtain existing information. In some cases, however, there may be a gap in data, or information necessary for the analysis may be lacking. A decision must be made on whether to create the information by conducting studies or field work. The lead agency and EIS team members should discuss the need for the information in order to make the final decision. If the information is necessary for an analysis of a significant issue, as opposed to that of an insignificant issue, the lead agency is more likely to decide that the research effort should proceed.

The minimum field work required for an EIS is a visit to the location or locations of the proposed project. Some EISs that are programmatic, whose proposal's effects cover a large region or several states, are not site-specific, and field visits may not be possible or may be only to known environmentally sensitive areas. Field studies may be necessary, for example, to document the presence of wildlife if the species are of concern and existing information is unavailable. In some cases, field work is seasonal or can only be done at certain times of the year. In this case, the EIS schedule (and possibly cost) may be substantially increased to accommodate the field work. Monitoring groundwater wells for volume and direction of flow, or presence of contaminants, also requires field work over a period of time. The time and amount of field work, as well as its methods, are planned and agreed upon by those who are evaluating the resource (the ID team members) and the responsible agencies (EIS lead agency and agencies with jurisdiction).

Where field visits to an area are not possible, or where the effects of a proposal are over a large area, aerial photographs are an option for analyzing environmental conditions (for those who have the skill to interpret and analyze aerial photographs). Computer modeling is another method for evaluating impacts; however, critics of computer models feel that some models have no relation to the "real world," can be misleading (how the model is set up and the choice of data used in the model influence the results), and are not required for the purposes of most EISs. If computer modeling is used, the analysts should provide an explanation of how the modeling and its results relate to the problems under analyses.

CREDIBILITY: ON WHAT ARE CONCLUSIONS BASED?

Conclusions regarding potential environmental impacts receive the most criticism from reviewers. The reasons are obvious: Conclusions of environmental evaluations determine whether impacts will be significant, and therefore whether an EIS will be prepared; whether certain impacts will receive detailed analyses; and whether mitigation measures are necessary and, if so, what type and to what degree impacts should be mitigated. A common criticism of reviewers regarding an author's conclusion is that it is not supported by facts presented in the author's analysis. Facts or information also must be credible—just as the author must have credibility by virtue of his or her qualifications—and the methodology for analyzing the information must be appropriate for the situation.

Criticism that *enough* information was not presented, and therefore that the conclusions are not supportable, may or may not be valid. The amount of information is not necessarily a factor in establishing the credibility of conclusions. The quality and relevance of the information are the most important factors, and the detail should be sufficient to support the conclusions. The detail and amount of data will vary, depending on the significance of the issue, the assumptions and methodology of the analysis, and any limitations of the data. Also, data presented in technical issues are frequently subject to interpretation by experts who may come to different conclusions regarding the potential environmental impacts of an action. In this case, the relevant data, including known differences in interpretation of the data, should be discussed in the EIS. An EIS author should explain why certain data were used and not others, and why another methodology or interpretation of data is not appropriate or not preferable for the analysis.

Establishing credibility in an analysis does not guarantee that everyone will agree with it. But reviewers should at least understand how authors came to their conclusions. Tell them what information was relied upon and how the information was used; and if you are aware of other information but chose not to use it, explain your reasons.

IMPACT MATRIX: TO USE OR NOT TO USE

An impact matrix shows the anticipated environmental impacts for each alternative in a graphic form. It is a tool used by many agencies to compare and contrast the environmental impacts of each alternative. An impact matrix would consist of, for example, project alternatives on one axis and environmental elements on the other axis. The ID team would develop a method of representing the impacts of each element under a particular alternative such as a clear circle for minor impact, a circle that is half black for moderate impact, and a circle that is full black for significant impact. Another method includes a numbering system that represents levels of environmental impact (e.g., level 1 might represent the least impact, and level 4 the highest).

Reviewers like impact matrixes because they show, in one location in an EIS, the differences in the expected impacts of each alternative. Reviewers do not have to wade through the EIS and read how the alternatives compare; instead, they can see it in one place in graphic form. A matrix simplifies the review of the impact analysis, which is at once its attraction and its downfall. In order to assign one value to an impact (e.g., low, medium, or high), the person who assigns the values:

■ Interprets the data
■ Evaluates the impacts
■ Includes or doesn't include mitigation measures
■ Assigns a subjective value—the value of the person preparing the matrix

All of the above evaluation processes are distilled into one symbol or value on a chart. The symbol is the conclusion of an analysis. All of the factors that are normally considered when evaluating the credibility of a conclusion cannot be shown. However, impacts are not black or white, they do not stand alone, and they are interactive. Conclusions regarding impacts may depend on certain circumstances (such as mitigation measures) or assumptions. An impact to a resource may be beneficial in one way but adverse in another. A matrix cannot show qualifiers or variances; at best it can show averages. A reviewer tends to accept the values as presented, which may be misleading because the reviewer does not know how the values were assigned.

A matrix does not show the condition of the existing environment, cannot indicate uncertainties from inadequate or limited data, cannot indicate environmental variability (e.g., the possibility of an extreme occurrence), cannot show probabilities or likelihood of an impact's occurrence, and cannot show indirect impacts. It is unlikely that the preparer of a matrix considers all of these factors.

Reviewers could refer to the appropriate chapter in an EIS to evaluate the credibility of the value assigned in a matrix. However, unless the author explained the process of distilling the analysis into one value, it would be difficult for a reviewer to understand how it was done. And in many cases, the author of a particular analysis does not assign the value in a matrix. An ID team member reviews the team members' analyses, makes value judgments, assigns the values, and prepares the matrix.

CUMULATIVE IMPACT ANALYSIS: THE BLACK HOLE[160]

The CEQ regulations define cumulative impact as

> . . . the impact on the environment which results from the incremental impact of the action when added to other past, present, and reasonably foreseeable future actions regardless of what agency (Federal or non-Federal) or person undertakes such other actions. Cumulative impacts can result from individually minor but collectively significant actions taking place over a period of time.[161]

---

[160]A black hole is a region of space, caused by the collapse of a star, with a gravitational field so intense that matter and light cannot escape it. It is also defined as a great void or an abyss (American Heritage Dictionary).

[161]40 CFR §1508.7.

The definition of cumulative impact essentially encompasses anything and everything, with no limits on timeframes or types of actions. There are no guidelines on how far into the past or into the future one should look. It does not matter whose actions they are or who has the authority to control the actions. There is no consideration of what type of information would be needed to conduct such an analysis or whether information would even be available.

Although it is easy to grasp the concept that unrelated actions may result in additive or cumulative environmental impacts, developing the scope, methodology, and conducting the cumulative impacts analysis is far from easy. The treatment of cumulative impacts may be uneven, with no method or consistency from one environmental element to another in an EIS. Each author analyzes cumulative impacts from his or her own understanding of what the analysis should entail. Regulations provide little guidance on cumulative impact analysis, and the problems associated with the analysis do not become apparent until EIS team members attempt to prepare the analysis.

Cumulative impact analysis, more than any other analysis in an EIS, is susceptible to criticism that not enough information was included because the definition has no limits, and no matter how much is included, more could always be included. Thus, during the preparation of EISs, internal reviewers continually add projects or actions, impacts or issues, and background information or other details to the scope of the effort until someone realizes that the analysis has become never-ending. Public reviewers are no less confused than an ID team about the purpose of a cumulative impacts analysis and often think the analysis is a separate EIS.

Cumulative Effect or Impact Versus Cumulative Actions

EIS preparers and reviewers frequently confuse cumulative effects or impacts analysis and cumulative actions analysis. A cumulative actions analysis includes actions related to the action proposed by the lead agency. A cumulative effect or impact analysis focuses on the impacts on environmental elements or resources, whereas a cumulative actions analysis focuses on whether actions are related and comprise a local, regional, or national program.[162]

---

[162]Kleppe v. Sierra Club (1976) (Fogleman, 1990).

Requirement for Cumulative Impact or Action Analysis

The CEQ regulations and the courts appear to have differing opinions regarding the requirement to address cumulative actions and impacts in EAs and EISs. CEQ regulations require EISs to address cumulative actions, while EAs do not have to address cumulative actions. Although the regulations do not require EAs to consider cumulative actions, some courts have considered them when determining whether an EA was adequate. Cumulative actions should be considered when making a threshold determination and when deciding whether actions are connected and, therefore, should be evaluated in the same EIS.

The regulations require both EISs and EAs to address cumulative impacts or effects. Some courts, however, have stated that cumulative impacts do not have to be considered in an EA because the requirement would place a burden on a federal agency's EA screening process that is equal to that required for an EIS.[163]

When determining the scope of an EIS, cumulative actions should be considered. When making a threshold determination and when preparing an EIS, cumulative actions and impacts should be included in the analyses.

Format

Cumulative impact analysis is part of an EIS, not a separate study or a separate EIS. A scope and methodology should be developed for the analysis similar to the analysis for an environmental element. However, the discussion of cumulative impacts can be within the discussion of impacts to the environmental element, rather than a separate chapter on cumulative impacts. Some agencies prefer to have the cumulative impact analysis as a separate chapter; however, this sometimes furthers the appearance that the analysis of cumulative impacts is separate from the analyses of environmental elements in an EIS.

An outline of an environmental element's sections that include cumulative impacts might look like the following:

■ Affected/existing environment
■ Environmental impacts/consequences

---

[163]Ringsred v. City of Duluth (1987) (Fogleman, 1990).

■ Mitigation measures
■ Unavoidable adverse impacts
■ Cumulative impacts

Methodology

The EIS team should develop a methodology for each team member to use in evaluating the cumulative impacts of his or her environmental element or resource. The scope (issues and geographic study area) of the analysis of cumulative impacts will normally be the same as for the overall EIS. In other words, when addressing cumulative impacts, evaluate the environmental impacts or issues that are within the scope of the EIS. The geographic area of study for analyzing cumulative impacts would be in the same study area as identified for the EIS. The scope might be different for cumulative impacts, however, if future actions would increase the area affected by the actions or result in additional types of impacts. Thus, the scope and methodology of the analysis depends on the type of proposed action and its area of influence. A methodology for analyzing cumulative impacts would include the following considerations:

■ Timeframe of past, present, and future impacts
■ Timeframe of past, present, and future actions
■ Types of actions
■ Identification of other actions and impacts
■ Evaluation of information (appropriateness for the study)

Relationship of Past and Present Timeframes

Actions and impacts are related when analyzing cumulative impacts. In order to evaluate past, present, and future impacts, one must know what past, present, and future actions caused or are causing the impacts. This is the first of several considerations regarding the timeframes of the analysis. For practical purposes, the baseline for the analysis normally is the present, not some date in the past, even though the regulations state "past, present, and future." The main reason is that the present conditions reflect the actions and impacts of the past. What we have today is a result of what we did in all of the yesterdays. Therefore, other than discussing past actions that may have resulted in current conditions as a topic of historic interest, there is no point in identifying a baseline or environmental conditions that existed 10, 50, or 100 years ago. Thus, the environmental baseline would be the same as for the overall EIS.

Similarly, there is no point in identifying past actions and their impacts, as separate from existing impacts, because past impacts are reflected in present conditions. Or, viewed another way, how would you separate present environmental impacts from those of the past? Today's environmental impacts are a continuation of impacts from the past. How would you draw the line of where the past impacts stopped and where the present impacts started? As another practical matter, unless EISs were prepared for past actions within the study area (limited to the past 25 years), there are few records of (1) what the environmental conditions were at the time of past actions and (2) what specific or quantified impacts resulted from those actions. Without such information, one is again left with the present conditions to start the analysis of cumulative impacts.

> **How would you draw the line of where the past impacts stopped, and the present impacts started?**

Relationship of Present and Future Timeframes

What timeframe is the EIS calling the "present?" Since an EIS often takes over a year to prepare, the present is not one day in time. Also, other actions to be included in the cumulative impacts analysis will be in varying stages that change over time. Actions may be

■ Proposed and may take weeks, months, or years to implement; or may only reach the proposal stage and never be implemented
■ Approved but not yet implemented
■ Under construction or implementation
■ Completed

Which stage of actions and their impacts will be included in the present (the baseline), and which in the future? An action that is proposed at the time the EIS started may be implemented by the time the EIS is completed (not all actions require EISs). How will you define "reasonably foreseeable future" actions? Will you continue to add actions as they are proposed to the cumulative impact analysis during the 1 to 2 years that many EISs require for completion? These are decisions that the EIS team must make before starting the analysis.

Impacts of present actions are normally considered those that are completed or under implementation. Proposed actions that have reached a permit application stage, and actions that are reasonably foreseeable, are future actions. The analysis assumes that the actions will be implemented, although there may be no assurance that the actions will even be approved much less obtain financing and other necessities for implementation.

Because the stages of actions, and their impacts, as well as the existing environmental conditions, may be changing during the time the EIS is under preparation, many agencies take a "snapshot in time" approach. You must "freeze" the moving picture in order to analyze it—or continually revise it. To take a snapshot in time, you might state that the existing environmental condition and "present" actions are those that existed as of a specified date. To identify "future" actions, include, for example, actions that were proposed during a specified period of time such as between January 1 and June 30 of 1996 if the Draft EIS is scheduled to be completed by August 31, 1996. The reason for setting a beginning date of January 1 for your specified period is to set a limit on how far back you will research records for proposed actions. Some actions may have been proposed years before and may not have progressed any further in the project planning process, yet an agency may still have the projects in its records whether or not the projects are still viable. The EIS team must decide on a time period that is reasonable for the circumstances.

A time period should be set for future impacts as well as for future actions. An EIS may be evaluating potential impacts of a lead agency's proposed action for 10 years into the future, as an example. The cumulative impacts analysis may also be within this timeframe for future impacts, or it may predict impacts further into the future depending on the impacts of future actions. How far should you go to anticipate impacts? Impacts should be considered if they are reasonably foreseeable, but not if they are highly speculative or indefinite.[164] Reasonably foreseeable impacts are often indirect impacts. An increase in growth or development is a reasonably foreseeable impact of constructing a new road to an area that has limited access. However, for a

---

[164]Sierra Club v. Marsh (1985); Trout Unlimited v. Morton (1974).

case involving expansion of an airport, the court ruled that an increase in tourism was a reasonably foreseeable impact, but increased growth of the area because of the increase in tourism was not a reasonably foreseeable impact. The increase in growth was one step removed from the proposed action and was considered too speculative.[165]

Types of Actions

A cumulative impacts analysis evaluates how the impacts of the lead agency's proposed action contribute to the impacts of other current and future actions. Therefore, the EIS team must consider what types of actions to include in the cumulative impacts analysis. An action that does not result in any of the impacts of the lead agency's proposed action should not be included in the analysis. As an example, if a proposed action would only cause noise and air traffic impacts, other actions that do not cause noise and air traffic impacts should not be included in the analysis.

Identification of Other Actions and Impacts

How would you identify other present and reasonably foreseeable future actions and their impacts? The first step is to identify the actions; the second step is to identify the impacts that resulted from those actions. One way to identify actions is to identify agencies with authority to approve proposals within the EIS's study area. Many proposals require construction or operation permits, or both. If the study area includes cities and counties, you could contact their planning departments and ask for information regarding permit applications for proposals, or pre-proposal meetings, that were received or took place during the analysis' specified time period. If the study area is within the jurisdiction of one federal land managing agency, that agency should know about all proposed actions within its area of responsibility. Some discussions with agency personnel will be necessary to determine if a proposal is really a proposal or if there is some likelihood that the action might take place, as opposed to rumors or wishful thinking. Reasonably foreseeable future actions are also those that have a reasonable likelihood of occurring, even if no one has proposed a specific action at the time the EIS is being prepared.

---

[165]Life of the Land v. Brinegar (1973) (Fogleman, 1990).

Impacts of actions that have taken place or are being implemented may be identified if environmental documentation, such as EAs or EISs, were prepared for those actions. Many actions, however, do not have such documentation. Impacts of future actions are even more difficult to identify because information regarding the actions' design, construction, and operation may not be known. Therefore, several assumptions regarding the impacts of proposed future actions are normally made in cumulative impact analyses:

■ The "footprint" of the action (if known) will be the (assumed) area of construction impact.
■ A proposed action will be constructed or implemented.
■ Mitigation measures required by regulations will be implemented (although specific mitigation measures may not be known).

Analyzing Cumulative Impacts

The analysis in an EIS is inherently cumulative. The basic methodology for impact analysis in an EIS is to identify the existing environment, including existing impacts, and to overlay (or add) features of the proposed action (its design, construction, and operation) on the existing environment to identify the proposed action's expected impacts. The existing environment, as discussed earlier, is representative of past and current impacts. Thus, the analysis in an EIS identifies the contribution of the proposed action's impacts to cumulative *existing* impacts (as reflected by the condition of the existing environment) in the study area. A cumulative impacts analysis picks up at this point, primarily to include other *future* actions that would contribute to cumulative impacts in the study area. If, however, the study area for cumulative impacts is larger than the study area for the overall EIS, present or existing actions and impacts would not have been addressed for the broader area and those actions and impacts would need to be identified and addressed.

The analysis in an EIS is inherently cumulative.

When evaluating cumulative impacts, the assumption is that impacts are additive over time. This is not entirely accurate because over time, impacts also can be subtracted or reduced. Mitigation, for example, subtracts adverse environmental impacts from the equation (e.g., replacement or creation of wetlands). Yet, for actions that did not have environmental documentation prepared, actual mitigation measures are not known and therefore are not considered when attempting to quantify cumulative impacts. Buildings may be demolished, and open space may be created; however, these types of actions may not be recorded with information sources. Agencies at the federal, state, and local level have various programs to benefit the environment, which include partnerships with private landowners to enhance, restore, or create wetlands and wildlife habitat.[166] Information regarding these actions would not be recorded in one place, such as an office for local building permits. Therefore, to do a reasonably thorough job of identifying actions for both sides of the cumulative impacts equation requires discussions with sources knowledgeable of the study area, in addition to researching records or files.

Limitations of Available Data

Agencies and the public have various expectations of a cumulative impacts analysis. Some expect an EIS on cumulative environmental impacts, while others have no idea of what the analysis should include. Regardless of the expectations of an EIS team when starting the analysis, the final product will be determined by the information that was available for the analysis. An EIS team may find, after identifying the available data, that the scope of the cumulative impacts analysis needs reevaluation to develop more realistic expectations.

Identifying Actions and Their Physical Impacts

Certain basic information regarding proposed actions is necessary in order to analyze their contribution to cumulative impacts. At a minimum, a description of the proposal sufficient to identify the area of impact, and types of impacts that would be expected, must be available. The specific location of the proposed action must also be known. If any of

---

[166]In 1993 the U.S. Fish and Wildlife Service, along with private landowners and their partners, restored over 44,703 acres of wetland, riparian, stream, and grassland habitats for federal trust species and other wildlife. Seventeen thousand sites comprising over 210,000 acres of wetlands and associated habitats, involving 10,900 landowner agreements, have been restored under the ecosystem approach for habitat restoration (Council on Environmental Quality, *Twenty-fourth Annual Report*, 1993).

this information is lacking, the impacts of the proposal cannot be identified and evaluated. With this information, the cumulative impacts analysis normally identifies very broad impacts, and it does not quantify impacts with specificity. The available information regarding other proposed actions is too limited to permit the same level of analysis as provided for the proposed action under evaluation in the EIS. This does not mean that the analysis is inadequate if the lead agency does not collect detailed information on proposals other than the proposed agency action. The courts have determined that impacts of other proposed actions do not have to receive detailed analyses.[167] Any speculative actions that are included in the analyses also do not require extensive evaluation.

Identifying Usable Data A cumulative impacts analysis is frequently for a region, or some other large study area, for which many agencies or organizations may have data on various topics. A common problem in obtaining data is that agencies or organizations do not use the same methods for collecting data. The assumptions or the basis for their studies may be different, their measurements may be based on different methods, the areas of analyses and timeframes for their studies may be different, and their conclusions may be inconsistent. This is understandable for agencies that have different purposes for their studies, but agencies with similar missions also have very different procedures for collecting data. Therefore, some data may not be appropriate because they are not compatible with the methodology or timeframes of the EIS. In many cases, however, most of the available information regarding a particular issue or topic is inconsistent or incomplete or does not definitively address an issue. For these reasons, a cumulative impact analysis frequently discusses trends suggested by the data and makes general observations, rather than quantify environmental impacts.

Identifying Mitigation Measures Generally, if mitigation measures are identified that will reduce a proposed action's impacts, those mitigation measures will also reduce the action's contribution to

---

[167]Fritiofson v. Alexander (1985).

cumulative impacts. In most cases, a project proponent is not required to mitigate the impacts caused by other actions. Cumulative impacts analysis takes a view of the "big picture," of which a proposed action may be a small part. The degree or extent of mitigation should correspond to the contribution of the project's impacts to the cumulative impacts in the study area. For some types of cumulative impacts, the only feasible mitigation may involve the adoption of ordinances or regulations rather than the imposition of conditions on a project-by-project basis.

TIPS

On Neutral Language

Beware of writing exposés as if you were a newspaper reporter; watch for tone, and avoid words that show bias.

To Avoid Rewriting Sections

Make a list of who is writing each section, and give the list to everyone involved so the team members can consult with each other on topics of mutual interest while preparing their analyses. This also makes it easier for team members to refer the reader to another section for additional information, rather than rewriting the information.

To Save Time and Effort

Have the preparers of each section write the summary section for his or her area of responsibility. It takes longer for another person to read someone's analysis, digest it, and then write a summary. And that person is more likely to make errors.

Make sure EIS authors have an outline of the summary, as well as the outline for the overall EIS, prior to starting work, and identify topics that overlap so team members can decide who will write about the topic and who will simply refer the reader to the appropriate section.

List references as you prepare your analysis rather than waiting until the analysis is finished. Remember to list personal communications.

If subconsultants are on the team, make sure they also receive all information regarding the project. Have subconsultants prepare "drop-in" sections, or sections that are written according to the guidelines and outlines provided to other team members. This saves time otherwise spent rewriting or reorganizing the subconsultants' work to be consistent with the rest of the EIS. If you've never worked with them before, don't assume that subconsultants' drop-in sections will be done properly, even if you've given them all the necessary information. Ask for their work earlier than the rest of the team to allow time for adjustments. And ask subconsultants for electronic copies of their work in a format compatible or easily translatable to your computer's word processing software (as well as hard copies).

Have subconsultants use "would" in their analysis rather than "will" so you don't have to change the words. There are exceptions to the rule, when "will" really is the right word, and it's easier for the authors to know the correct word to use.

To Make EISs More Useful to Decisionmakers

Present the conclusions of the alternatives analysis separately, copy them from the EIS so that they are easily accessible to the decisionmaker. The agency will need this information to discuss the agency's preferences among alternatives based on relevant factors including economic and technical considerations and agency statutory missions as required in the record of decision.

To Make Mitigation Measures More Useful

Have each team member list "potential" mitigation measures, and give them to the EIS team leader. The team leader can ask the project proponent if he or she will commit to any of the suggested mitigation measures and make them part of the proposed action. A mitigation measure that is part of the proposed project, and therefore a commitment, is more meaningful than a mitigation measure that is only a possibility.

Prepare a separate list of mitigation measures discussed in an EIS so that agencies may decide which to include as conditions to approvals or otherwise commit to mitigation, as well as any monitoring program (including providing results of monitoring to the public) to ensure mitigation measures are enforced.

Prepare a list of mitigation measures that an agency has committed to implement, and attach the list to all subsequent contracts, agreements, or other instructions for individuals who will be taking the proposal through the phases of project implementation (e.g., design, construction, and operation).

To Illustrate Cumulative Impacts

Because impacts are difficult to quantify, maps are very useful tools in cumulative impacts analyses for visually showing the extent of existing impacts and for showing areas of future impacts. Graphs and charts are also useful for showing trends in changing environmental conditions, demographics, revenues, or other socioeconomic impacts.

CHAPTER SUMMARY OF TIPS

◼ To avoid analyzing a "moving target" of changing environmental conditions, take a "snapshot in time" approach for defining the existing environment.

◼ To identify existing environmental conditions, including existing environmental impacts, visit the area within the EIS's study area. Also, look for any off-site sources of environmental impacts.

◼ To ensure that field work and analysis of impacts cover a proposed project's area of impact, stake or flag the study area for any field work and flag the footprint of construction areas (buildings, roads, utility lines).

◼ To emphasize the alternatives section of an EIS, the lead agency must evaluate certain parts of the discussion that are within the agency's purview, such as the agency's preference with regard to alternatives, identification of any alternatives that would be "reasonable" yet outside the jurisdiction of the lead agency, and "essential considerations of national policy."

◼ To avoid delays in an EIS process, the ID team and lead agency staff should identify, discuss, and resolve any differences in perspective early in the preparation of the EIS's foundation and framework.

◼ To maintain an EIS's focus on significant issues:

Ensure that an EIS's assumptions and scope receive internal agreement early in the process.
Thoroughly define at the beginning of an EIS process the proposed action and its significant effects and issues.

■ To ensure evaluation of the specific characteristics of a proposal, rather than generic projects and assumed impacts, know the details related to each feature of the action—where it would be located, how it would be sited or situated, how it will be designed, how it will be operated, and other details. Visit similar projects if possible.

■ To develop specific mitigation measures, identify specific project impacts.

■ To describe the significance of a potential impact, place environmental impacts in a common context so that the reader can tell if the issue is of any concern. Explain the significance of an impact.

■ To describe project-specific environmental impacts, include project and regulatory mitigation measures applicable to the proposal in the evaluation.

■ To convey accurate meaning:

Use precise, plain, and unbiased language, and show (with visual aids such as maps, graphs, or other drawings) how something works or how you came to your conclusions.

Explain rationale, and provide information from other sections in the description of impacts (rather than assuming a reader has deduced the rationale from information presented in previous sections).

Define terms that are not commonly known, and explain the significance of facts or observations if, without the explanation, the significance would only be clear to someone with your specialty.

■ To evaluate all aspects of the no-action alternative, identify ongoing (existing) impacts and reasonably foreseeable future impacts that might occur if the proposed project were not implemented.

CHAPTER

13

REDUCING TIME, EFFORT, AND PAPERWORK

It is better to be ten percent effective in achieving a worthwhile goal than one hundred percent efficient in doing something worthless.

Sid Taylor

INTRODUCTION

In terms of understanding the EIS process—so that everyone involved in it works more efficiently and effectively—this entire book is about reducing time, effort, and paperwork. This applies to public reviewers as well as lead agencies and project proponents. Specific time- and effort-saving methodologies can be used in preparing EISs, including:

■ Integrating the EIS process into early planning
■ Combining EISs with other documents such as general plans or land use and natural resource plans
■ Preparing joint EISs
■ Using EAs and FONSIs (and their state equivalents) when a project will not have significant effects on the environment (e.g., build mitigation into a project)
■ Establishing cooperating agency relationships
■ Adopting other agencies' analyses

- Incorporating information by reference
- Conducting scoping
- Setting time and page limits
- Analyzing only significant issues
- Omitting extraneous background information
- Tiering

This chapter summarizes time- and effort-saving tips from previous chapters. The CEQ regulations, state EIS regulations, and specific agency guidelines address many of the same suggestions. Reducing time, effort, and paperwork does not mean that corners will be cut in an EIS process, but that unnecessary or redundant effort and paperwork will be eliminated, thereby streamlining the process (streamlining environmental and planning processes is discussed in Chapter 3).

INTEGRATING THE EIS PROCESS INTO EARLY PLANNING

Project delays resulting for the environmental review process are commonly caused because environmental review did not start until late in a proposal's planning process. Design and budget were close to being finalized before the proposal received environmental review. Because the EIS process is one that looks for problems, it often finds them. If problems are identified early, there will be time to identify solutions and make adjustments in a proposal. Therefore, an agency should integrate its planning process with environmental review as soon as a proposed action is sufficiently defined. Private individuals and organizations should discuss private proposals with agencies who will be reviewing the proposals at the earliest possible time in their project planning process.

> **Because the EIS process is one that looks for problems, it often finds them.**

Choosing the right time to start an EIS process involves balancing of competing factors. EISs and EAs, as well as equivalent state documents, should be prepared as early as possible in the planning process to enable environmental consideration to influence the proposed action's design, but late enough to provide sufficient project information for environmental analysis (the "EIS Catch-22" described in Chapter 3). In order to prepare environmental documentation early in a planning process, the proposed action or project may only have preliminary design information. If the information is not sufficient to identify and evaluate potential environmental impacts, the analyses cannot begin. Project sponsors must develop and provide sufficient details of a proposal to permit analysis of impacts. The amount of detail necessary to describe a proposal will depend on the type of proposal, the circumstances surrounding the action (such as the existing environment and environmental sensitivities), and whether additional environmental documentation will be prepared later in the process (see "Tiering," below).

COMBINING EISs WITH OTHER DOCUMENTS

Environmental documents should be coordinated and, when possible, combined with agency planning, review, and project approval processes. Land use and natural resource management plans are planning documents. Planning documents and the planning process may be combined with EISs and the EIS process. Agencies such as the U.S. Forest Service and Bureau of Land Management routinely combine their planning and EIS review processes. For federal plans, combining agency planning documents with environmental documents is easier for agency-proposed actions than for actions proposed by someone other than the lead agency. This is because an agency is likely to know what actions it is contemplating for the future and, thus, can include those actions in its planning documents. An agency may know about private actions that require long lead time and planning (e.g., large development proposals) and could, theoretically, include those proposals in its planning document. If an action sponsored by an entity other than the lead agency is controversial, however, the drawback to including the action in the agency's combined planning and EIS document is that the agency may increase the potential for litigation of its EIS (and thereby delay implementation of its actions) by including the other action. In contrast to federal plans, local land use plans primarily control development on privately owned land, and an EIS would, for example, evaluate the types of private proposals that may fall within certain zoning classifications.

PREPARING JOINT EISs

Proposed actions that require substantial involvement and approval of more than one agency, and that require EISs under those agencies' regulations, can be evaluated in one EIS that combines the procedural and substantive requirements of the agencies. For example, a state and a federal EIS can be evaluated in one EIS that is prepared jointly by the state and federal agencies. Preparing one EIS eliminates duplication of effort and reduces time and paperwork. The agencies cooperate with each other to combine their respective EIS processes so that one process will meet the requirements of both agencies. The CEQ regulations go further to suggest that the planning processes of agencies be combined, where possible, to eliminate duplication of effort.

Combining EIS and planning processes, while not technically difficult, requires a substantial amount of planning and coordination to develop the scope of the EIS, the schedule, and milestones and define the respective roles of the agencies. If a consultant will be assisting in the preparation of the EIS, the agencies should designate one point of contact with the consultant. Agencies often sign memorandums of agreement or understanding to formalize their understanding of roles and how the joint EIS process will work. Agency cooperation and mutual support are prerequisites for the success of a joint EIS. Misunderstandings or disagreements, especially if late in the process, can cause serious delay rather than the expedited process initially sought. If the parties can come to agreement on the basic assumptions for the EIS as well as for the process (discussed in Chapter 8), misunderstandings may be avoided and differences may be identified and resolved early.

USING EAs AND FONSIs AND THEIR STATE EQUIVALENTS

Proposed actions that are designed, sited, or otherwise modified to reduce potential environmental impacts below a threshold of significance do not require preparation of an EIS. If a proposal's mitigation measures are commitments of the proposal, or conditions of approval, or both, and the proposal will not result in significant impacts, the purposes of the regulations will be met and the time, effort, and expense of an EIS will not be necessary. State EIS regulations encourage this approach. However, the CEQ is more cautious and states that if a project was proposed with mitigation measures that would reduce potential environmental impacts below a significant level, an EA and FONSI would be appropriate. However, if a proposal as initially proposed would result in significant impacts, but mitigation measures identified after the proposal was submitted would reduce those impacts below a significant level, the CEQ's opinion is that the possibility of mitigation measures is not sufficient, and an EIS should be prepared. If scoping results in a redefinition of a project as a result of mitigation proposals, an agency may alter its previous decision to do an EIS as long as the project sponsor resubmits the entire proposal, and the EA and FONSI are available for 30 days of review and comment.[168]

Under the CEQ regulations, a FONSI must be made available to the public; however, an agency is not precluded from implementing its proposal for a period of time after the FONSI has been distributed (such as for a ROD), unless an agency's internal regulations provide for a waiting period before an action may be taken. In certain circumstances, an agency must make the FONSI available for public review for 30 days before the agency makes its final determination whether to prepare an EIS and before the action may begin. The circumstances are as follows:

■ The proposed action is, or is closely similar to, one which normally requires the preparation of an EIS, or
■ The nature of the proposed action is one without precedent.[169]

ESTABLISHING COOPERATING AGENCY RELATIONSHIPS

Agencies with jurisdiction by law or expertise, including state and local agencies by agreement (and American Indian tribes when effects are on a reservation), may become cooperating agencies in the preparation of an EIS. Cooperating agencies participate early in an EIS process and in scoping, assist in providing information and preparing environmental analyses, or make staff available to enhance the lead agency's interdisciplinary capability. Cooperation and involvement of other agencies saves time and effort by ensuring that environmental issues are addressed early in an EIS process and with agency cooperation, rather than discovering agency objections late in the process.

---

[168]Council on Environmental Quality, *Forty Most Asked Questions Concerning CEQ's National Environmental Policy Act Regulations*, 1981.
[169]40 CFR §1501.4(e)(2).

ADOPTING OTHER AGENCIES' ANALYSES

An agency may adopt another agency's Draft or Final EIS if the EIS meets the standards for an adequate EIS under the adopting agency's regulations. A federal agency may adopt another federal agency's EIS (or portions of an EIS), and state agencies may adopt federal EISs or other state agencies' EISs. If a federal cooperating agency independently reviews a federal lead agency's EIS and is satisfied with it, the agency may adopt the EIS of a lead agency without recirculating it for public review. If an action covered by another agency's EIS and the proposed action are substantially the same, the adopting agency does not have to recirculate another agency's EIS except as a final statement. Otherwise the adopting agency recirculates the EIS as a draft statement.

INCORPORATING INFORMATION BY REFERENCE

The purpose for incorporating information by reference is to reduce an EIS's bulk. The incorporated information must be cited in an EIS and its content briefly described. If information is generally available, it does not need to be repeated if it is adequately cited in the EIS. Other information, including supporting studies and technical references, also need not be included in an EIS as long as it is identified, available, and accessible. Proprietary data, for example, cannot be incorporated by reference.

CONDUCTING SCOPING

The scoping process determines the scope of issues to be addressed in an EIS. Scoping includes input from the public to identify the significant issues which are the focus of an EIS, to identify insignificant issues that will be eliminated from detailed study, and to narrow the discussion of these issues to a brief presentation of why they will not have a significant effect on the environment. Scoping identifies significant issues early in an EIS process so the EIS team does not spend more time and effort on insignificant issues than is required to establish that they are not significant. Scoping also identifies information sources that the EIS team may not have known about, and it saves the team members time and effort in finding information.

ANALYZING ONLY SIGNIFICANT ISSUES

State and federal EIS regulations emphasize the importance of identifying and focusing on significant issues. Significant issues are of the most interest to the public and agency decisionmakers; therefore, by not cluttering the analyses with insignificant issues, EIS reviewers can more easily compare and contrast the issues by alternatives. Deleting insignificant issues from detailed study reduces the bulk of the EIS and allows the available time and resources to be spent on the issues that really matter. However, this is not as easy as it sounds because "significance" is subjective. The first problem is in deciding which impacts are significant and which are not. EIS reviewers will disagree on the significance of an issue because of differences in their perspectives. The next area of potential controversy is regarding how much analysis to provide for a significant issue so that it is considered "adequate." Some EIS reviewers will say that the analysis is inadequate, while others will say it is too detailed. Generally, reviewers who are specialists or who have an interest in a subject tend to expect more detail, while those who are not interested in a subject expect less detail. There are no clear guidelines regarding the amount of analysis that is required in an EIS because adequacy depends on the facts and circumstances of a particular case. A "reasonably thorough" analysis is required. Exhaustive analysis is not. Also, see "Focus on Significant Impacts" in Chapter 12.

> **By not cluttering the analyses with insignificant issues, EIS reviewers can more easily compare and contrast the issues by alternatives.**

SETTING TIME AND PAGE LIMITS

The CEQ regulations (and some state guidelines) state that the text of Final EISs shall normally be less than 150 pages, or 300 pages for proposals of unusual scope or complexity. As described in one section of the CEQ regulations, a Final EIS is to contain most of the analysis that was in the Draft EIS.[170] However, the regulations state, in an-

---

[170]40 CFR §1502.7 states "The text of final environmental impact statements (e.g., paragraphs (d) through (g) of §1502.10) shall normally be less than 150 pages. . . ."

other section, that Final EISs do not have to include the Draft EIS or a revised Draft EIS.[171] Therefore, most federal agencies consider the 150- to 300-page guideline when preparing Draft EISs as well as Final EISs.

An applicant for a proposed action, state and local agencies, or members of the public may request a federal lead agency to set time limits. A federal lead agency may set time limits provided that the limits are consistent with NEPA's purposes and other considerations including the potential for environmental harm, size of the proposed action, and number of persons and agencies affected. In practice, few agencies set page or time limits (and if they do, the limits aren't enforced) because of a concern that the limits might constrain the EIS process. Any time constraints for preparation of an EIS are usually because of factors such as the proposed action's schedule or limitations of available funding.

OMITTING EXTRANEOUS BACKGROUND INFORMATION

Background information is usually presented in the description of the existing or affected environment, and it frequently accounts for the majority of the bulk of an EIS. The CEQ regulations express clearly their opinion of background information:

> The descriptions shall be no longer than is necessary to understand the effects of the alternatives. Data and analyses in a statement shall be commensurate with the importance of the impact, with less important material summarized, consolidated, or simply referenced. Agencies shall avoid useless bulk in statements and shall concentrate effort and attention on important issues. Verbose descriptions of the affected environment are themselves no measure of the adequacy of an environmental impact statement.[172]

TIERING

EIS regulations permit the analyses of a proposal's environmental effects to be accomplished in phases or tiers. Some agencies have the responsibility for actions that are programs, or actions that are very

---

[171]40 CFR §1500.4(m) and §1503.4(c)
[172]40 CFR §1502.15.

broad in their application. These actions affect a large area or region rather than a particular site. Specific actions or projects, or the specific features of those actions (e.g., size, shape, location, operation) within the region, may not be known at the time the program action is being contemplated. Yet, an agency is required to prepare an EIS for its program or broad action without knowing the details regarding the specific actions that may follow. When an agency prepares a comprehensive EIS, or programmatic EIS for a broad program, it may prepare supplemental or site-specific EISs for a second tier of EISs under the programmatic EIS. Alternatively, depending on the circumstances, the agency may prepare EAs as a second tier, or a third tier under a site-specific EIS.

Tiering eliminates repetitive discussions of the same issues and focuses on the actual issues "ripe" for decision at each level of environmental review. When tiering, an agency does not need to repeat analyses contained in a programmatic EIS but may summarize the analyses in an EIS of narrower scope, a site-specific EIS, or an EA and incorporate the broader analyses by reference. Tiered and programmatic EISs are also discussed in Chapter 7.

THE PRODUCTION PROCESS

Time, effort, and paperwork are factors in the production process of an EIS, as are the technical analyses. The production process is the part of an EIS preparation process that pulls the sections of an EIS together, giving it form as a document. The EIS project manager or ID team leader orchestrates this process to ensure the accomplishment of all required tasks.

In addition to the ID team, the process of preparing EISs includes the clerical staff, editors, word processors, graphic artists, receptionists, and anyone else who can be coerced into performing odd tasks and working long hours, weekends, and sometimes holidays to meet production deadlines. Technical staff (planners, architects, engineers, and scientists) also work long hours in the effort to meet deadlines. Delays in obtaining information, internal reviews from agencies, last minute changes in the project, and numerous other factors can destroy a well-planned schedule. But if the deadline can't be budged, it is usually the production staff who receive the most pressure because they are closest in the EIS preparation process to the deadline. However, there are some things that can be done internally at the beginning of the process that will save time and effort later. One approach is to streamline the internal EIS preparation process by identifying, in as much detail as possible, staff and resources that will be required and how much time will be necessary for each step of production, then integrating the production process into an EIS's scope of work.

Typically, production of an EIS begins *after* the technical staff have completed their analyses for the EIS. Any production problems, therefore, are not identified until late in an EIS process's schedule. If the production process and staff were integrated as part of the EIS team from the start of the process, production issues could be identified and resolved to expedite the process before an EIS is in production.

The scope of work and cost estimate of an EIS (prepared before work on the EIS began) should have identified the expected number of copies of the EIS and appendices, number of pages, type of paper (e.g., its weight, whether to use recycled paper), number of figures and any that exceed 8.5 × 11 inches, color figures, photos, dividers, tabs, type of binders or fasteners, and any special formatting features. The production staff and the rest of the EIS team members should discuss the production staff's requirements to identify time and staff resources required for production to accomplish its tasks. The following are some examples.

■ The word processor for an EIS should discuss with the EIS team any word processing issues related to the project.

■ If a client wants to have an EIS printed in double columns, or some other special format, discuss any time or resource requirements with the word processor. Some options may take longer to produce an EIS or may require certain skills that the word processor may not have.

■ All members of the EIS team (including subconsultants) should use the same word processing software if possible. Although translation programs are available, certain symbols, tables, and other features sometimes do not translate and have to be retyped.

■ All EIS team members should be following the same format for their sections in the EIS to reduce the need for the editor to reformat, and the word processor to retype, text.

- All EIS team members should be using the same style manual.
- The graphic artist and EIS team members should discuss issues and options regarding the figures in an EIS. All team members should be using the same base maps, when possible, to expedite the production of figures and provide consistency in the EIS.
- If copies of the EIS will be made by a print shop, make sure they are aware of the EIS's schedule so their staff, as well as the type and color of paper, are available when needed. EISs usually present large jobs for print shops that must plan for extra staff and order supplies in advance.

TIPS

To Reduce the Amount of Paper Used in EISs

All EISs Print double-sided copies, and make judicious use of the white space on a page. A three-inch margin is a waste of space; however, crowding text on a page will cause the reader eye fatigue.

Final EISs How do you conserve paper if you get hundreds of comment letters? One comment letter can be 30 pages or more, with one hundred or more specific comments.[173] Some suggestions to save space and paper include the following:

- Comments to a Draft EIS and responses to comments must be included in a Final EIS, but may be summarized if the comments are voluminous.[174]
- If a letter is signed by numerous individuals (as if a petition), it only requires one response.
- Public hearing or meeting comments on a Draft EIS often repeat written comments. Don't repeat responses already included in response to written comments. Prepare a table that summarizes and numbers comments and responses, and refer readers to the appropriately numbered responses to written comments.[175]

---

[173]This doesn't necessarily mean that the Draft EIS was inadequate; most often a response to comment is "see page 00 of the Draft EIS" because the reader either missed the information or didn't read the entire EIS.

[174]40 CFR §1503.4(b).

[175]Each letter would receive an identifying number, and each comment within the letter would also be numbered, so that a reference to a previous response might read: "Please see response to Comment Letter 48, comment 22."

■ The same comments may be made by different people, or the same people may send more than one letter with the same comment. Instead of repeating a response to identical comments, refer readers to the first time the response was given.

■ Reduce comment letters and agency response pages to half size, duplicate side-by-side.

Public comment letters come in all sizes and shapes, typed, and handwritten. The letters range from carefully thought-out to rambling and incoherent.

■ For rambling or incoherent letters, the difficulty lies in discerning what point or concern the commentor is attempting to make. If the point is not evident, the most appropriate response may be to say that the comment is unclear in its concerns and therefore a response cannot be provided. Some agency staff feel compelled to guess the point of the comment in order to provide a response, which can, in turn, result in some rambling responses such as "If the commentor is referring to . . ., then . . .; however, if the commentor means to say that . . ., then. . . . On the other hand, if what the commentor really meant to say is . . ., then. . . ."

Other Environmentally Friendly Tips[176]

■ Use recycled paper; but the cost may be higher, so make sure the budget includes the use of recycled paper.

■ Send summaries of the EIS if it is very long, except to agencies and others who require the full EIS.

■ Send appendices to technical reviewing agencies, or those who request appendices, not to everyone on the mailing list.

CHAPTER SUMMARY OF TIPS

■ To permit analysis of impacts early in a planning process, develop, as soon as possible, descriptions of a proposal sufficient to begin an EIS process.

■ To avoid misunderstandings, late discovery of differences of opinion, and delay in a joint EIS process, the agencies involved should

---

[176]Save trees, energy, and money.

come to agreement (including in-house agreement) on the framework and foundation of an EIS before work on the EIS is begun.

■ To reduce a proposal's potential environmental impacts to below the threshold of significance, design, site, or otherwise modify a proposal, or commit to mitigation measures that would reduce the proposals's impacts to insignificant levels, or both.

■ To identify and address environmental issues early in an EIS process, and thereby avoid delaying the process, maximize cooperation and involvement of other agencies with jurisdiction by law or expertise over a proposed action.

■ To avoid delays in an EIS's production, identify and resolve production issues early in an EIS process by including the production staff in the kickoff meeting and progress meetings.

CHAPTER 14

REVIEWING EISs

The industrial concept of man as master of an inexhaustible world
for his convenience is giving way to an ecological concept of man as
caretaker of limited resources that must be protected.

President Gerald Ford

INTRODUCTION

Environmental impact statements (EISs) are prepared in two separate stages: a Draft EIS and a Final EIS. The names of the stages are somewhat misleading. Many internal versions and reviews are normally required before a final Draft EIS and a final Final EIS are ready for public distribution and review. Thus, a "Draft EIS" is actually quite complete in terms of the amount of data and analysis that went into it; it is not a "first cut" that the public sees. The Draft EIS is called a "Draft" because it is not "Final" until the public has reviewed it, and any required changes are made in response to those public comments.

After public review of a Draft EIS, the ID team analyzes the comments to decide whether changes to the EIS are necessary. The purpose of a Final EIS is to respond to public comments on the Draft EIS. If public comments simply require clarification or other minor responses such as minor corrections or modifications to data, or responses that simply explain the reason further response is not warranted, the entire Draft EIS need not be rewritten and reissued to the public. If the Final EIS consists of responses to comments (and not a rewritten Draft EIS), "the EIS" is comprised of the Draft EIS and Final EIS. You would need both documents to get all of the information.

In preparing a Final EIS, the decision of whether to modify and reissue the Draft EIS is made by the lead agency. Some agencies simply prefer rewriting it because they feel it is easier to find all the information in one document than to have to look at two volumes[177] (the Draft and the Final). It is a tradeoff between saving time, effort, and paperwork (as encouraged by regulations) or having the convenience of only one document—the Final EIS—to refer to when the process is done.

The Reviewers

Reviewers of EISs include (1) internal reviewers, those who are working on an EIS, or lead agency staff members who review preliminary EISs for accuracy and consistency with agency programs and (2) external or public reviewers. The public reviews Draft EISs that have been internally reviewed several times before the documents are considered ready for public distribution. Public reviewers include individuals, organizations, companies, agencies, and American Indian tribes. Some lead agencies refer to other agencies and "the public" as separate groups. However, in this book, the public includes federal, state, and local agencies who review Draft EISs during the public review period, as well as individual citizens and organizations.

IN-HOUSE REVIEW OF PRELIMINARY EISs

EISs are written and rewritten by an ID team before they are ready for public review. At each stage of an EIS, for the Draft EIS and for the Final EIS, two or more internal reviews may be necessary (six internal reviews per document is not uncommon). The internal documents might be: First Preliminary Draft EIS, Second Preliminary Draft EIS; and First Preliminary Final EIS, Second Preliminary Final EIS (or Pre-

---

[177]EISs often are multiple volumes; this example assumes the simplest case of one volume each for the Draft and Final EIS.

liminary Draft EIS #1, #2, and so on). During internal review, a dozen or more people may be reviewing the EIS and suggesting changes. The different versions of EISs need to be labeled on each page, or each page should have a date, to ensure that changes are being made to the correct version and to compare or verify changes between versions.

Internal reviews of preliminary EISs are often by agency personnel who were not involved in the development of the proposed action and any assumptions and who were not involved in agreements made during meetings between the project proponent, lead agency, and those directly involved in the EIS. Therefore, it is important to provide notes to reviewers that describe the scope of analyses and the methodology, including any agreed-to limitations or constraints of the analyses. Otherwise, reviewers may have expectations of the analyses which are contrary to the understanding of the ID team. To have effective in-house reviews, all reviewers of the document must have the same understanding of the proposal and the EIS's foundation and framework as those who are preparing the document.

> **To have effective in-house reviews, all reviewers of the document must have the same understanding of the EIS's foundation and framework as those who are preparing the document.**

At a minimum, reviewers should have a copy of the EIS's scope of work, purpose and need (including objectives) for the proposed action, and the description of the proposed action and alternatives. These descriptions are usually at the beginning of an EIS, but some reviewers are provided only the section of the EIS that relates to their technical area of expertise (the wildlife section or the water quality section), while other reviewers will ignore the sections of an EIS that do not directly discuss their technical area of responsibility. Thus, many review comments are that information is lacking, when it was provided in sections other than the ones read by the reviewer; or that the scope of the analyses was inadequate when the reviewer's agency specified the scope. The agency EIS project manager should provide information describing the foundation and framework of the analyses. The project manager will receive more relevant and focused review comments, and, even though reviewers had to spend some time in the beginning reading background information, they will save time in the long run.

ROLES OF REVIEWERS

The following section describes roles from the perspective of major reviewers: consultant, lead agency, reviewing agencies, and citizens and groups.

Consultant's Perspective

Consultants prepare EISs for federal, state, and local agencies under the agencies' direction and in accordance with applicable regulations. As discussed above, members of the consultant team review and revise the EIS prior to sending a preliminary EIS to the lead agency for review. After the lead agency's staff have reviewed and commented on a preliminary EIS, the consultant team reviews the comments, discusses them with the agency project manager for clarification, and makes revisions to the document. This process of review and revision, by the consultant and by the agency, will take place several times before an EIS is distributed for public review.

After the Draft EIS has been publicly reviewed and comments have been sent to the lead agency, the EIS team and project manager for the lead agency discuss the public review comments to determine appropriate ways for responding to the comments in the Final EIS. The preliminary Final EIS receives the same type of internal reviews and revisions by the consultant and lead agency as took place with the preliminary Draft EIS.

When responding to the lead agency's review comments, a consultant team attempts to find ways to make the requested changes. This raises some basic questions for the consultant to consider. First, are the changes, especially those that require additional information, within the scope of work for the EIS? If the agency is requesting a change that is outside the EIS's scope of work such as adding issues or topics, or completely changing an EIS's format, the requested changes may require a contract change order. Second, do any of the requested changes conflict with previous agreements or understandings, or conflict with other requested changes? If the consultant does not evaluate the requested changes for potential conflicts before starting to make the changes in an EIS, one agency staffer's request may supersede another agency staffer's request by virtue of being the most recent request the consultant sees. To find such conflicts (and discuss them with the lead agency), a consultant can make changes by hand on a hard copy of the EIS, so that requested changes and potential conflicts can be seen on the hard copy before changes are actually made on the document in the computer. Although it is faster to make changes directly to the document in the computer, it is not advisable because the consultant may not be able to remember which items were changed by previous agency comments. For similar reasons, the agency comments should not be given directly to a word processor to make changes to the EIS. The consultant team leader should make changes on a hard copy of the latest version of the EIS (cut and tape a copy of the agency comments to save handwriting time) and give the annotated copy to the word processor. This annotated copy of the EIS can be useful later if someone at the lead agency questions the reason a change was made.

If the lead agency's project manager would compile all of the agency's comments on one hard copy (not a computer disk) of the latest version of the EIS, rather than provide lists of comments to the consultant, the revision process would be more efficient. This approach, however, takes more time for the lead agency's project manager than passing to the consultant the task of organizing the comments and finding conflicting agency direction.

Lead Agency's Perspective

Lead agencies review preliminary EISs prepared by consultants to ensure that the EISs meet regulatory requirements and the standards of the agency. The lead agency normally designates a project manager who is the point of contact for the consultant team, and he or she provides guidance and direction to the team. The agency project manager receives the preliminary EIS from the consultant, distributes it to other reviewers within the agency, and requests comments by a certain date. The agency project manager should discuss and evaluate comments from fellow agency reviewers before giving the comments to the consultant, but in many cases the comments are given to the consultant without any evaluation. The consultant must then evaluate the comments, and if some comments conflict or are unclear, he must discuss the comments with the agency project manager.

If a lead agency prepares an EIS without the assistance of a consultant, many of the agency personnel, who would have been reviewers, will most likely be on the EIS team as preparers of the EIS. Whether prepared by consultant or in-house, EISs frequently are also reviewed by an agency's "chain of command." Depending on the number of layers in the bureaucracy, and the level within the bureaucracy that has authority to approve an EIS, an EIS may require review and approval at several higher levels. Internal review could take months to complete, particularly if more than one specialist must review the EIS at any given level in the hierarchy. If possible, send informal advance copies to contacts at the various levels in the hierarchy to expedite the review and approval process. Also, see "In-House Review of Preliminary EISs," above, for additional suggestions on expediting internal review.

Obtaining Public Review Comments To understand public perceptions and encourage commentors to express their concerns, a lead agency must make an effort to explain the process and provide guidance to the public so the public knows what type of information will be useful to the agency. When requesting public review comments, agencies usually state what is in the regulations. However, regulations do not convey enough information of what is desired of the commentor. Regulations are understandable to people who are already familiar with the process and know the regulations' intent. "We invite you to comment on the Draft EIS" provides little guidance. Tell reviewers what their roles are, what types of comments are being solicited, and how the comments will be useful to the agency (see Chapter 11 for more information on requesting the public's input).

Anticipate the public's concerns when requesting their participation. Give some background or information regarding those concerns to the public when sending out public notices. Let them know what issues the lead agency has already identified, or how the proposed action addresses those concerns. These points should be statements of fact, not possibilities or promises that sound as though the agency is trying to sell the project to the public.

Reviewing Agencies' Perspective

Agencies that review EISs frequently have jurisdiction over a proposed action by law or special expertise. Jurisdiction may exist because of a proposal's location, the type of proposed action, or potential to affect a resource or sensitive area. Some agencies have permitting authority over a proposal; that is, the agencies can issue or deny a permit. In general, the role of permitting agencies is to regulate and control private or public activities. Therefore, agency reviewers tend to review the project (as though an EIS was the project sponsor's application for a permit or license) rather than review the EIS. An EIS should contain information sufficient to understand the potential impacts of a project and the adequacy of mitigation measures, but the level of information may not be to the level of detail that an agency may need to issue a permit. Proposed projects may require many permits and approvals from several agencies, but an EIS is not required to provide all the information that those agencies will need to decide whether to issue permits.

> **Agency reviewers tend to review the project . . . rather than the EIS.**

A comment letter from a reviewing agency, signed by someone with authority to represent the agency, represents the concerns or positions of that agency even if employees within the agency have varying opinions. Comments by individual agency employees that are contrary to the agency's comments do not have to be addressed in an EIS.[178]

Agency reviewers who are part of a large bureaucracy should check to see who else within the bureaucracy may be commenting on the EIS. Ideally, all comments should be coordinated and should be submitted through a single spokesperson. Many agencies have procedures for coordinated review of EISs. States have clearinghouses so that state agencies who review EISs send comments through their clearinghouse. It is not uncommon, however, for agencies to provide conflicting comments to a lead agency because of differences in agency programs and goals.

Citizen's and Group's Perspective

As discussed in Chapter 11, the public is invited to participate at several points in the EIS process. However, a public review process

---

[178]Sierra Club v. United States Department of Transportation (1987); Union of Concerned Scientists v. AEC (1974); No Oilport! v. Carter (1981) (Fogleman, 1990).

normally is provided only for a Draft EIS. A Final EIS is not publicly reviewed in the same manner as a Draft EIS because a Final EIS is publicly distributed, but review comments are not solicited. The purpose of a Final EIS is for the public to see how the lead agency responded to public review comments on a Draft EIS.

A lead agency will normally provide a Draft EIS to those who request it, to agencies with jurisdiction, and to those who own property adjacent to the site of a proposed action. Public notification of the availability of a Draft EIS is provided by various means such as in newspapers, radio and TV, or notices posted on a proposed project site. To receive a copy of a Draft EIS, an interested party should contact the person whose name and phone number is on the public notice.

In some cases, individuals and organizations may request that their names and addresses be placed on a lead agency's mailing list if they are aware of a pending action, and they request notices regarding the action or type of action. Many agencies have public affairs officers or public information specialists who may be able to provide interested parties with options on the types of actions for which they may receive notifications, or they may provide contacts within the agency for additional information. Another option is to become a member of a group that is on an agency's mailing list such as community groups or environmental groups.

If an individual attends a public scoping meeting for an EIS, he or she may request a copy of the EIS at the meeting, although the agency will most likely ask meeting attendees to place their names and addresses on a list to obtain copies of the EIS and any public notices. If interested parties review a Draft EIS and send comments to the lead agency, they may or may not receive a letter from the lead agency in response to their letters. Most agencies respond to comments in a Final EIS, although some agencies will send letters to commentors as well. If interested parties do not comment on a Draft EIS, they may not receive a copy of the Final EIS, unless they notify the lead agency that they wish to receive the Final EIS. Therefore, to receive a copy of a Final EIS, interested parties may provide their name and address at the Draft EIS public hearing, or they may call the contact at the lead agency and ask that a copy of the Final EIS be mailed when it becomes available. A Final EIS may take several months to prepare after the comment period on a Draft EIS has closed. When the Final EIS is available, the lead agency will distribute copies of the EIS and will issue public notices through the public media and through mailings to those on its mailing list.

ADEQUACY OF ANALYSES

A common concern of EIS reviewers is how one determines if the analysis in an EIS is adequate, particularly if the reviewers are not experts in the subjects under analysis. Since EISs are reviewed by a wide variety of agencies and individuals, chances are very good that some of the reviewers of the EIS will have the expertise to technically evaluate the analysis. If the authors of an EIS present the information in nontechnical terms, most reviewers should be able to understand the information, including the analyses' rationale and conclusions. Thus, most reviewers should be able to comment on the technical subjects in an EIS, even if the technical methodology requires training to understand.

During public review, attention should focus on the description of the proposed action's purpose and need, the scope of the EIS, and the overall methodology of the EIS. These are the premises on which the EIS is built, and they are described in Chapter 8. Review the EIS's description of the proposed action's purpose and need to see if the agency has clearly described the need for the action as well as its purposes and objectives. The alternatives that are developed for evaluation in an EIS should be based on this discussion, and therefore it is very important.

The scope of an EIS includes the range of alternatives, the significant issues, and the geographic study area that is evaluated in an EIS. Review the EIS to determine if the project alternatives are logical in terms of fulfilling the project's basic purpose and need. Alternatives (other than the no-action alternative) do not have to completely meet a project's objectives, but must reasonably meet those objectives. Alternatives should present a range of environmental impacts as well, so the agency decisionmaker is informed about the degree of environmental impact that is expected to occur for the various project alternatives. Every possible alternative does not have to be evaluated, but the EIS should provide a reasonable range or selection of alternatives. When evaluating alternatives, consider the following:

- Did the agency clearly present how it identified alternatives for analyses in the EIS?
- Are the features of each alternative explained in terms of how an alternative will or will not meet the purpose and need of the proposed action?
- Does each alternative provide a different array of environmental impacts in terms of types of impact, or degree of impact, when compared with the other alternatives?
- Is the environmentally preferred alternative identified (even though the agency does not have to choose it)?
- For Federal EISs, is the agency's preferred alternative identified in a Final EIS unless prohibited by law (the preferred alternative may be identified in a Draft EIS)?
- Does the EIS present fair and consistent analyses of environmental impacts for each *reasonable* alternative, and does it present the analyses in comparative form (compare impacts of an alternative to the impacts of other alternatives such as the no-action alternative and the agency's preferred alternative)?

An EIS analyzes the environmental impacts that would be caused by a proposed action if the action were implemented. The focus, or emphasis of the EIS, however, should be on the significant issues (discussed in Chapter 12 under "Focus on Significant Impacts"). When evaluating the EIS's treatment of significant issues, consider the following:

- Does the EIS identify which issues are considered significant, which are considered insignificant, and the reasons the issues are significant or insignificant?
- Do the significant issues receive thorough analyses?
- Are mitigation measures addressed for each significant environmental impact?
- Does the EIS explain the reasons an environmental impact cannot be mitigated?
- Are the significant impacts described with as much specificity as possible, given the preliminary nature of the proposed action or available information?
- Does the EIS present all sides of any controversy regarding significant environmental impacts?

The geographic study area for the evaluation of environmental impacts should be identified in an EIS. The study area is the area in which the environmental effects of a proposed action are expected to be felt. An EIS may have an overall study area, and the specific subjects of analyses in the EIS (e.g., air, water, traffic) may also have study areas in addition to the overall study area. When evaluating the study area of an EIS, consider the following:

■ Are the study area boundaries clearly defined? For example, is it legally defined (township, range, and section) and shown on a map?
■ Is the the study area appropriate? For example, does it include the areas that would be significantly affected by the proposed action?
■ Does the study area include the areas that would receive impacts as well as the area in which the impacts would be created?
■ Does the EIS explain the reason for any differences in study areas within the EIS?

Review and evaluate the methodology and mitigation measures in an EIS. An EIS should describe the methodologies used in preparing the analyses of the various issues, and in preparing the overall EIS if the methodology is unusual or complicated (e.g., legislative EISs, joint EISs). Mitigation measures are actions that may be taken to reduce or avoid environmental impacts (discussed in Chapter 12). Although a lead agency is not required to adopt the mitigation measures identified in an EIS, the conclusions in the analyses of impacts may be based on assumptions that certain mitigation measures would be implemented. An EIS should disclose

■ Whether mitigation measures are likely to be implemented (especially if they are not required by regulations)
■ The effectiveness of mitigation measures
■ Whether the conclusions regarding the type and degree of anticipated impacts take mitigation measures into consideration

Most of the items discussed above will be discussed throughout the EIS, within the analyses for each subject or topic. Therefore, when evaluating an EIS for adequacy one cannot find a type of information in only one location in an EIS. The various sections are interrelated, and any weaknesses in the analyses are identified by evaluating how the agency defined the proposed action, developed alternatives, and analyzed the environmental impacts of those alternatives from the EIS's beginning to its end.

Specificity of Comments

Because the public's comments are read and considered by the EIS team, commentors should make their comments clear and understandable. What are you requesting the agency to do? Make comments a requested action, such as a request that an error be corrected, or that the agency include certain issues in an EIS, or that the agency consider an additional alternative (specify what the issues or alternatives might be and why they should be analyzed in the EIS). Keep in mind that the EIS team is going to respond to comments. If you say that "the project is bad" or "the project will result in unacceptable impacts," the only possible response, in the absence of something more specific, will be "comment noted." If you say the project is beneficial and that you support it, the response also will be "comment noted."

To provide comments on an EIS, one must read the document. Many reviewers comment on a proposed action without reading the EIS. To comment on the conclusions or methodology used in the EIS, you should read more than just the summary. The summary will not contain the detail necessary to understand how the analysis was done and how the conclusions were reached. Make comments based on your understanding of the project and its potential impacts, not, for example, something from the news media. Keep in mind that EISs are written at an early planning stage of a proposed project and sometimes information about the project and therefore its impacts is not available. However, the project does have to be sufficiently defined to allow meaningful analyses. Also, for many projects, detailed project design drawings and specifications will be required in the final planning stages of the project by other agencies who must approve or permit the project (e.g., local construction permits). Over time, project designs and information in EISs have become progressively more detailed, even if not "final."

If you disagree with an EIS's conclusions regarding the environmental impacts, state why you disagree. Do you feel the conclusions were supported by the analysis? If not, why not? If you feel the alternatives were inadequate, what reasonable alternative do you think was missed? Why should it be included? If you think the methodology that was used in the EIS was not appropriate, state why it was inappropriate, what other methodology you would prefer, and why your methodology is better than the one used.[179] If you feel that the analysis was inadequate, state why it was inadequate. What was not included that should have been included? Why should that information or analyses be included? Remember that agencies are directed to *not* write encyclopedic EISs.[180] Large amounts of background data usually are unnecessary. Detailed analyses should be provided only for those impacts that would be significant. If your comments are that the EIS needs more analyses or more information, state how much more is required, and consider whether the information you are requesting is necessary to allow the lead agency decisionmaker to make an informed decision. An informed decision normally does not require exhaustive studies or multiple pages of environmental resource inventories, such as species lists.

Reasonability of Comments

Comments will be much more effective if any concerns are clearly described. If, for example, you don't want a sanitary landfill or a detention facility in your neighborhood, explain your concerns regarding such facilities. Some people view such facilities as sources of jobs and as beneficial for the economy. If your concern is safety and security, focus on the discussions in the EIS that address the project's proposed methods of ensuring security. Identify any weaknesses of the proposal, and make suggestions, if possible, on what specific actions would make you feel more secure. If your point is that the facility be placed somewhere else, identify other locations you feel are more appropriate. If an EIS states that an alternative site has to be within a certain distance or travel time, from a certain point for logistical reasons, keep that in mind when suggesting alternative locations. One may always disagree with the rationale or criteria of a proposed project. But make suggested alternatives reasonable in terms of the objectives of the proposed project.

As an example, an EIS was prepared on a proposal for construction of a police shooting range. Neighbors around the proposed site

---

[179]40 CFR §1503.3(b).
[180]40 CFR §1500.4(b); §1502.1; §1502.2(a); §1502.15.

objected to the project because of potential noise from the facility. A suggested alternative was that the police use a shooting range at a military base that would require almost two hours travel time, one way. The travel time to the military range exceeded (doubled) the EIS's maximum travel time criteria for evaluation of alternatives. Another suggested alternative was that the site be used for an elementary school or park. However, an EIS evaluates a proposed action and alternative ways of meeting the objectives of that action, not actions that do not meet the objectives of the proposal (with the exception of the no-action alternative).

CHAPTER SUMMARY OF TIPS

■ To ensure that changes are being made to the correct internal version of an EIS, place the version name or number (e.g., fourth preliminary EIS, or preliminary EIS #4), along with the date of revision, on each page.

■ To have effective in-house reviews, receive relevant and focused review comments, and save time in the long run, an agency project manager should ensure that all agency reviewers understand an EIS's scope, methodology, assumptions, and agreed-to limitations (its foundation and framework).

■ To find conflicts in agency staff directions for changes to an EIS, a consultant can note changes on a hard copy of the EIS (cut and tape a copy of the agency comments to save handwriting time) so that all of the requested revisions are visible prior to making changes to the document in the computer.

■ To expedite revisions of internal EISs and reduce potential for incorrect changes, rather than provide lists of comments to the consultant, a lead agency's project manager should compile all agency staff's comments on one hard copy (not a computer disk) of the EIS.

■ To expedite the review and approval process in a bureaucracy, send informal advance copies of EISs to contacts at the various levels in the hierarchy.

■ To provide effective comments on EISs:

Request the lead agency to take specific actions in an EIS (e.g., correct errors, add significant issues) rather than ask questions or state opinions.

Make comments as specific as possible regarding subjects such as significant issues, alternatives, mitigation measures, the rationale for analyses, and methodology in the EIS.

Make suggested alternatives reasonable in terms of meeting the basic purpose and objectives of a proposed action.

Consider whether requests for additional information or analyses in an EIS are necessary in order for the decisionmaker to make an informed decision.

Provide the reasons for your requests.

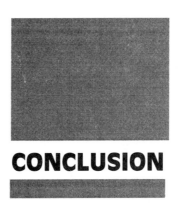

CONCLUSION

The environmental review process will only meet the goals and purposes for which the laws were enacted if participants in the process, agencies, and the public understand their roles and support the goals of the process. The need to make the environmental impact statement (EIS) process more effective so that participants recognize the value of the process is evident as the business community, as well as private citizens, question the requirement for the cost and time spent conducting environmental review. This book provides suggestions for the following:

Agencies

- To make an extra effort to inform the public about their roles in the process in order to obtain more meaningful and effective public involvement and input for a decisionmaker's consideration
- To understand the differences in your roles as lead agencies, reviewing agencies (of EISs), cooperating agencies, and permitting or licensing agencies
- To prepare EISs that comply with regulations and that are readable and useful to the public and decisionmakers
- To effectively manage interdisciplinary teams and the EIS process, including public meetings

■ To streamline environmental review and project review (reduce cost, delay, and paperwork) by combining environmental and planning processes

■ To use the EIS process to design better projects—projects that meet mission requirements and minimize impacts to the environment and society

The Public (Individuals, Organizations, Developers, Special Interest Groups, Reviewing Agencies)

■ To understand your roles in an EIS or planning process, or both

■ To understand how decisions are made in a lead agency's hierarchy and to know who the decisionmaker for a proposed action will be

■ To make comments and input on an EIS more effective by

Reading an EIS and understanding the purpose and objectives of a proposed action

Providing comments that are specific as to the type of concern and that request a lead agency to, for example, make a certain change or correct errors identified by the commentor in an EIS

Providing suggestions for reasonable alternatives (alternatives that meet the underlying purpose and objectives of a proposal)

Project Sponsors

■ To understand your role in the EIS process in relation to the lead agency and consultant

■ To carefully and thoroughly define the proposed action

■ To start the environmental review process early in project planning

■ To include environmental considerations in design, operation, and other characteristics of the proposed project

■ To propose rational projects that do not push the envelope of agencies' authorities

COOPERATION

An EIS process seeks cooperative working relationships, yet the nature of most proposals that require EISs is that they are controversial, and they polarize rather than bring participants together. To reduce polarization of participants, agencies and members of the public alike must make the effort to identify and communicate with specific participants for a specific proposal. "The government," or agencies within the (federal, state, or local) government are comprised of people. "The public" includes agencies (other than the lead agency), organizations, businesses, and citizens, which are similarly comprised of people. Generally, an approach of people dealing with each other is more effective than organizations presenting their positions to other organizations who also have their positions. An EIS process will most likely include people at odds with each other no matter what is done; however, litigation may be less frequent if the participants are making good faith efforts to identify, discuss, and understand the specific concerns and positions of everyone involved.

BIBLIOGRAPHY

Advisory Council on Historic Preservation, and GSA Interagency Training Center. 1993. *Introduction to Federal Projects and Historic Preservation Law, Participant's Course Book*. A Section 106 Training Publication.

Bregman, Jacob I., and Kenneth M. Mackenthun. 1992. *Environmental Impact Statements*. Lewis Publishers, Boca Raton, FL.

California Governor's Office of Planning and Research. 1995. *California Environmental Quality Act Statutes and Guidelines*. Sacramento, CA.

Corwin, Ruthann, et al. 1975. *Environmental Impact Assessment*. Freeman, Cooper & Company, San Francisco, CA.

Council on Environmental Quality. 1993. *Environmental Quality: The Twenty-fourth Annual Report*. Washington, D.C.

Council on Environmental Quality. 1991. *Environmental Quality: The Twenty-second Annual Report*. Washington, D.C.

Council on Environmental Quality. 1976. *Environmental Quality: The Seventh Annual Report*. Washington, D.C.

Erickson, Paul A. 1994. *A Practical Guide to Environmental Impact Assessment*. Academic Press, San Diego, CA.

Fogleman, Valerie M. 1990. *Guide to the National Environmental Policy Act: Interpretations, Applications, and Compliance.* Quorum Books, Westport, CT.

Jain, R. K. 1993. *Environmental Assessment.* McGraw-Hill, New York.

Knight, Jeffrey. 1980. "Seventies into Eighties," in *The New Environmental Handbook.* Garrett DeBell, editor. Friends of the Earth, San Francisco, CA.

Laurence, Peter J. 1986. *The Peter Pyramid, or, Will We Ever Get the Point?* William Morrow and Company, New York.

Mandelker, Daniel R. 1995. *NEPA Law and Litigation*, 2nd ed., Release #3. Clark Boardman Callaghan, Deerfield, IL.

McHarg, Ian L. 1969. *Design With Nature.* Doubleday/Natural History Press, Doubleday & Company, Garden City, NY.

Merriam-Webster. 1993. *Webster's Tenth New Collegiate Dictionary.* Merriam-Webster, Inc., Springfield, MA.

Lee, N., et al., editors. 1995. *EIA Newsletter 10.* EIA Centre, Department of Planning and Landscape, University of Manchester, UK.

Lee, N., et al., editors. 1995. *EIA Legislation and Regulations in the EU.* EIA Leaflet Series #5. EIA Centre, Department of Planning and Landscape, University of Manchester, UK.

Lee, N., et al., editors. 1993. *Five Year Review of the Implementation of the EIA Directive.* EIA Leaflet Series #14. EIA Centre, Department of Planning and Landscape, University of Manchester, UK.

Rodgers, William H., Jr. 1977. *Handbook on Environmental Law.* West Publishing Co., St. Paul, MN.

Roe, Dilys, Barry Dalal-Clayton, and Ross Hughes. 1995. *A Directory of Impact Assessment Guidelines.* International Institute for Environment and Development, London, UK.

Rubenson, David, et al. 1994. *Marching to Different Drummers: Evolution of the Army's Environmental Program.* RAND, Santa Monica, CA.

Rubenson, David, et al. 1992. *Two Shades of Green: Environmental Protection and Combat Training.* RAND, Santa Monica, CA.

Southern California Association of Governments. 1995. "Open Space and Conservation," *Regional Comprehensive Plan and Guide.* Los Angeles, CA.

Tipler, Paul A. 1976. *Physics.* Worth Publishers, New York.

U.S. Army Corps of Engineers. 1995. *Auburn Thoroughbred Horse Racing Facility, Final Federal Environmental Impact Statement (NEPA), Volume I, Revised EIS*. Seattle, WA.

U.S. Army Corps of Engineers. 1988. *Environmental Quality; Procedures for Implementing the National Environmental Policy Act (NEPA); Final Rule*. 33 CFR Parts 230 and 325.

U.S. Bureau of Land Management. 1988. *National Environmental Policy Act Handbook*. BLM Handbook H-1790-1. Washington, D.C.

U.S. Bureau of Land Management. 1994. *Guidelines for Assessing and Documenting Cumulative Impacts*. Information Bulletin 94-310. Washington, D.C.

U.S. Bureau of Land Management. 1993. *Environmental Analysis Handbook*. California State Office, Sacramento, CA.

U.S. Environmental Protection Agency. 1984. *Policy and Procedures for the Review of Federal Actions Impacting the Environment*. Washington, D.C.

U.S. Forest Service. 1992. *National Environmental Policy Act; Revised Policy and Procedures*. Federal Register, Volume 57, Number 182.

U.S. Navy. 1994. "Procedures for Implementing the National Environmental Policy Act (NEPA)," *Environmental and Natural Resources Program Manual*. OPNAVINST 5090.1B, Office of the Chief of Naval Operations, Washington, D.C.

U.S. Navy. 1989. *Proposed Expansion and Realignment of the Marine Corps Base Camp Lejeune, Onslow County, North Carolina, Draft Environmental Impact Statement*. Atlantic Division, Naval Facilities Engineering Command, Norfolk, VA.

U.S. Office of the Federal Register, National Archives and Records Administration. 1994. *The United States Government Manual 1994/1995*. U.S. Government Printing Office, Washington, D.C.

Washington Department of Ecology. no date. *State Environmental Policy Act, SEPA Handbook; SEPA Rules*. WAC 197-11; SEPA, RCW 43.21C, Olympia, WA.

Washington Department of Ecology. 1995. *Improving the Permit Process: Integration, Tracking, and Assistance*. Publication #95-253. Olympia, WA.

Washington Department of Ecology. 1994. *Permit Handbook: Commonly Required Environmental Permits for Washington State*. Publication # 90-29. Olympia, WA.

Washington, State of. 1995. *House Bill 1724: An Act Relating to Implementing the Recommendations of the Governor's Task Force on Regulatory Reform on Integrating Growth Management Planning and Environmental Review.* Olympia, WA.

Wellington, Joseph A. 1989. "A Primer on Environmental Law for the Naval Services," *Naval Law Review.* Naval Justice School, Newport, RI.

Wellington, Joseph A. 1988. *Naval Readiness, Operational Training, and Environmental Protection: Achieving an Appropriate Balance Between Competing National Interests.* Unpublished.

Wellington, Joseph A. 1987. *Extending the Environmental Mandate Beyond the Decision to Proceed with a Federal Project: NEPA, CEQ and the Duty to Mitigate.* A thesis submitted to the faculty of the National Law Center of the George Washington University in partial satisfaction of the requirements for the degree of Master of Laws.

NATIONAL ENVIRONMENTAL
POLICY ACT

THE NATIONAL ENVIRONMENTAL
POLICY ACT OF 1969, AS AMENDED*

An Act to establish a national policy for the environment, to provide for the establishment of a Council on Environmental Quality, and for other purposes.

Be it enacted by the Senate and House of Representatives of the United States of America in Congress assembled, That this Act may be cited as the "National Environmental Policy Act of 1969."

PURPOSE   [42 USC § 4321]

SEC. 2. The purposes of this Act are: To declare a national policy which will encourage productive and enjoyable harmony between man and his environment; to promote efforts which will prevent or eliminate damage to the environment and biosphere and stimulate the health and welfare of man; to enrich the understanding of the ecological systems and natural resources important to the Nation; and to establish a Council on Environmental Quality.

TITLE I

DECLARATION OF NATIONAL ENVIRONMENTAL POLICY   42 USC § 4331

SEC. 101. (a) The Congress, recognizing the profound impact of man's activity on the interrelations of all components of the natural environment, particularly the profound influences of population growth, high-density urbanization, industrial expansion, resource exploitation, and new and expanding technological advances and recognizing further the critical importance of restoring and maintaining environmental quality to the overall welfare and development of man, declares that it is the continuing policy of the Federal Government, in cooperation with State and local governments, and other concerned public and private organizations, to use all practicable means and measures, including financial and technical assistance, in a manner calculated to foster and promote the general welfare, to create and maintain conditions under which man and nature can exist in productive harmony, and fulfill the social, economic, and other requirements of present and future generations of Americans.

(b) In order to carry out the policy set forth in this Act, it is the continuing responsibility of the Federal Government to use all practicable means, consistent with other essential considerations of national policy, to improve and coordinate Federal plans, functions, programs, and resources to the end that the Nation may—

(1) fulfill the responsibilities of each generation as trustee of the environment for succeeding generations;

(2) assure for all Americans safe, healthful, productive, and esthetically and culturally pleasing surroundings;

(3) attain the widest range of beneficial uses of the environment without degradation, risk to health or safety, or other undesirable and unintended consequences;

(4) preserve important historic, cultural, and natural aspects of our national heritage, and maintain, wherever possible, an environment which supports diversity, and variety of individual choice;

(5) achieve a balance between population and resource use which will permit high standards of living and a wide sharing of life's amenities; and

(6) enhance the quality of renewable resources and approach the maximum attainable recycling of depletable resources.

---

*Pub. L. 91–190, 42 U.S.C. 4321–4347, January 1, 1970, as amended by Pub. L. 94–52, July 3, 1975, and Pub. L. 94–83, August 9, 1975.

ENVIRONMENTAL JUSTICE.

(c) The Congress recognizes that each person should enjoy a healthful environment and that each person has a responsibility to contribute to the preservation and enhancement of the environment.

[42 USC § 4332]

SEC. 102. The Congress authorizes and directs that, to the fullest extent possible: (1) the policies, regulations, and public laws of the United States shall be interpreted and administered in accordance with the policies set forth in this Act, and (2) all agencies of the Federal Government shall—

102(2)A

(A) Utilize a systematic, interdisciplinary approach which will insure the integrated use of the natural and social sciences and the environmental design arts in planning and in decisionmaking which may have an impact on man's environment; *HIRE DIFFERENT GROUPS*

(B) Identify and develop methods and procedures, in consultation with the Council on Environmental Quality established by title II of this Act, which will insure that presently unquantified environmental amenities and values may be given appropriate consideration in decision-making along with economic and technical considerations;

WAS A WHOLE TITLED SECTION

(C) Include in every recommendation or report on proposals for legislation and other major Federal actions significantly affecting the quality of the human environment, a detailed statement by the responsible official on— *EIS ACTION*

(i) The environmental impact of the proposed action,

(ii) Any adverse environmental effects which cannot be avoided should the proposal be implemented,

(iii) Alternatives to the proposed action,

(iv) The relationship between local short-term uses of man's environment and the maintenance and enhancement of long-term productivity, and *TRADE OFFS*

(v) Any irreversible and irretrievable commitments of resources which would be involved in the proposed action should it be implemented.

Prior to making any detailed statement, the responsible Federal official shall consult with and obtain the comments of any Federal agency which has jurisdiction by law or special expertise with respect to any environmental impact involved. Copies of such statement and the comments and views of the appropriate Federal, State, and local agencies, which are authorized to develop and enforce environmental standards, shall be made available to the President, the Council on Environmental Quality and to the public as provided by section 552 of title 5, United States Code, and shall accompany the proposal through the existing agency review processes; *DRAFT EIS FOIA*

Ammendment

(D) Any detailed statement required under subparagraph (c) after January 1, 1970, for any major Federal action funded under a program of grants to States shall not be deemed to be legally insufficient solely by reason of having been prepared by a State agency or official, if:

(i) the State agency or official has statewide jurisdiction and has the responsibility for such action,

(ii) the responsible Federal official furnishes guidance and participates in such preparation,

(iii) the responsible Federal official independently evaluates such statement prior to its approval and adoption, and

(iv) after January 1, 1976, the responsible Federal official provides early notification to, and solicits the views of, any other State or any Federal land management entity of any action or any alternative thereto which may have significant impacts upon such State or affected Federal land management entity and, if there is any disagreement on such impacts, prepares a written assessment of such impacts and views for incorporation into such detailed statement.

The procedures in this subparagraph shall not relieve the Federal official of his responsibilities for the scope, objectivity, and content of the entire statement or of any other responsibility under this Act; and further, this subparagraph does not affect the legal sufficiency of statements prepared by State agencies with less than statewide jurisdiction.

(E) Study, develop, and describe appropriate alternatives to recommended courses of action in any proposal which involves unresolved conflicts concerning alternative uses of available resources;

(F) Recognize the worldwide and long-range character of environmental problems and, where consistent with the foreign policy of the United States, lend appropriate support to initiatives, resolutions, and programs designed to maximize international cooperation in anticipating and preventing a decline in the quality of mankind's world environment;

(G) Make available to States, counties, municipalities, institutions, and individuals, advice and information useful in restoring, maintaining, and enhancing the quality of the environment;

(H) Initiate and utilize ecological information in the planning and development of resource-oriented projects; and

(I) Assist the Council on Environmental Quality established by title II of this Act.

42 USC §4333 SEC. 103. All agencies of the Federal Government shall review their present statutory authority, administrative regulations, and current policies and procedures for the purpose of determining whether there are any deficiencies or inconsistencies therein which prohibit full compliance with the purposes and provisions of this Act and shall propose to the President not later than July 1, 1971, such measures as may be necessary to bring their authority and policies into conformity with the intent, purposes, and procedures set forth in this Act.

42 USC §4334 SEC. 104. Nothing in section 102 or 103 shall in any way affect the specific statutory obligations of any Federal agency (1) to comply with criteria or standards of environmental quality, (2) to coordinate or consult with any other Federal or State agency, or (3) to act, or refrain from acting contingent upon the recommendations or certification of any other Federal or State agency.

42 USC §4335 SEC. 105. The policies and goals set forth in this Act are supplementary to those set forth in existing authorizations of Federal agencies.

TITLE II

COUNCIL ON ENVIRONMENTAL QUALITY   42 USC § 4341

PRES. REPORT SEC. 201. The President shall transmit to the Congress annually beginning July 1, 1970, an Environmental Quality Report (hereinafter referred to as the "report") which shall set forth (1) the status and condition of the major natural, manmade, or altered environmental classes of the Nation, including, but not limited to, the air, the aquatic, including marine, estuarine, and fresh water, and the terrestrial environment, including, but not limited to, the forest, dryland, wetland, range, urban, suburban and rural environment; (2) current and foreseeable trends in the quality, management and utilization of such environments and the effects of those trends on the social, economic, and other requirements of the Nation; (3) the adequacy of available natural resources for fulfilling human and economic requirements of the Nation in the light of expected population pressures; (4) a review of the programs and activities (including regulatory activities) of the Federal Government, the State and local governments, and nongovernmental entities or individuals with particular reference to their effect on the environment and on the conservation, development and utilization of natural resources; and (5) a program for remedying the deficiencies of existing programs and activities, together with recommendations for legislation.

COUNCIL SEC. 202. There is created in the Executive Office of the President a Coun- 4342 cil on Environmental Quality (hereinafter referred to as the "Council"). The Council shall be composed of three members who shall be appointed by the President to serve at his pleasure, by and with the advice and consent of the Senate. The President shall designate one of the members of the Council to serve as Chairman. Each member shall be a person who, as a result of his training, experience, and attainments, is exceptionally well qualified to analyze and interpret environmental trends and information of all kinds; to appraise programs and activities of the Federal Government in the light of the policy set forth in title I of this Act; to be conscious of and responsive to the scientific, economic, social, esthetic, and cultural needs and interests of the Nation; and to formulate and recommend national policies to promote the improvement of the quality of the environment.

SEC. 203. The Council may employ such officers and employees as may be necessary to carry out its functions under this Act. In addition, the Council may employ and fix the compensation of such experts and consultants as may be necessary for the carrying out of its functions under this Act, in accordance with section 3109 of title 5, United States Code (but without regard to the last sentence thereof). 4343

SEC. 204. It shall be the duty and function of the Council— 4344

(1) to assist and advise the President in the preparation of the Environmental Quality Report required by section 201 of this title;

(2) to gather timely and authoritative information concerning the conditions and trends in the quality of the environment both current and prospective, to analyze and interpret such information for the purpose of determining whether such conditions and trends are interfering, or are likely to interfere, with the achievement of the policy set forth in title I of this Act, and to compile and submit to the President studies relating to such conditions and trends;

(3) to review and appraise the various programs and activities of the Federal Government in the light of the policy set forth in title I of this Act for the purpose of determining the extent to which such programs and activities are contributing to the achievement of such policy, and to make recommendations to the President with respect thereto;

(4) to develop and recommend to the President national policies to foster and promote the improvement of environmental quality to meet the conservation, social, economic, health, and other requirements and goals of the Nation;

(5) to conduct investigations, studies, surveys, research, and analyses relating to ecological systems and environmental quality;

(6) to document and define changes in the natural environment, including the plant and animal systems, and to accumulate necessary data and other information for a continuing analysis of these changes or trends and an interpretation of their underlying causes;

(7) to report at least once each year to the President on the state and condition of the environment; and

(8) to make and furnish such studies, reports thereon, and recommendations with respect to matters of policy and legislation as the President may request.

SEC. 205. In exercising its powers, functions, and duties under this Act, the Council shall— 4345

(1) Consult with the Citizens' Advisory Committee on Environmental Quality established by Executive Order No. 11472, dated May 29, 1969, and with such representatives of science, industry, agriculture, labor, conservation organizations, State and local governments and other groups, as it deems advisable; and

(2) Utilize, to the fullest extent possible, the services, facilities and information (including statistical information) of public and private agencies and organizations, and individuals, in order that duplication of effort and expense may be avoided, thus assuring that the Council's activities will not unnecessarily overlap or conflict with similar activities authorized by law and performed by established agencies.

SEC. 206. Members of the Council shall serve full time and the Chairman of the Council shall be compensated at the rate provided for Level II of the Executive Schedule Pay Rates (5 U.S.C. 5313). The other members of the Council shall be compensated at the rate provided for Level IV of the Executive Schedule Pay Rates (5 U.S.C. 5315). 4346

SEC. 207. The Council may accept reimbursements from any private nonprofit organization or from any department, agency, or instrumentality of the Federal Government, any State, or local government, for the reasonable travel expenses incurred by an officer or employee of the Council in connection with his attendance at any conference, seminar, or similar meeting conducted for the benefit of the Council. 4348a

SEC. 208. The Council may make expenditures in support of its international activities, including expenditures for: (1) international travel; (2) activities in implementation of international agreements; and (3) the sup- 4348b port of international exchange programs in the United States and in foreign countries.

4347   SEC. 209. There are authorized to be appropriated to carry out the provisions of this chapter not to exceed $300,000 for fiscal year 1970, $700,000 for fiscal year 1971, and $1,000,000 for each fiscal year thereafter.

APPENDIX B

COUNCIL ON ENVIRONMENTAL QUALITY REGULATIONS

Council on Environmental Quality
Executive Office of the President

REGULATIONS

For Implementing The Procedural Provisions Of The

NATIONAL
ENVIRONMENTAL
POLICY ACT

Reprint
40 CFR Parts 1500–1508
(as of July 1, 1986)

CONTENTS

For further information, contact:
General Counsel
Council on Environmental Quality
Executive Office of the President
722 Jackson Pl. N.W.
Washington, D.C. 20503
(202) 395-5754

For sale by the Superintendent of Documents, Congressional Sales Office
U.S. Government Printing Office, Washington, DC 20402

TABLE OF CONTENTS

2

PART 1500—PURPOSE, POLICY, AND MANDATE

Sec.
1500.1 Purpose.
1500.2 Policy.
1500.3 Mandate.
1500.4 Reducing paperwork.
1500.5 Reducing delay.
1500.6 Agency authority.

AUTHORITY: NEPA, the Environmental Quality Improvement Act of 1970, as amended (42 U.S.C. 4371 *et seq.*), sec. 309 of the Clean Air Act, as amended (42 U.S.C. 7609) and E.O. 11514, Mar. 5, 1970, as amended by E.O. 11991, May 24, 1977).

SOURCE: 43 FR 55990, Nov. 28, 1978, unless otherwise noted.

§ 1500.1 Purpose.

(a) The National Environmental Policy Act (NEPA) is our basic national charter for protection of the environment. It establishes policy, sets goals (section 101), and provides means (section 102) for carrying out the policy. Section 102(2) contains "action-forcing" provisions to make sure that federal agencies act according to the letter and spirit of the Act. The regulations that follow implement section 102(2). Their purpose is to tell federal agencies what they must do to comply with the procedures and achieve the goals of the Act. The President, the federal agencies, and the courts share responsibility for enforcing the Act so as to achieve the substantive requirements of section 101.

(b) NEPA procedures must insure that environmental information is available to public officials and citizens before decisions are made and before actions are taken. The information must be of high quality. Accurate scientific analysis, expert agency comments, and public scrutiny are essential to implementing NEPA. Most important, NEPA documents must concentrate on the issues that are truly significant to the action in question, rather than amassing needless detail.

(c) Ultimately, of course, it is not better documents but better decisions that count. NEPA's purpose is not to generate paperwork—even excellent paperwork—but to foster excellent action. The NEPA process is intended to help public officials make decisions that are based on understanding of environmental consequences, and take actions that protect, restore, and enhance the environment. These regulations provide the direction to achieve this purpose.

§ 1500.2 Policy.

Federal agencies shall to the fullest extent possible:

(a) Interpret and administer the policies, regulations, and public laws of the United States in accordance with the policies set forth in the Act and in these regulations.

(b) Implement procedures to make the NEPA process more useful to decisionmakers and the public; to reduce paperwork and the accumulation of extraneous background data; and to emphasize real environmental issues and alternatives. Environmental impact statements shall be concise, clear, and to the point, and shall be supported by evidence that agencies have made the necessary environmental analyses.

(c) Integrate the requirements of NEPA with other planning and environmental review procedures required by law or by agency practice so that all such procedures run concurrently rather than consecutively.

(d) Encourage and facilitate public involvement in decisions which affect the quality of the human environment.

(e) Use the NEPA process to identify and assess the reasonable alternatives to proposed actions that will avoid or minimize adverse effects of these actions upon the quality of the human environment.

(f) Use all practicable means, consistent with the requirements of the Act and other essential considerations of national policy, to restore and enhance the quality of the human environment and avoid or minimize any possible adverse effects of their actions upon the quality of the human environment.

§ 1500.3 Mandate.

Parts 1500 through 1508 of this title provide regulations applicable to and binding on all Federal agencies for implementing the procedural provisions

3

of the National Environmental Policy Act of 1969, as amended (Pub. L. 91-190, 42 U.S.C. 4321 et seq.) (NEPA or the Act) except where compliance would be inconsistent with other statutory requirements. These regulations are issued pursuant to NEPA, the Environmental Quality Improvement Act of 1970, as amended (42 U.S.C. 4371 et seq.) section 309 of the Clean Air Act, as amended (42 U.S.C. 7609) and Executive Order 11514, Protection and Enhancement of Environmental Quality (March 5, 1970, as amended by Executive Order 11991, May 24, 1977). These regulations, unlike the predecessor guidelines, are not confined to sec. 102(2)(C) (environmental impact statements). The regulations apply to the whole of section 102(2). The provisions of the Act and of these regulations must be read together as a whole in order to comply with the spirit and letter of the law. It is the Council's intention that judicial review of agency compliance with these regulations not occur before an agency has filed the final environmental impact statement, or has made a final finding of no significant impact (when such a finding will result in action affecting the environment), or takes action that will result in irreparable injury. Furthermore, it is the Council's intention that any trivial violation of these regulations not give rise to any independent cause of action.

§ 1500.4 **Reducing paperwork.**

Agencies shall reduce excessive paperwork by:

(a) Reducing the length of environmental impact statements (§ 1502.2(c)), by means such as setting appropriate page limits (§§ 1501.7(b)(1) and 1502.7).

(b) Preparing analytic rather than encyclopedic environmental impact statements (§ 1502.2(a)).

(c) Discussing only briefly issues other than significant ones (§ 1502.2(b)).

(d) Writing environmental impact statements in plain language (§ 1502.8).

(e) Following a clear format for environmental impact statements (§ 1502.10).

(f) Emphasizing the portions of the environmental impact statement that are useful to decisionmakers and the public (§§ 1502.14 and 1502.15) and reducing emphasis on background material (§ 1502.16).

(g) Using the scoping process, not only to identify significant environmental issues deserving of study, but also to deemphasize insignificant issues, narrowing the scope of the environmental impact statement process accordingly (§ 1501.7).

(h) Summarizing the environmental impact statement (§ 1502.12) and circulating the summary instead of the entire environmental impact statement if the latter is unusually long (§ 1502.19).

(i) Using program, policy, or plan environmental impact statements and tiering from statements of broad scope to those of narrower scope, to eliminate repetitive discussions of the same issues (§§ 1502.4 and 1502.20).

(j) Incorporating by reference (§ 1502.21).

(k) Integrating NEPA requirements with other environmental review and consultation requirements (§ 1502.25).

(l) Requiring comments to be as specific as possible (§ 1503.3).

(m) Attaching and circulating only changes to the draft environmental impact statement, rather than rewriting and circulating the entire statement when changes are minor (§ 1503.4(c)).

(n) Eliminating duplication with State and local procedures, by providing for joint preparation (§ 1506.2), and with other Federal procedures, by providing that an agency may adopt appropriate environmental documents prepared by another agency (§ 1506.3).

(o) Combining environmental documents with other documents (§ 1506.4).

(p) Using categorical exclusions to define categories of actions which do not individually or cumulatively have a significant effect on the human environment and which are therefore exempt from requirements to prepare an environmental impact statement (§ 1508.4).

(q) Using a finding of no significant impact when an action not otherwise excluded will not have a significant effect on the human environment and is therefore exempt from requirements to prepare an environmental impact statement (§ 1508.13).

[43 FR 55990, Nov. 29, 1978; 44 FR 873, Jan. 3, 1979]

§ 1500.5 **Reducing delay.**

Agencies shall reduce delay by:

(a) Integrating the NEPA process into early planning (§ 1501.2).

(b) Emphasizing interagency cooperation before the environmental impact statement is prepared, rather than submission of adversary comments on a completed document (§ 1501.6).

(c) Insuring the swift and fair resolution of lead agency disputes (§ 1501.5).

(d) Using the scoping process for an early identification of what are and what are not the real issues (§ 1501.7).

(e) Establishing appropriate time limits for the environmental impact statement process (§§ 1501.7(b)(2) and 1501.8).

(f) Preparing environmental impact statements early in the process (§ 1502.5).

(g) Integrating NEPA requirements with other environmental review and consultation requirements (§ 1502.25).

(h) Eliminating duplication with State and local procedures by providing for joint preparation (§ 1506.2) and with other Federal procedures by providing that an agency may adopt appropriate environmental documents prepared by another agency (§ 1506.3).

(i) Combining environmental documents with other documents (§ 1506.4).

(j) Using accelerated procedures for proposals for legislation (§ 1506.8).

(k) Using categorical exclusions to define categories of actions which do not individually or cumulatively have a significant effect on the human environment (§ 1508.4) and which are therefore exempt from requirements to prepare an environmental impact statement.

(l) Using a finding of no significant impact when an action not otherwise excluded will not have a significant effect on the human environment (§ 1508.13) and is therefore exempt from requirements to prepare an environmental impact statement.

§ 1500.6 **Agency authority.**

Each agency shall interpret the provisions of the Act as a supplement to its existing authority and as a mandate to view traditional policies and missions in the light of the Act's national environmental objectives. Agencies shall review their policies, procedures, and regulations accordingly and revise them as necessary to insure full compliance with the purposes and provisions of the Act. The phrase "to the fullest extent possible" in section 102 means that each agency of the Federal Government shall comply with that section unless existing law applicable to the agency's operations expressly prohibits or makes compliance impossible.

PART 1501—NEPA AND AGENCY PLANNING

Sec.
1501.1 Purpose.
1501.2 Apply NEPA early in the process.
1501.3 When to prepare an environmental assessment.
1501.4 Whether to prepare an environmental impact statement.
1501.5 Lead agencies.
1501.6 Cooperating agencies.
1501.7 Scoping.
1501.8 Time limits.

AUTHORITY: NEPA, the Environmental Quality Improvement Act of 1970, as amended (42 U.S.C. 4371 *et seq.*), sec. 309 of the Clean Air Act, as amended (42 U.S.C. 7609, and E.O. 11514 (Mar. 5, 1970, as amended by E.O. 11991, May 24, 1977).

SOURCE: 43 FR 55992, Nov. 29, 1978, unless otherwise noted.

§ 1501.1 **Purpose.**

The purposes of this part include:

(a) Integrating the NEPA process into early planning to insure appropriate consideration of NEPA's policies and to eliminate delay.

(b) Emphasizing cooperative consultation among agencies before the environmental impact statement is prepared rather than submission of adversary comments on a completed document.

(c) Providing for the swift and fair resolution of lead agency disputes.

(d) Identifying at an early stage the significant environmental issues deserving of study and deemphasizing insignificant issues, narrowing the scope of the environmental impact statement accordingly.

(e) Providing a mechanism for putting appropriate time limits on the environmental impact statement process.

§ 1501.2 **Apply NEPA early in the process.**

Agencies shall integrate the NEPA process with other planning at the earliest possible time to insure that planning and decisions reflect environmental values, to avoid delays later in the process, and to head off potential conflicts. Each agency shall:

(a) Comply with the mandate of section 102(2)(A) to "utilize a systematic, interdisciplinary approach which will insure the integrated use of the natural and social sciences and the environmental design arts in planning and in decisionmaking which may have an impact on man's environment," as specified by § 1507.2.

(b) Identify environmental effects and values in adequate detail so they can be compared to economic and technical analyses. Environmental documents and appropriate analyses shall be circulated and reviewed at the same time as other planning documents.

(c) Study, develop, and describe appropriate alternatives to recommended courses of action in any proposal which involves unresolved conflicts concerning alternative uses of available resources as provided by section 102(2)(E) of the Act.

(d) Provide for cases where actions are planned by private applicants or other non-Federal entities before Federal involvement so that:

(1) Policies or designated staff are available to advise potential applicants of studies or other information foreseeably required for later Federal action.

(2) The Federal agency consults early with appropriate State and local agencies and Indian tribes and with interested private persons and organizations when its own involvement is reasonably foreseeable.

(3) The Federal agency commences its NEPA process at the earliest possible time.

§ 1501.3 **When to prepare an environmental assessment.**

(a) Agencies shall prepare an environmental assessment (§ 1508.9) when necessary under the procedures adopted by individual agencies to supplement these regulations as described in § 1507.3. An assessment is not necessary if the agency has decided to prepare an environmental impact statement.

(b) Agencies may prepare an environmental assessment on any action at any time in order to assist agency planning and decisionmaking.

§ 1501.4 **Whether to prepare an environmental impact statement.**

In determining whether to prepare an environmental impact statement the Federal agency shall:

(a) Determine under its procedures supplementing these regulations (described in § 1507.3) whether the proposal is one which:

(1) Normally requires an environmental impact statement, or

(2) Normally does not require either an environmental impact statement or an environmental assessment (categorical exclusion).

(b) If the proposed action is not covered by paragraph (a) of this section, prepare an environmental assessment (§ 1508.9). The agency shall involve environmental agencies, applicants, and the public, to the extent practicable, in preparing assessments required by § 1508.9(a)(1).

(c) Based on the environmental assessment make its determination whether to prepare an environmental impact statement.

(d) Commence the scoping process (§ 1501.7), if the agency will prepare an environmental impact statement.

(e) Prepare a finding of no significant impact (§ 1508.13), if the agency determines on the basis of the environmental assessment not to prepare a statement.

(1) The agency shall make the finding of no significant impact available

to the affected public as specified in § 1506.6.

(2) In certain limited circumstances, which the agency may cover in its procedures under § 1507.3, the agency shall make the finding of no significant impact available for public review (including State and areawide clearinghouses) for 30 days before the agency makes its final determination whether to prepare an environmental impact statement and before the action may begin. The circumstances are:

(i) The proposed action is, or is closely similar to, one which normally requires the preparation of an environmental impact statement under the procedures adopted by the agency pursuant to § 1507.3, or

(ii) The nature of the proposed action is one without precedent.

§ 1501.5 Lead agencies.

(a) A lead agency shall supervise the preparation of an environmental impact statement if more than one Federal agency either:

(1) Proposes or is involved in the same action; or

(2) Is involved in a group of actions directly related to each other because of their functional interdependence or geographical proximity.

(b) Federal, State, or local agencies, including at least one Federal agency, may act as joint lead agencies to prepare an environmental impact statement (§ 1506.2).

(c) If an action falls within the provisions of paragraph (a) of this section the potential lead agencies shall determine by letter or memorandum which agency shall be the lead agency and which shall be cooperating agencies. The agencies shall resolve the lead agency question so as not to cause delay. If there is disagreement among the agencies, the following factors (which are listed in order of descending importance) shall determine lead agency designation:

(1) Magnitude of agency's involvement.

(2) Project approval/disapproval authority.

(3) Expertise concerning the action's environmental effects.

(4) Duration of agency's involvement.

(5) Sequence of agency's involvement.

(d) Any Federal agency, or any State or local agency or private person substantially affected by the absence of lead agency designation, may make a written request to the potential lead agencies that a lead agency be designated.

(e) If Federal agencies are unable to agree on which agency will be the lead agency or if the procedure described in paragraph (c) of this section has not resulted within 45 days in a lead agency designation, any of the agencies or persons concerned may file a request with the Council asking it to determine which Federal agency shall be the lead agency.

A copy of the request shall be transmitted to each potential lead agency. The request shall consist of:

(1) A precise description of the nature and extent of the proposed action.

(2) A detailed statement of why each potential lead agency should or should not be the lead agency under the criteria specified in paragraph (c) of this section.

(f) A response may be filed by any potential lead agency concerned within 20 days after a request is filed with the Council. The Council shall determine as soon as possible but not later than 20 days after receiving the request and all responses to it which Federal agency shall be the lead agency and which other Federal agencies shall be cooperating agencies.

[43 FR 55992, Nov. 29, 1978; 44 FR 873, Jan. 3, 1979]

§ 1501.6 Cooperating agencies.

The purpose of this section is to emphasize agency cooperation early in the NEPA process. Upon request of the lead agency, any other Federal agency which has jurisdiction by law shall be a cooperating agency. In addition any other Federal agency which has special expertise with respect to any environmental issue, which should be addressed in the statement may be a cooperating agency upon request of the lead agency. An agency may request the lead agency to designate it a cooperating agency.

(a) The lead agency shall:

(1) Request the participation of each cooperating agency in the NEPA process at the earliest possible time.

(2) Use the environmental analysis and proposals of cooperating agencies with jurisdiction by law or special expertise, to the maximum extent possible consistent with its responsibility as lead agency.

(3) Meet with a cooperating agency at the latter's request.

(b) Each cooperating agency shall:

(1) Participate in the NEPA process at the earliest possible time.

(2) Participate in the scoping process (described below in § 1501.7).

(3) Assume on request of the lead agency responsibility for developing information and preparing environmental analyses including portions of the environmental impact statement concerning which the cooperating agency has special expertise.

(4) Make available staff support at the lead agency's request to enhance the latter's interdisciplinary capability.

(5) Normally use its own funds. The lead agency shall, to the extent available funds permit, fund those major activities or analyses it requests from cooperating agencies. Potential lead agencies shall include such funding requirements in their budget requests.

(c) A cooperating agency may in response to a lead agency's request for assistance in preparing the environmental impact statement (described in paragraph (b) (3), (4), or (5) of this section) reply that other program commitments preclude any involvement or the degree of involvement requested in the action that is the subject of the environmental impact statement. A copy of this reply shall be submitted to the Council.

§ 1501.7 Scoping.

There shall be an early and open process for determining the scope of issues to be addressed and for identifying the significant issues related to a proposed action. This process shall be termed scoping. As soon as practicable after its decision to prepare an environmental impact statement and

before the scoping process the lead agency shall publish a notice of intent (§ 1508.22) in the FEDERAL REGISTER except as provided in § 1507.3(e).

(a) As part of the scoping process the lead agency shall:

(1) Invite the participation of affected Federal, State, and local agencies, any affected Indian tribe, the proponent of the action, and other interested persons (including those who might not be in accord with the action on environmental grounds), unless there is a limited exception under § 1507.3(c). An agency may give notice in accordance with § 1506.6.

(2) Determine the scope (§ 1508.25) and the significant issues to be analyzed in depth in the environmental impact statement.

(3) Identify and eliminate from detailed study the issues which are not significant or which have been covered by prior environmental review (§ 1506.3), narrowing the discussion of these issues in the statement to a brief presentation of why they will not have a significant effect on the human environment or providing a reference to their coverage elsewhere.

(4) Allocate assignments for preparation of the environmental impact statement among the lead and cooperating agencies, with the lead agency retaining responsibility for the statement.

(5) Indicate any public environmental assessments and other environmental impact statements which are being or will be prepared that are related to but are not part of the scope of the impact statement under consideration.

(6) Identify other environmental review and consultation requirements so the lead and cooperating agencies may prepare other required analyses and studies concurrently with, and integrated with, the environmental impact statement as provided in § 1502.25.

(7) Indicate the relationship between the timing of the preparation of environmental analyses and the agency's tentative planning and decisionmaking schedule.

(b) As part of the scoping process the lead agency may:

(1) Set page limits on environmental documents (§ 1502.7).

(2) Set time limits (§ 1501.8).

(3) Adopt procedures under § 1507.3 to combine its environmental assessment process with its scoping process.

(4) Hold an early scoping meeting or meetings which may be integrated with any other early planning meeting the agency has. Such a scoping meeting will often be appropriate when the impacts of a particular action are confined to specific sites.

(c) An agency shall revise the determinations made under paragraphs (a) and (b) of this section if substantial changes are made later in the proposed action, or if significant new circumstances or information arise which bear on the proposal or its impacts.

§ 1501.8 Time limits.

Although the Council has decided that prescribed universal time limits for the entire NEPA process are too inflexible, Federal agencies are encouraged to set time limits appropriate to individual actions (consistent with the time intervals required by § 1506.10). When multiple agencies are involved the reference to agency below means lead agency.

(a) The agency shall set time limits if an applicant for the proposed action requests them: *Provided,* That the limits are consistent with the purposes of NEPA and other essential considerations of national policy.

(b) The agency may:

(1) Consider the following factors in determining time limits:

(i) Potential for environmental harm.

(ii) Size of the proposed action.

(iii) State of the art of analytic techniques.

(iv) Degree of public need for the proposed action, including the consequences of delay.

(v) Number of persons and agencies affected.

(vi) Degree to which relevant information is known and if not known the time required for obtaining it.

(vii) Degree to which the action is controversial.

(viii) Other time limits imposed on the agency by law, regulations, or executive order.

(2) Set overall time limits or limits for each constituent part of the NEPA process, which may include:

(i) Decision on whether to prepare an environmental impact statement (if not already decided).

(ii) Determination of the scope of the environmental impact statement.

(iii) Preparation of the draft environmental impact statement.

(iv) Review of any comments on the draft environmental impact statement from the public and agencies.

(v) Preparation of the final environmental impact statement.

(vi) Review of any comments on the final environmental impact statement.

(vii) Decision on the action based in part on the environmental impact statement.

(3) Designate a person (such as the project manager or a person in the agency's office with NEPA responsibilities) to expedite the NEPA process.

(c) State or local agencies or members of the public may request a Federal Agency to set time limits.

PART 1502—ENVIRONMENTAL IMPACT STATEMENT

Sec.
1502.23 Cost-benefit analysis.
1502.24 Methodology and scientific accuracy.
1502.25 Environmental review and consultation requirements.

AUTHORITY: NEPA, the Environmental
Quality Improvement Act of 1970, as
amended (42 U.S.C. 4371 *et seq.*), sec. 309 of
the Clean Air Act, as amended (42 U.S.C.
7609), and E.O. 11514 (Mar. 5, 1970, as
amended by E.O. 11991, May 24, 1977).

SOURCE: 43 FR 55994, Nov. 29, 1978, unless
otherwise noted.

§ 1502.1 Purpose.

The primary purpose of an environmental impact statement is to serve as
an action-forcing device to insure that
the policies and goals defined in the
Act are infused into the ongoing programs and actions of the Federal Government. It shall provide full and fair
discussion of significant environmental impacts and shall inform decisionmakers and the public of the reasonable alternatives which would avoid or
minimize adverse impacts or enhance
the quality of the human environment. Agencies shall focus on significant environmental issues and alternatives and shall reduce paperwork and
the accumulation of extraneous background data. Statements shall be concise, clear, and to the point, and shall
be supported by evidence that the
agency has made the necessary environmental analyses. An environmental
impact statement is more than a disclosure document. It shall be used by
Federal officials in conjunction with
other relevant material to plan actions
and make decisions.

§ 1502.2 Implementation.

To achieve the purposes set forth in
§ 1502.1 agencies shall prepare environmental impact statements in the
following manner:

(a) Environmental impact statements shall be analytic rather than
encyclopedic.

(b) Impacts shall be discussed in pro-
portion to their significance. There
shall be only brief discussion of other
than significant issues. As in a finding
of no significant impact, there should
be only enough discussion to show
why more study is not warranted.

(c) Environmental impact statements shall be kept concise and shall
be no longer than absolutely necessary
to comply with NEPA and with these
regulations. Length should vary first
with potential environmental problems and then with project size.

(d) Environmental impact statements shall state how alternatives considered in it and decisions based on it
will or will not achieve the requirements of sections 101 and 102(1) of the
Act and other environmental laws and
policies.

(e) The range of alternatives discussed in environmental impact statements shall encompass those to be
considered by the ultimate agency
decisionmaker.

(f) Agencies shall not commit resources prejudicing selection of alternatives before making a final decision
(§ 1506.1).

(g) Environmental impact statements shall serve as the means of assessing the environmental impact of
proposed agency actions, rather than
justifying decisions already made.

§ 1502.3 Statutory requirements for state-
ments.

As required by sec. 102(2)(C) of
NEPA environmental impact statements (§ 1508.11) are to be included in
every recommendation or report.

On proposals (§ 1508.23).

For legislation and (§ 1508.17).

Other major Federal actions
(§ 1508.18).

Significantly (§ 1508.27).

Affecting (§§ 1508.3, 1508.8).

The quality of the human environment (§ 1508.14).

§ 1502.4 Major Federal actions requiring
the preparation of environmental
impact statements.

(a) Agencies shall make sure the proposal which is the subject of an environmental impact statement is properly defined. Agencies shall use the criteria for scope (§ 1508.25) to determine
which proposal(s) shall be the subject
of a particular statement. Proposals or
parts of proposals which are related to
each other closely enough to be, in
effect, a single course of action shall

be evaluated in a single impact statement.

(b) Environmental impact statements may be prepared, and are sometimes required, for broad Federal actions such as the adoption of new agency programs or regulations (§ 1508.18). Agencies shall prepare statements on broad actions so that they are relevant to policy and are timed to coincide with meaningful points in agency planning and decisionmaking.

(c) When preparing statements on broad actions (including proposals by more than one agency), agencies may find it useful to evaluate the proposal(s) in one of the following ways:

(1) Geographically, including actions occurring in the same general location, such as body of water, region, or metropolitan area.

(2) Generically, including actions which have relevant similarities, such as common timing, impacts, alternatives, methods of implementation, media, or subject matter.

(3) By stage of technological development including federal or federally assisted research, development or demonstration programs for new technologies which, if applied, could significantly affect the quality of the human environment. Statements shall be prepared on such programs and shall be available before the program has reached a stage of investment or commitment to implementation likely to determine subsequent development or restrict later alternatives.

(d) Agencies shall as appropriate employ scoping (§ 1501.7), tiering (§ 1502.20), and other methods listed in §§ 1500.4 and 1500.5 to relate broad and narrow actions and to avoid duplication and delay.

§ 1502.5 Timing.

An agency shall commence preparation of an environmental impact statement as close as possible to the time the agency is developing or is presented with a proposal (§ 1508.23) so that preparation can be completed in time for the final statement to be included in any recommendation or report on the proposal. The statement shall be prepared early enough so that it can serve practically as an important contribution to the decisionmaking process and will not be used to rationalize or justify decisions already made (§§ 1500.2(c), 1501.2, and 1502.2). For instance:

(a) For projects directly undertaken by Federal agencies the environmental impact statement shall be prepared at the feasibility analysis (go-no go) stage and may be supplemented at a later stage if necessary.

(b) For applications to the agency appropriate environmental assessments or statements shall be commenced no later than immediately after the application is received. Federal agencies are encouraged to begin preparation of such assessments or statements earlier, preferably jointly with applicable State or local agencies.

(c) For adjudication, the final environmental impact statement shall normally precede the final staff recommendation and that portion of the public hearing related to the impact study. In appropriate circumstances the statement may follow preliminary hearings designed to gather information for use in the statements.

(d) For informal rulemaking the draft environmental impact statement shall normally accompany the proposed rule.

§ 1502.6 Interdisciplinary preparation.

Environmental impact statements shall be prepared using an inter-disciplinary approach which will insure the integrated use of the natural and social sciences and the environmental design arts (section 102(2)(A) of the Act). The disciplines of the preparers shall be appropriate to the scope and issues identified in the scoping process (§ 1501.7).

§ 1502.7 Page limits.

The text of final environmental impact statements (e.g., paragraphs (d) through (g) of § 1502.10) shall normally be less than 150 pages and for proposals of unusual scope or complexity shall normally be less than 300 pages.

§ 1502.8 **Writing.**

Environmental impact statements shall be written in plain language and may use appropriate graphics so that decisionmakers and the public can readily understand them. Agencies should employ writers of clear prose or editors to write, review, or edit statements, which will be based upon the analysis and supporting data from the natural and social sciences and the environmental design arts.

§ 1502.9 **Draft, final, and supplemental statements.**

Except for proposals for legislation as provided in § 1506.8 environmental impact statements shall be prepared in two stages and may be supplemented.

(a) Draft environmental impact statements shall be prepared in accordance with the scope decided upon in the scoping process. The lead agency shall work with the cooperating agencies and shall obtain comments as required in Part 1503 of this chapter. The draft statement must fulfill and satisfy to the fullest extent possible the requirements established for final statements in section 102(2)(C) of the Act. If a draft statement is so inadequate as to preclude meaningful analysis, the agency shall prepare and circulate a revised draft of the appropriate portion. The agency shall make every effort to disclose and discuss at appropriate points in the draft statement all major points of view on the environmental impacts of the alternatives including the proposed action.

(b) Final environmental impact statements shall respond to comments as required in Part 1503 of this chapter. The agency shall discuss at appropriate points in the final statement any responsible opposing view which was not adequately discussed in the draft statement and shall indicate the agency's response to the issues raised.

(c) Agencies:

(1) Shall prepare supplements to either draft or final environmental impact statements if:

(i) The agency makes substantial changes in the proposed action that are relevant to environmental concerns; or

(ii) There are significant new circumstances or information relevant to environmental concerns and bearing on the proposed action or its impacts.

(2) May also prepare supplements when the agency determines that the purposes of the Act will be furthered by doing so.

(3) Shall adopt procedures for introducing a supplement into its formal administrative record, if such a record exists.

(4) Shall prepare, circulate, and file a supplement to a statement in the same fashion (exclusive of scoping) as a draft and final statement unless alternative procedures are approved by the Council.

§ 1502.10 **Recommended format.**

Agencies shall use a format for environmental impact statements which will encourage good analysis and clear presentation of the alternatives including the proposed action. The following standard format for environmental impact statements should be followed unless the agency determines that there is a compelling reason to do otherwise:

(a) Cover sheet.

(b) Summary.

(c) Table of contents.

(d) Purpose of and need for action.

(e) Alternatives including proposed action (sections 102(2)(C)(iii) and 102(2)(E) of the Act).

(f) Affected environment.

(g) Environmental consequences (especially sections 102(2)(C) (i), (ii), (iv), and (v) of the Act).

(h) List of preparers.

(i) List of Agencies, Organizations, and persons to whom copies of the statement are sent.

(j) Index.

(k) Appendices (if any).

If a different format is used, it shall include paragraphs (a), (b), (c), (h), (i), and (j), of this section and shall include the substance of paragraphs (d), (e), (f), (g), and (k) of this section, as further described in §§ 1502.11 through 1502.18, in any appropriate format.

§ 1502.11 Cover sheet.

The cover sheet shall not exceed one page. It shall include:

(a) A list of the responsible agencies including the lead agency and any co-operating agencies.

(b) The title of the proposed action that is the subject of the statement (and if appropriate the titles of related cooperating agency actions), together with the State(s) and county(ies) (or other jurisdiction if applicable) where the action is located.

(c) The name, address, and telephone number of the person at the agency who can supply further information.

(d) A designation of the statement as a draft, final, or draft or final supplement.

(e) A one paragraph abstract of the statement.

(f) The date by which comments must be received (computed in cooperation with EPA under § 1506.10).

The information required by this section may be entered on Standard Form 424 (in items 4, 6, 7, 10, and 18).

§ 1502.12 Summary.

Each environmental impact statement shall contain a summary which adequately and accurately summarizes the statement. The summary shall stress the major conclusions, areas of controversy (including issues raised by agencies and the public), and the issues to be resolved (including the choice among alternatives). The summary will normally not exceed 15 pages.

§ 1502.13 Purpose and need.

The statement shall briefly specify the underlying purpose and need to which the agency is responding in proposing the alternatives including the proposed action.

§ 1502.14 Alternatives including the proposed action.

This section is the heart of the environmental impact statement. Based on the information and analysis presented in the sections on the Affected Environment (§ 1502.15) and the Environmental Consequences (§ 1502.16), it should present the environmental impacts of the proposal and the alternatives in comparative form, thus sharply defining the issues and providing a clear basis for choice among options by the decisionmaker and the public. In this section agencies shall:

(a) Rigorously explore and objectively evaluate all reasonable alternatives, and for alternatives which were eliminated from detailed study, briefly discuss the reasons for their having been eliminated.

(b) Devote substantial treatment to each alternative considered in detail including the proposed action so that reviewers may evaluate their comparative merits.

(c) Include reasonable alternatives not within the jurisdiction of the lead agency.

(d) Include the alternative of no action.

(e) Identify the agency's preferred alternative or alternatives, if one or more exists, in the draft statement and identify such alternative in the final statement unless another law prohibits the expression of such a preference.

(f) Include appropriate mitigation measures not already included in the proposed action or alternatives.

§ 1502.15 Affected environment.

The environmental impact statement shall succinctly describe the environment of the area(s) to be affected or created by the alternatives under consideration. The descriptions shall be no longer than is necessary to understand the effects of the alternatives. Data and analyses in a statement shall be commensurate with the importance of the impact, with less important material summarized, consolidated, or simply referenced. Agencies shall avoid useless bulk in statements and shall concentrate effort and attention on important issues. Verbose descriptions of the affected environment are themselves no measure of the adequacy of an environmental impact statement.

§ 1502.16 Environmental consequences.

This section forms the scientific and analytic basis for the comparisons under § 1502.14. It shall consolidate

13

the discussions of those elements required by sections 102(2)(C) (i), (ii), (iv), and (v) of NEPA which are within the scope of the statement and as much of section 102(2)(C)(iii) as is necessary to support the comparisons. The discussion will include the environmental impacts of the alternatives including the proposed action, any adverse environmental effects which cannot be avoided should the proposal be implemented, the relationship between short-term uses of man's environment and the maintenance and enhancement of long-term productivity, and any irreversible or irretrievable commitments of resources which would be involved in the proposal should it be implemented. This section should not duplicate discussions in § 1502.14. It shall include discussions of:

(a) Direct effects and their significance (§ 1508.8).

(b) Indirect effects and their significance (§ 1508.8).

(c) Possible conflicts between the proposed action and the objectives of Federal, regional, State, and local (and in the case of a reservation, Indian tribe) land use plans, policies and controls for the area concerned. (See § 1506.2(d).)

(d) The environmental effects of alternatives including the proposed action. The comparisons under § 1502.14 will be based on this discussion.

(e) Energy requirements and conservation potential of various alternatives and mitigation measures.

(f) Natural or depletable resource requirements and conservation potential of various alternatives and mitigation measures.

(g) Urban quality, historic and cultural resources, and the design of the built environment, including the reuse and conservation potential of various alternatives and mitigation measures.

(h) Means to mitigate adverse environmental impacts (if not fully covered under § 1502.14(f)).

[43 FR 55994, Nov. 29, 1978; 44 FR 873, Jan. 3, 1979]

§ 1502.17 List of preparers.

The environmental impact statement shall list the names, together with their qualifications (expertise, experience, professional disciplines), of the persons who were primarily responsible for preparing the environmental impact statement or significant background papers, including basic components of the statement (§§ 1502.6 and 1502.8). Where possible the persons who are responsible for a particular analysis, including analyses in background papers, shall be identified. Normally the list will not exceed two pages.

§ 1502.18 Appendix.

If an agency prepares an appendix to an environmental impact statement the appendix shall:

(a) Consist of material prepared in connection with an environmental impact statement (as distinct from material which is not so prepared and which is incorporated by reference (§ 1502.21)).

(b) Normally consist of material which substantiates any analysis fundamental to the impact statement.

(c) Normally be analytic and relevant to the decision to be made.

(d) Be circulated with the environmental impact statement or be readily available on request.

§ 1502.19 Circulation of the environmental impact statement.

Agencies shall circulate the entire draft and final environmental impact statements except for certain appendices as provided in § 1502.18(d) and unchanged statements as provided in § 1503.4(c). However, if the statement is unusually long, the agency may circulate the summary instead, except that the entire statement shall be furnished to:

(a) Any Federal agency which has jurisdiction by law or special expertise with respect to any environmental impact involved and any appropriate Federal, State or local agency authorized to develop and enforce environmental standards.

(b) The applicant, if any.

(c) Any person, organization, or agency requesting the entire environmental impact statement.

(d) In the case of a final environmental impact statement any person, organization, or agency which submitted substantive comments on the draft.

If the agency circulates the summary and thereafter receives a timely request for the entire statement and for additional time to comment, the time for that requestor only shall be extended by at least 15 days beyond the minimum period.

§ 1502.20 Tiering.

Agencies are encouraged to tier their environmental impact statements to eliminate repetitive discussions of the same issues and to focus on the actual issues ripe for decision at each level of environmental review (§ 1508.28). Whenever a broad environmental impact statement has been prepared (such as a program or policy statement) and a subsequent statement or environmental assessment is then prepared on an action included within the entire program or policy (such as a site specific action) the subsequent statement or environmental assessment need only summarize the issues discussed in the broader statement and incorporate discussions from the broader statement by reference and shall concentrate on the issues specific to the subsequent action. The subsequent document shall state where the earlier document is available. Tiering may also be appropriate for different stages of actions. (Section 1508.28).

§ 1502.21 Incorporation by reference.

Agencies shall incorporate material into an environmental impact statement by reference when the effect will be to cut down on bulk without impeding agency and public review of the action. The incorporated material shall be cited in the statement and its content briefly described. No material may be incorporated by reference unless it is reasonably available for inspection by potentially interested persons within the time allowed for comment. Material based on proprietary data which is itself not available for review and comment shall not be incorporated by reference.

§ 1502.22 Incomplete or unavailable information.

When an agency is evaluating reasonably foreseeable significant adverse effects on the human environment in an environmental impact statement and there is incomplete or unavailable information, the agency shall always make clear that such information is lacking.

(a) If the incomplete information relevant to reasonably foreseeable significant adverse impacts is essential to a reasoned choice among alternatives and the overall costs of obtaining it are not exorbitant, the agency shall include the information in the environmental impact statement.

(b) If the information relevant to reasonably foreseeable significant adverse impacts cannot be obtained because the overall costs of obtaining it are exorbitant or the means to obtain it are not known, the agency shall include within the environmental impact statement: (1) A statement that such information is incomplete or unavailable; (2) a statement of the relevance of the incomplete or unavailable information to evaluating reasonably foreseeable significant adverse impacts on the human environment; (3) a summary of existing credible scientific evidence which is relevant to evaluating the reasonably foreseeable significant adverse impacts on the human environment, and (4) the agency's evaluation of such impacts based upon theoretical approaches or research methods generally accepted in the scientific community. For the purposes of this section, "reasonably foreseeable" includes impacts which have catastrophic consequences, even if their probability of occurrence is low, provided that the analysis of the impacts is supported by credible scientific evidence, is not based on pure conjecture, and is within the rule of reason.

(c) The amended regulation will be applicable to all environmental impact statements for which a Notice of Intent (40 CFR 1508.22) is published in the FEDERAL REGISTER on or after May 27, 1986. For environmental impact statements in progress, agencies may choose to comply with the requirements of either the original or amended regulation.

[51 FR 15625, Apr. 25, 1986]

§ 1502.23 Cost-benefit analysis.

If a cost-benefit analysis relevant to the choice among environmentally different alternatives is being considered for the proposed action, it shall be incorporated by reference or appended to the statement as an aid in evaluating the environmental consequences. To assess the adequacy of compliance with section 102(2)(B) of the Act the statement shall, when a cost-benefit analysis is prepared, discuss the relationship between that analysis and any analyses of unquantified environmental impacts, values, and amenities. For purposes of complying with the Act, the weighing of the merits and drawbacks of the various alternatives need not be displayed in a monetary cost-benefit analysis and should not be when there are important qualitative considerations. In any event, an environmental impact statement should at least indicate those considerations, including factors not related to environmental quality, which are likely to be relevant and important to a decision.

§ 1502.24 Methodology and scientific accuracy.

Agencies shall insure the professional integrity, including scientific integrity, of the discussions and analyses in environmental impact statements. They shall identify any methodologies used and shall make explicit reference by footnote to the scientific and other sources relied upon for conclusions in the statement. An agency may place discussion of methodology in an appendix.

§ 1502.25 Environmental review and consultation requirements.

(a) To the fullest extent possible, agencies shall prepare draft environmental impact statements concurrently with and integrated with environmental impact analyses and related surveys and studies required by the Fish and Wildlife Coordination Act (16 U.S.C. 661 et seq.), the National Historic Preservation Act of 1966 (16 U.S.C. 470 et seq.), the Endangered Species Act of 1973 (16 U.S.C. 1531 et seq.), and other environmental review laws and executive orders.

(b) The draft environmental impact statement shall list all Federal permits, licenses, and other entitlements which must be obtained in implementing the proposal. If it is uncertain whether a Federal permit, license, or other entitlement is necessary, the draft environmental impact statement shall so indicate.

PART 1503—COMMENTING

Sec.
1503.1 Inviting comments.
1503.2 Duty to comment.
1503.3 Specificity of comments.
1503.4 Response to comments.

AUTHORITY: NEPA, the Environmental Quality Improvement Act of 1970, as amended (42 U.S.C. 4371 *et seq.*), sec. 309 of the Clean Air Act, as amended (42 U.S.C. 7609), and E.O. 11514 (Mar. 5, 1970, as amended by E.O. 11991, May 24, 1977).

SOURCE: 43 FR 55997, Nov. 29, 1978, unless otherwise noted.

§ 1503.1 Inviting comments.

(a) After preparing a draft environmental impact statement and before preparing a final environmental impact statement the agency shall:

(1) Obtain the comments of any Federal agency which has jurisdiction by law or special expertise with respect to any environmental impact involved or which is authorized to develop and enforce environmental standards.

(2) Request the comments of:

(i) Appropriate State and local agencies which are authorized to develop and enforce environmental standards;

(ii) Indian tribes, when the effects may be on a reservation; and

(iii) Any agency which has requested that it receive statements on actions of the kind proposed.

Office of Management and Budget Circular A-95 (Revised), through its system of clearinghouses, provides a means of securing the views of State and local environmental agencies. The clearinghouses may be used, by mutual agreement of the lead agency and the clearinghouse, for securing

State and local reviews of the draft environmental impact statements.

(3) Request comments from the applicant, if any.

(4) Request comments from the public, affirmatively soliciting comments from those persons or organizations who may be interested or affected.

(b) An agency may request comments on a final environmental impact statement before the decision is finally made. In any case other agencies or persons may make comments before the final decision unless a different time is provided under § 1506.10.

§ 1503.2 Duty to comment.

Federal agencies with jurisdiction by law or special expertise with respect to any environmental impact involved and agencies which are authorized to develop and enforce environmental standards shall comment on statements within their jurisdiction, expertise, or authority. Agencies shall comment within the time period specified for comment in § 1506.10. A Federal agency may reply that it has no comment. If a cooperating agency is satisfied that its views are adequately reflected in the environmental impact statement, it should reply that it has no comment.

§ 1503.3 Specificity of comments.

(a) Comments on an environmental impact statement or on a proposed action shall be as specific as possible and may address either the adequacy of the statement or the merits of the alternatives discussed or both.

(b) When a commenting agency criticizes a lead agency's predictive methodology, the commenting agency should describe the alternative methodology which it prefers and why.

(c) A cooperating agency shall specify in its comments whether it needs additional information to fulfill other applicable environmental reviews or consultation requirements and what information it needs. In particular, it shall specify any additional information it needs to comment adequately on the draft statement's analysis of significant site-specific effects associated with the granting or approving by that cooperating agency of neces-

sary Federal permits, licenses, or entitlements.

(d) When a cooperating agency with jurisdiction by law objects to or expresses reservations about the proposal on grounds of environmental impacts, the agency expressing the objection or reservation shall specify the mitigation measures it considers necessary to allow the agency to grant or approve applicable permit, license, or related requirements or concurrences.

§ 1503.4 Response to comments.

(a) An agency preparing a final environmental impact statement shall assess and consider comments both individually and collectively, and shall respond by one or more of the means listed below, stating its response in the final statement. Possible responses are to:

(1) Modify alternatives including the proposed action.

(2) Develop and evaluate alternatives not previously given serious consideration by the agency.

(3) Supplement, improve, or modify its analyses.

(4) Make factual corrections.

(5) Explain why the comments do not warrant further agency response, citing the sources, authorities, or reasons which support the agency's position and, if appropriate, indicate those circumstances which would trigger agency reappraisal or further response.

(b) All substantive comments received on the draft statement (or summaries thereof where the response has been exceptionally voluminous), should be attached to the final statement whether or not the comment is thought to merit individual discussion by the agency in the text of the statement.

(c) If changes in response to comments are minor and are confined to the responses described in paragraphs (a) (4) and (5) of this section, agencies may write them on errata sheets and attach them to the statement instead of rewriting the draft statement. In such cases only the comments, the responses, and the changes and not the final statement need be circulated (§ 1502.19). The entire document with a new cover sheet shall be filed as the final statement (§ 1506.9).

PART 1504—PREDECISION REFERRALS TO THE COUNCIL OF PROPOSED FEDERAL ACTIONS DETERMINED TO BE ENVIRONMENTALLY UNSATISFACTORY

Sec.
1504.1 Purpose.
1504.2 Criteria for referral.
1504.3 Procedure for referrals and response.

AUTHORITY: NEPA, the Environmental Quality Improvement Act of 1970, as amended (42 U.S.C. 4371 *et seq.*), sec. 309 of the Clean Air Act, as amended (42 U.S.C. 7609), and E.O. 11514 (Mar. 5, 1970, as amended by E.O. 11991, May 24, 1977).

SOURCE: 43 FR 55998, Nov. 29, 1978, unless otherwise noted.

§ 1504.1 Purpose.

(a) This part establishes procedures for referring to the Council Federal interagency disagreements concerning proposed major Federal actions that might cause unsatisfactory environmental effects. It provides means for early resolution of such disagreements.

(b) Under section 309 of the Clean Air Act (42 U.S.C. 7609), the Administrator of the Environmental Protection Agency is directed to review and comment publicly on the environmental impacts of Federal activities, including actions for which environmental impact statements are prepared. If after this review the Administrator determines that the matter is "unsatisfactory from the standpoint of public health or welfare or environmental quality," section 309 directs that the matter be referred to the Council (hereafter "environmental referrals").

(c) Under section 102(2)(C) of the Act other Federal agencies may make similar reviews of environmental impact statements, including judgments on the acceptability of anticipated environmental impacts. These reviews must be made available to the President, the Council and the public.

§ 1504.2 Criteria for referral.

Environmental referrals should be made to the Council only after concerted, timely (as early as possible in the process), but unsuccessful attempts to resolve differences with the lead agency. In determining what environmental objections to the matter are appropriate to refer to the Council, an agency should weigh potential adverse environmental impacts, considering:

(a) Possible violation of national environmental standards or policies.

(b) Severity.

(c) Geographical scope.

(d) Duration.

(e) Importance as precedents.

(f) Availability of environmentally preferable alternatives.

§ 1504.3 Procedure for referrals and response.

(a) A Federal agency making the referral to the Council shall:

(1) Advise the lead agency at the earliest possible time that it intends to refer a matter to the Council unless a satisfactory agreement is reached.

(2) Include such advice in the referring agency's comments on the draft environmental impact statement, except when the statement does not contain adequate information to permit an assessment of the matter's environmental acceptability.

(3) Identify any essential information that is lacking and request that it be made available at the earliest possible time.

(4) Send copies of such advice to the Council.

(b) The referring agency shall deliver its referral to the Council not later than twenty-five (25) days after the final environmental impact statement has been made available to the Environmental Protection Agency, commenting agencies, and the public. Except when an extension of this period has been granted by the lead agency, the Council will not accept a referral after that date.

(c) The referral shall consist of:

(1) A copy of the letter signed by the head of the referring agency and delivered to the lead agency informing the lead agency of the referral and the reasons for it, and requesting that no action be taken to implement the matter until the Council acts upon the

referral. The letter shall include a copy of the statement referred to in (c)(2) of this section.

(2) A statement supported by factual evidence leading to the conclusion that the matter is unsatisfactory from the standpoint of public health or welfare or environmental quality. The statement shall:

(i) Identify any material facts in controversy and incorporate (by reference if appropriate) agreed upon facts,

(ii) Identify any existing environmental requirements or policies which would be violated by the matter,

(iii) Present the reasons why the referring agency believes the matter is environmentally unsatisfactory,

(iv) Contain a finding by the agency whether the issue raised is of national importance because of the threat to national environmental resources or policies or for some other reason,

(v) Review the steps taken by the referring agency to bring its concerns to the attention of the lead agency at the earliest possible time, and

(vi) Give the referring agency's recommendations as to what mitigation alternative, further study, or other course of action (including abandonment of the matter) are necessary to remedy the situation.

(d) Not later than twenty-five (25) days after the referral to the Council the lead agency may deliver a response to the Council, and the referring agency. If the lead agency requests more time and gives assurance that the matter will not go forward in the interim, the Council may grant an extension. The response shall:

(1) Address fully the issues raised in the referral.

(2) Be supported by evidence.

(3) Give the lead agency's response to the referring agency's recommendations.

(e) Interested persons (including the applicant) may deliver their views in writing to the Council. Views in support of the referral should be delivered not later than the referral. Views in support of the response shall be delivered not later than the response.

(f) Not later than twenty-five (25) days after receipt of both the referral and any response or upon being informed that there will be no response

(unless the lead agency agrees to a longer time), the Council may take one or more of the following actions:

(1) Conclude that the process of referral and response has successfully resolved the problem.

(2) Initiate discussions with the agencies with the objective of mediation with referring and lead agencies.

(3) Hold public meetings or hearings to obtain additional views and information.

(4) Determine that the issue is not one of national importance and request the referring and lead agencies to pursue their decision process.

(5) Determine that the issue should be further negotiated by the referring and lead agencies and is not appropriate for Council consideration until one or more heads of agencies report to the Council that the agencies' disagreements are irreconcilable.

(6) Publish its findings and recommendations (including where appropriate a finding that the submitted evidence does not support the position of an agency).

(7) When appropriate, submit the referral and the response together with the Council's recommendation to the President for action.

(g) The Council shall take no longer than 60 days to complete the actions specified in paragraph (f) (2), (3), or (5) of this section.

(h) When the referral involves an action required by statute to be determined on the record after opportunity for agency hearing, the referral shall be conducted in a manner consistent with 5 U.S.C. 557(d) (Administrative Procedure Act).

[43 FR 55998, Nov. 29, 1978; 44 FR 873, Jan. 3, 1979]

PART 1505—NEPA AND AGENCY DECISIONMAKING

AUTHORITY: NEPA, the Environmental Quality Improvement Act of 1970, as amended (42 U.S.C. 4371 et seq.), sec. 309 of the Clean Air Act, as amended (42 U.S.C.

7609), and E.O. 11514 (Mar. 5, 1970, as amended by E.O. 11991, May 24, 1977).

SOURCE: 43 FR 55999, Nov. 29, 1978, unless otherwise noted.

§ 1505.1 Agency decisionmaking procedures.

Agencies shall adopt procedures (§ 1507.3) to ensure that decisions are made in accordance with the policies and purposes of the Act. Such procedures shall include but not be limited to:

(a) Implementing procedures under section 102(2) to achieve the requirements of sections 101 and 102(1).

(b) Designating the major decision points for the agency's principal programs likely to have a significant effect on the human environment and assuring that the NEPA process corresponds with them.

(c) Requiring that relevant environmental documents, comments, and responses be part of the record in formal rulemaking or adjudicatory proceedings.

(d) Requiring that relevant environmental documents, comments, and responses accompany the proposal through existing agency review processes so that agency officials use the statement in making decisions.

(e) Requiring that the alternatives considered by the decisionmaker are encompassed by the range of alternatives discussed in the relevant environmental documents and that the decisionmaker consider the alternatives described in the environmental impact statement. If another decision document accompanies the relevant environmental documents to the decisionmaker, agencies are encouraged to make available to the public before the decision is made any part of that document that relates to the comparison of alternatives.

§ 1505.2 Record of decision in cases requiring environmental impact statements.

At the time of its decision (§ 1506.10) or, if appropriate, its recommendation to Congress, each agency shall prepare a concise public record of decision. The record, which may be integrated into any other record prepared by the agency, including that required by

OMB Circular A-95 (Revised), part I, sections 6 (c) and (d), and part II, section 5(b)(4), shall:

(a) State what the decision was.

(b) Identify all alternatives considered by the agency in reaching its decision, specifying the alternative or alternatives which were considered to be environmentally preferable. An agency may discuss preferences among alternatives based on relevant factors including economic and technical considerations and agency statutory missions. An agency shall identify and discuss all such factors including any essential considerations of national policy which were balanced by the agency in making its decision and state how those considerations entered into its decision.

(c) State whether all practicable means to avoid or minimize environmental harm from the alternative selected have been adopted, and if not, why they were not. A monitoring and enforcement program shall be adopted and summarized where applicable for any mitigation.

§ 1505.3 Implementing the decision.

Agencies may provide for monitoring to assure that their decisions are carried out and should do so in important cases. Mitigation (§ 1505.2(c)) and other conditions established in the environmental impact statement or during its review and committed as part of the decision shall be implemented by the lead agency or other appropriate consenting agency. The lead agency shall:

(a) Include appropriate conditions in grants, permits or other approvals.

(b) Condition funding of actions on mitigation.

(c) Upon request, inform cooperating or commenting agencies on progress in carrying out mitigation measures which they have proposed and which were adopted by the agency making the decision.

(d) Upon request, make available to the public the results of relevant monitoring.

20

PART 1506—OTHER REQUIREMENTS OF NEPA

Sec.

AUTHORITY: NEPA, the Environmental Quality Improvement Act of 1970, as amended (42 U.S.C. 4371 *et seq.*), sec. 309 of the Clean Air Act, as amended (42 U.S.C. 7609), and E.O. 11514 (Mar. 5, 1970, as amended by E.O. 11991, May 24, 1977).

SOURCE: 43 FR 56000, Nov. 29, 1978, unless otherwise noted.

§ 1506.1 Limitations on actions during NEPA process.

(a) Until an agency issues a record of decision as provided in § 1505.2 (except as provided in paragraph (c) of this section), no action concerning the proposal shall be taken which would:

(1) Have an adverse environmental impact; or

(2) Limit the choice of reasonable alternatives.

(b) If any agency is considering an application from a non-Federal entity, and is aware that the applicant is about to take an action within the agency's jurisdiction that would meet either of the criteria in paragraph (a) of this section, then the agency shall promptly notify the applicant that the agency will take appropriate action to insure that the objectives and procedures of NEPA are achieved.

(c) While work on a required program environmental impact statement is in progress and the action is not covered by an existing program statement, agencies shall not undertake in the interim any major Federal action covered by the program which may significantly affect the quality of the human environment unless such action:

(1) Is justified independently of the program;

(2) Is itself accompanied by an adequate environmental impact statement; and

(3) Will not prejudice the ultimate decision on the program. Interim action prejudices the ultimate decision on the program when it tends to determine subsequent development or limit alternatives.

(d) This section does not preclude development by applicants of plans or designs or performance of other work necessary to support an application for Federal, State or local permits or assistance. Nothing in this section shall preclude Rural Electrification Administration approval of minimal expenditures not affecting the environment (*e.g.* long leadtime equipment and purchase options) made by non-governmental entities seeking loan guarantees from the Administration.

§ 1506.2 Elimination of duplication with State and local procedures.

(a) Agencies authorized by law to cooperate with State agencies of statewide jurisdiction pursuant to section 102(2)(D) of the Act may do so.

(b) Agencies shall cooperate with State and local agencies to the fullest extent possible to reduce duplication between NEPA and State and local requirements, unless the agencies are specifically barred from doing so by some other law. Except for cases covered by paragraph (a) of this section, such cooperation shall to the fullest extent possible include:

(1) Joint planning processes.

(2) Joint environmental research and studies.

(3) Joint public hearings (except where otherwise provided by statute).

(4) Joint environmental assessments.

(c) Agencies shall cooperate with State and local agencies to the fullest extent possible to reduce duplication between NEPA and comparable State and local requirements, unless the agencies are specifically barred from doing so by some other law. Except for cases covered by paragraph (a) of this section, such cooperation shall to the fullest extent possible include joint environmental impact statements. In such cases one or more Federal agencies and one or more State or local agencies shall be joint lead agencies. Where State laws or local ordinances have environmental impact statement requirements in addition to but not in conflict with those in NEPA, Federal agencies shall cooperate in fulfilling these requirements as well as those of Federal laws so that one document will comply with all applicable laws.

(d) To better integrate environmental impact statements into State or local planning processes, statements shall discuss any inconsistency of a proposed action with any approved State or local plan and laws (whether or not federally sanctioned). Where an inconsistency exists, the statement should describe the extent to which the agency would reconcile its proposed action with the plan or law,

§ 1506.3 Adoption.

(a) An agency may adopt a Federal draft or final environmental impact statement or portion thereof provided that the statement or portion thereof meets the standards for an adequate statement under these regulations.

(b) If the actions covered by the original environmental impact statement and the proposed action are substantially the same, the agency adopting another agency's statement is not required to recirculate it except as a final statement. Otherwise the adopting agency shall treat the statement as a draft and recirculate it (except as provided in paragraph (c) of this section).

(c) A cooperating agency may adopt without recirculating the environmental impact statement of a lead agency when, after an independent review of the statement, the cooperating agency concludes that its comments and suggestions have been satisfied.

(d) When an agency adopts a statement which is not final within the agency that prepared it, or when the action it assesses is the subject of a referral under Part 1504, or when the statement's adequacy is the subject of a judicial action which is not final, the agency shall so specify.

§ 1506.4 Combining documents.

Any environmental document in compliance with NEPA may be combined with any other agency document to reduce duplication and paperwork.

§ 1506.5 Agency responsibility.

(a) *Information.* If an agency requires an applicant to submit environmental information for possible use by the agency in preparing an environmental impact statement, then the agency should assist the applicant by outlining the types of information required. The agency shall independently evaluate the information submitted and shall be responsible for its accuracy. If the agency chooses to use the information submitted by the applicant in the environmental impact statement, either directly or by reference, then the names of the persons responsible for the independent evaluation shall be included in the list of preparers (§ 1502.17). It is the intent of this paragraph that acceptable work not be redone, but that it be verified by the agency.

(b) *Environmental assessments.* If an agency permits an applicant to prepare an environmental assessment, the agency, besides fulfilling the requirements of paragraph (a) of this section, shall make its own evaluation of the environmental issues and take responsibility for the scope and content of the environmental assessment.

(c) *Environmental impact statements.* Except as provided in §§ 1506.2 and 1506.3 any environmental impact statement prepared pursuant to the requirements of NEPA shall be prepared directly by or by a contractor selected by the lead agency or where appropriate under § 1501.6(b), a cooperating agency. It is the intent of these regulations that the contractor be chosen solely by the lead agency, or by the lead agency in cooperation with cooperating agencies, or where appropriate by a cooperating agency to avoid any conflict of interest. Contractors shall execute a disclosure statement prepared by the lead agency, or where appropriate the cooperating agency, specifying that they have no financial or other interest in the outcome of the project. If the document is prepared by contract, the responsible Federal official shall furnish guidance and participate in the preparation and shall independently evaluate the statement prior to its approval and take responsibility for its scope and contents. Nothing in this section is intended to prohibit any agency from requesting any person to submit information to it or to prohibit any person from submitting information to any agency.

§ 1506.6 Public involvement.

Agencies shall:

(a) Make diligent efforts to involve the public in preparing and implementing their NEPA procedures.

(b) Provide public notice of NEPA-related hearings, public meetings, and the availability of environmental documents so as to inform those persons and agencies who may be interested or affected.

(1) In all cases the agency shall mail notice to those who have requested it on an individual action.

(2) In the case of an action with effects of national concern notice shall include publication in the FEDERAL REGISTER and notice by mail to national organizations reasonably expected to be interested in the matter and may include listing in the *102 Monitor.* An agency engaged in rulemaking may provide notice by mail to national organizations who have requested that notice regularly be provided. Agencies shall maintain a list of such organizations.

(3) In the case of an action with effects primarily of local concern the notice may include:

(i) Notice to State and areawide clearinghouses pursuant to OMB Circular A-95 (Revised).

(ii) Notice to Indian tribes when effects may occur on reservations.

(iii) Following the affected State's public notice procedures for comparable actions.

(iv) Publication in local newspapers (in papers of general circulation rather than legal papers).

(v) Notice through other local media.

(vi) Notice to potentially interested community organizations including small business associations.

(vii) Publication in newsletters that may be expected to reach potentially interested persons.

(viii) Direct mailing to owners and occupants of nearby or affected property.

(ix) Posting of notice on and off site in the area where the action is to be located.

(c) Hold or sponsor public hearings or public meetings whenever appropriate or in accordance with statutory requirements applicable to the agency. Criteria shall include whether there is:

(1) Substantial environmental controversy concerning the proposed action or substantial interest in holding the hearing.

(2) A request for a hearing by another agency with jurisdiction over the action supported by reasons why a hearing will be helpful. If a draft environmental impact statement is to be considered at a public hearing, the agency should make the statement available to the public at least 15 days in advance (unless the purpose of the hearing is to provide information for the draft environmental impact statement).

(d) Solicit appropriate information from the public.

(e) Explain in its procedures where interested persons can get information or status reports on environmental impact statements and other elements of the NEPA process.

(f) Make environmental impact statements, the comments received, and any underlying documents available to the public pursuant to the provisions of the Freedom of Information Act (5 U.S.C. 552), without regard to the exclusion for interagency memoranda where such memoranda transmit comments of Federal agencies on the environmental impact of the proposed action. Materials to be made available to the public shall be provided to the public without charge to the extent practicable, or at a fee which is not more than the actual costs of reproducing copies required to be sent to other Federal agencies, including the Council.

§ 1506.7 Further guidance.

The Council may provide further guidance concerning NEPA and its procedures including:

(a) A handbook which the Council may supplement from time to time, which shall in plain language provide guidance and instructions concerning the application of NEPA and these regulations.

(b) Publication of the Council's Memoranda to Heads of Agencies.

(c) In conjunction with the Environmental Protection Agency and the publication of the 102 Monitor, notice of:

(1) Research activities;

(2) Meetings and conferences related to NEPA; and

(3) Successful and innovative procedures used by agencies to implement NEPA.

§ 1506.8 Proposals for legislation.

(a) The NEPA process for proposals for legislation (§ 1508.17) significantly affecting the quality of the human environment shall be integrated with the legislative process of the Congress. A legislative environmental impact statement is the detailed statement required by law to be included in a recommendation or report on a legislative proposal to Congress. A legislative environmental impact statement shall be considered part of the formal transmittal of a legislative proposal to Congress; however, it may be transmitted to Congress up to 30 days later in order to allow time for completion of an accurate statement which can serve as the basis for public and Congressional debate. The statement must be available in time for Congressional hearings and deliberations.

(b) Preparation of a legislative environmental impact statement shall conform to the requirements of these regulations except as follows:

(1) There need not be a scoping process.

(2) The legislative statement shall be prepared in the same manner as a draft statement, but shall be considered the "detailed statement" required by statute; *Provided,* That when any of the following conditions exist both the draft and final environmental impact statement on the legislative proposal shall be prepared and circulated as provided by §§ 1503.1 and 1506.10.

(i) A Congressional Committee with jurisdiction over the proposal has a rule requiring both draft and final environmental impact statements.

(ii) The proposal results from a study process required by statute (such as those required by the Wild and Scenic Rivers Act (16 U.S.C. 1271 et seq.) and the Wilderness Act (16 U.S.C. 1131 et seq.)).

(iii) Legislative approval is sought for Federal or federally assisted construction or other projects which the agency recommends be located at specific geographic locations. For proposals requiring an environmental impact statement for the acquisition of space by the General Services Administration, a draft statement shall accompany the Prospectus or the 11(b) Report of Building Project Surveys to Congress, and a final statement shall be completed before site acquisition.

(iv) The agency decides to prepare draft and final statements.

(c) Comments on the legislative statement shall be given to the lead agency which shall forward them along with its own responses to the Congressional committees with jurisdiction.

§ 1506.9 Filing requirements.

Environmental impact statements together with comments and responses shall be filed with the Environmental Protection Agency, attention Office of Federal Activities (A-104), 401 M Street SW., Washington, D.C. 20460. Statements shall be filed with EPA no earlier than they are also transmitted to commenting agencies and made available to the public. EPA shall deliver one copy of each statement to the Council, which shall satisfy the requirement of availability to the President. EPA may issue guidelines to agencies to implement its responsibilities under this section and § 1506.10.

§ 1506.10 Timing of agency action.

(a) The Environmental Protection Agency shall publish a notice in the FEDERAL REGISTER each week of the environmental impact statements filed during the preceding week. The minimum time periods set forth in this section shall be calculated from the date of publication of this notice.

(b) No decision on the proposed action shall be made or recorded under § 1505.2 by a Federal agency until the later of the following dates:

(1) Ninety (90) days after publication of the notice described above in paragraph (a) of this section for a draft environmental impact statement.

(2) Thirty (30) days after publication of the notice described above in paragraph (a) of this section for a final environmental impact statement.

An exception to the rules on timing may be made in the case of an agency decision which is subject to a formal internal appeal. Some agencies have a formally established appeal process which allows other agencies or the public to take appeals on a decision and make their views known, after publication of the final environmental impact statement. In such cases, where a real opportunity exists to alter the decision, the decision may be made and recorded at the same time the environmental impact statement is published. This means that the period for appeal of the decision and the 30-day period prescribed in paragraph (b)(2) of this section may run concurrently. In such cases the environmental impact statement shall explain the timing and the public's right of appeal. An agency engaged in rule-making under the Administrative Procedure Act or other statute for the purpose of protecting the public health or safety, may waive the time period in paragraph (b)(2) of this section and publish a decision on the final rule simultaneously with publication of the notice of the availability of the final environmental impact statement as described in paragraph (a) of this section.

(c) If the final environmental impact statement is filed within ninety (90) days after a draft environmental impact statement is filed with the Environmental Protection Agency, the minimum thirty (30) day period and the minimum ninety (90) day period may run concurrently. However, subject to paragraph (d) of this section agencies shall allow not less than 45 days for comments on draft statements.

(d) The lead agency may extend prescribed periods. The Environmental Protection Agency may upon a showing by the lead agency of compelling reasons of national policy reduce the prescribed periods and may upon a showing by any other Federal agency of compelling reasons of national policy also extend prescribed periods, but only after consultation with the lead agency. (Also see § 1507.3(d).) Failure to file timely comments shall not be a sufficient reason for extending a period. If the lead agency does not concur with the extension of time, EPA may not extend it for more than 30 days. When the Environmental Protection Agency reduces or extends any period of time it shall notify the Council.

[43 FR 56000, Nov. 29, 1978; 44 FR 874, Jan. 3, 1979]

§ 1506.11 Emergencies.

Where emergency circumstances make it necessary to take an action with significant environmental impact without observing the provisions of these regulations, the Federal agency taking the action should consult with the Council about alternative arrangements. Agencies and the Council will limit such arrangements to actions necessary to control the immediate impacts of the emergency. Other actions remain subject to NEPA review.

§ 1506.12 Effective date.

The effective date of these regulations is July 30, 1979, except that for agencies that administer programs that qualify under section 102(2)(D) of the Act or under sec. 104(h) of the Housing and Community Development Act of 1974 an additional four months shall be allowed for the State or local agencies to adopt their implementing procedures.

(a) These regulations shall apply to the fullest extent practicable to ongoing activities and environmental documents begun before the effective date. These regulations do not apply to an environmental impact statement or supplement if the draft statement was filed before the effective date of these regulations. No completed environmental documents need be redone by reasons of these regulations. Until these regulations are applicable, the Council's guidelines published in the FEDERAL REGISTER of August 1, 1973, shall continue to be applicable. In cases where these regulations are applicable the guidelines are superseded. However, nothing shall prevent an agency from proceeding under these regulations at an earlier time.

(b) NEPA shall continue to be applicable to actions begun before January 1, 1970, to the fullest extent possible.

PART 1507—AGENCY COMPLIANCE

Sec.
1507.1 Compliance.
1507.2 Agency capability to comply.
1507.3 Agency procedures.

AUTHORITY: NEPA, the Environmental Quality Improvement Act of 1970, as amended (42 U.S.C. 4371 *et seq.*), sec. 309 of the Clean Air Act, as amended (42 U.S.C. 7609), and E.O. 11514 (Mar. 5, 1970, as amended by E.O. 11991, May 24, 1977).

SOURCE: 43 FR 56002, Nov. 29, 1978, unless otherwise noted.

§ 1507.1 **Compliance.**

All agencies of the Federal Government shall comply with these regulations. It is the intent of these regulations to allow each agency flexibility in adapting its implementing procedures authorized by § 1507.3 to the requirements of other applicable laws.

§ 1507.2 **Agency capability to comply.**

Each agency shall be capable (in terms of personnel and other resources) of complying with the requirements enumerated below. Such compliance may include use of other's resources, but the using agency shall itself have sufficient capability to evaluate what others do for it. Agencies shall:

(a) Fulfill the requirements of section 102(2)(A) of the Act to utilize a systematic, interdisciplinary approach which will insure the integrated use of the natural and social sciences and the environmental design arts in planning and in decisionmaking which may have an impact on the human environment. Agencies shall designate a

person to be responsible for overall review of agency NEPA compliance.

(b) Identify methods and procedures required by section 102(2)(B) to insure that presently unquantified environmental amenities and values may be given appropriate consideration.

(c) Prepare adequate environmental impact statements pursuant to section 102(2)(C) and comment on statements in the areas where the agency has jurisdiction by law or special expertise or is authorized to develop and enforce environmental standards.

(d) Study, develop, and describe alternatives to recommended courses of action in any proposal which involves unresolved conflicts concerning alternative uses of available resources. This requirement of section 102(2)(E) extends to all such proposals, not just the more limited scope of section 102(2)(C)(iii) where the discussion of alternatives is confined to impact statements.

(e) Comply with the requirements of section 102(2)(H) that the agency initiate and utilize ecological information in the planning and development of resource-oriented projects.

(f) Fulfill the requirements of sections 102(2)(F), 102(2)(G), and 102(2)(I), of the Act and of Executive Order 11514, Protection and Enhancement of Environmental Quality, Sec. 2.

§ 1507.3 **Agency procedures.**

(a) Not later than eight months after publication of these regulations as finally adopted in the FEDERAL REGISTER, or five months after the establishment of an agency, whichever shall come later, each agency shall as necessary adopt procedures to supplement these regulations. When the agency is a department, major subunits are encouraged (with the consent of the department) to adopt their own procedures. Such procedures shall not paraphrase these regulations. They shall confine themselves to implementing procedures. Each agency shall consult with the Council while developing its procedures and before publishing them in the FEDERAL REGISTER for comment. Agencies with similar programs should consult with each other and the Council to coordinate their procedures, especially for programs requesting similar information from applicants. The procedures shall be adopted only after an opportunity for public review and after review by the Council for conformity with the Act and these regulations. The Council shall complete its review within 30 days. Once in effect they shall be filed with the Council and made readily available to the public. Agencies are encouraged to publish explanatory guidance for these regulations and their own procedures. Agencies shall continue to review their policies and procedures and in consultation with the Council to revise them as necessary to ensure full compliance with the purposes and provisions of the Act.

(b) Agency procedures shall comply with these regulations except where compliance would be inconsistent with statutory requirements and shall include:

(1) Those procedures required by §§ 1501.2(d), 1502.9(c)(3), 1505.1, 1506.6(e), and 1508.4.

(2) Specific criteria for and identification of those typical classes of action:

(i) Which normally do require environmental impact statements.

(ii) Which normally do not require either an environmental impact statement or an environmental assessment (categorical exclusions (§ 1508.4)).

(iii) Which normally require environmental assessments but not necessarily environmental impact statements.

(c) Agency procedures may include specific criteria for providing limited exceptions to the provisions of these regulations for classified proposals. They are proposed actions which are specifically authorized under criteria established by an Executive Order or statute to be kept secret in the interest of national defense or foreign policy and are in fact properly classified pursuant to such Executive Order or statute. Environmental assessments and environmental impact statements which address classified proposals may be safeguarded and restricted from public dissemination in accordance with agencies' own regulations applicable to classified information. These documents may be organized so that classified portions can be included as annexes, in order that the unclassified portions can be made available to the public.

(d) Agency procedures may provide for periods of time other than those presented in § 1506.10 when necessary to comply with other specific statutory requirements.

(e) Agency procedures may provide that where there is a lengthy period between the agency's decision to prepare an environmental impact statement and the time of actual preparation, the notice of intent required by § 1501.7 may be published at a reasonable time in advance of preparation of the draft statement.

PART 1508—TERMINOLOGY AND INDEX

AUTHORITY: NEPA, the Environmental Quality Improvement Act of 1970, as amended (42 U.S.C. 4371 *et seq.*), sec. 309 of the Clean Air Act, as amended (42 U.S.C. 7609), and E.O. 11514 (Mar. 5, 1970, as amended by E.O. 11991, May 24, 1977).

SOURCE: 43 FR 56003, Nov. 29, 1978, unless otherwise noted.

§ 1508.1 Terminology.

The terminology of this part shall be uniform throughout the Federal Government.

§ 1508.2 Act.

"Act" means the National Environmental Policy Act, as amended (42 U.S.C. 4321, et seq.) which is also referred to as "NEPA."

§ 1508.3 Affecting.

"Affecting" means will or may have an effect on.

§ 1508.4 Categorical exclusion.

"Categorical exclusion" means a category of actions which do not individually or cumulatively have a significant effect on the human environment and which have been found to have no such effect in procedures adopted by a Federal agency in implementation of these regulations (§ 1507.3) and for which, therefore, neither an environmental assessment nor an environmental impact statement is required. An agency may decide in its procedures or otherwise, to prepare environmental assessments for the reasons stated in § 1508.9 even though it is not required to do so. Any procedures under this section shall provide for extraordinary circumstances in which a normally excluded action may have a significant environmental effect.

§ 1508.5 Cooperating agency.

"Cooperating agency" means any Federal agency other than a lead agency which has jurisdiction by law or special expertise with respect to any environmental impact involved in a proposal (or a reasonable alternative) for legislation or other major Federal action significantly affecting the quality of the human environment. The selection and responsibilities of a cooperating agency are described in § 1501.6. A State or local agency of similar qualifications or, when the effects are on a reservation, an Indian Tribe, may by agreement with the lead agency become a cooperating agency.

§ 1508.6 Council.

"Council" means the Council on Environmental Quality established by Title II of the Act.

§ 1508.7 Cumulative impact.

"Cumulative impact" is the impact on the environment which results from the incremental impact of the action when added to other past, present, and reasonably foreseeable future actions regardless of what agency (Federal or non-Federal) or person undertakes such other actions. Cumulative impacts can result from individually minor but collectively significant actions taking place over a period of time.

§ 1508.8 Effects.

"Effects" include:

(a) Direct effects, which are caused by the action and occur at the same time and place.

(b) Indirect effects, which are caused by the action and are later in time or farther removed in distance, but are still reasonably foreseeable. Indirect effects may include growth inducing effects and other effects related to induced changes in the pattern of land use, population density or growth rate, and related effects on air and water and other natural systems, including ecosystems.

Effects and impacts as used in these regulations are synonymous. Effects includes ecological (such as the effects on natural resources and on the components, structures, and functioning of affected ecosystems), aesthetic, historic, cultural, economic, social, or health, whether direct, indirect, or cumulative. Effects may also include those resulting from actions which may have both beneficial and detrimental effects, even if on balance that the agency believes that the effect will be beneficial.

§ 1508.9 Environmental assessment.

"Environmental assessment":

(a) Means a concise public document for which a Federal agency is responsible that serves to:

(1) Briefly provide sufficient evidence and analysis for determining

whether to prepare an environmental impact statement or a finding of no significant impact.

(2) Aid an agency's compliance with the Act when no environmental impact statement is necessary.

(3) Facilitate preparation of a statement when one is necessary.

(b) Shall include brief discussions of the need for the proposal, of alternatives as required by section 102(2)(E), of the environmental impacts of the proposed action and alternatives, and a listing of agencies and persons consulted.

§ 1508.10 Environmental document.

"Environmental document" includes the documents specified in § 1508.9 (environmental assessment), § 1508.11 (environmental impact statement), § 1508.13 (finding of no significant impact), and § 1508.22 (notice of intent).

§ 1508.11 Environmental impact statement.

"Environmental impact statement" means a detailed written statement as required by section 102(2)(C) of the Act.

§ 1508.12 Federal agency.

"Federal agency" means all agencies of the Federal Government. It does not mean the Congress, the Judiciary, or the President, including the performance of staff functions for the President in his Executive Office. It also includes for purposes of these regulations States and units of general local government and Indian tribes assuming NEPA responsibilities under section 104(h) of the Housing and Community Development Act of 1974.

§ 1508.13 Finding of no significant impact.

"Finding of no significant impact" means a document by a Federal agency briefly presenting the reasons why an action, not otherwise excluded (§ 1508.4), will not have a significant effect on the human environment and for which an environmental impact statement therefore will not be prepared. It shall include the environmental assessment or a summary of it and shall note any other environmental documents related to it

(§ 1501.7(a)(5)). If the assessment is included, the finding need not repeat any of the discussion in the assessment but may incorporate it by reference.

§ 1508.14 Human environment.

"Human environment" shall be interpreted comprehensively to include the natural and physical environment and the relationship of people with that environment. (See the definition of "effects" (§ 1508.8).) This means that economic or social effects are not intended by themselves to require preparation of an environmental impact statement. When an environmental impact statement is prepared and economic or social and natural or physical environmental effects are interrelated, then the environmental impact statement will discuss all of these effects on the human environment.

§ 1508.15 Jurisdiction by law.

"Jurisdiction by law" means agency authority to approve, veto, or finance all or part of the proposal.

§ 1508.16 Lead agency.

"Lead agency" means the agency or agencies preparing or having taken primary responsibility for preparing the environmental impact statement.

§ 1508.17 Legislation.

"Legislation" includes a bill or legislative proposal to Congress developed by or with the significant cooperation and support of a Federal agency, but does not include requests for appropriations. The test for significant cooperation is whether the proposal is in fact predominantly that of the agency rather than another source. Drafting does not by itself constitute significant cooperation. Proposals for legislation include requests for ratification of treaties. Only the agency which has primary responsibility for the subject matter involved will prepare a legislative environmental impact statement.

§ 1508.18 Major Federal action.

"Major Federal action" includes actions with effects that may be major and which are potentially subject to

Federal control and responsibility. Major reinforces but does not have a meaning independent of significantly (§ 1508.27). Actions include the circumstance where the responsible officials fail to act and that failure to act is reviewable by courts or administrative tribunals under the Administrative Procedure Act or other applicable law as agency action.

(a) Actions include new and continuing activities, including projects and programs entirely or partly financed, assisted, conducted, regulated, or approved by federal agencies; new or revised agency rules, regulations, plans, policies, or procedures; and legislative proposals (§§ 1506.8, 1508.17). Actions do not include funding assistance solely in the form of general revenue sharing funds, distributed under the State and Local Fiscal Assistance Act of 1972, 31 U.S.C. 1221 et seq., with no Federal agency control over the subsequent use of such funds. Actions do not include bringing judicial or administrative civil or criminal enforcement actions.

(b) Federal actions tend to fall within one of the following categories:

(1) Adoption of official policy, such as rules, regulations, and interpretations adopted pursuant to the Administrative Procedure Act, 5 U.S.C. 551 et seq.; treaties and international conventions or agreements; formal documents establishing an agency's policies which will result in or substantially alter agency programs.

(2) Adoption of formal plans, such as official documents prepared or approved by federal agencies which guide or prescribe alternative uses of federal resources, upon which future agency actions will be based.

(3) Adoption of programs, such as a group of concerted actions to implement a specific policy or plan; systematic and connected agency decisions allocating agency resources to implement a specific statutory program or executive directive.

(4) Approval of specific projects, such as construction or management activities located in a defined geographic area. Projects include actions approved by permit or other regulatory decision as well as federal and federally assisted activities.

§ 1508.19 **Matter.**

"Matter" includes for purposes of Part 1504:

(a) With respect to the Environmental Protection Agency, any proposed legislation, project, action or regulation as those terms are used in section 309(a) of the Clean Air Act (42 U.S.C. 7609).

(b) With respect to all other agencies, any proposed major federal action to which section 102(2)(C) of NEPA applies.

§ 1508.20 **Mitigation.**

"Mitigation" includes:

(a) Avoiding the impact altogether by not taking a certain action or parts of an action.

(b) Minimizing impacts by limiting the degree or magnitude of the action and its implementation.

(c) Rectifying the impact by repairing, rehabilitating, or restoring the affected environment.

(d) Reducing or eliminating the impact over time by preservation and maintenance operations during the life of the action.

(e) Compensating for the impact by replacing or providing substitute resources or environments.

§ 1508.21 **NEPA process.**

"NEPA process" means all measures necessary for compliance with the requirements of section 2 and Title I of NEPA.

§ 1508.22 **Notice of intent.**

"Notice of intent" means a notice that an environmental impact statement will be prepared and considered. The notice shall briefly:

(a) Describe the proposed action and possible alternatives.

(b) Describe the agency's proposed scoping process including whether, when, and where any scoping meeting will be held.

(c) State the name and address of a person within the agency who can answer questions about the proposed action and the environmental impact statement.

§ 1508.23 Proposal.

"Proposal" exists at that stage in the development of an action when an agency subject to the Act has a goal and is actively preparing to make a decision on one or more alternative means of accomplishing that goal and the effects can be meaningfully evaluated. Preparation of an environmental impact statement on a proposal should be timed (§ 1502.5) so that the final statement may be completed in time for the statement to be included in any recommendation or report on the proposal. A proposal may exist in fact as well as by agency declaration that one exists.

§ 1508.24 Referring agency.

"Referring agency" means the federal agency which has referred any matter to the Council after a determination that the matter is unsatisfactory from the standpoint of public health or welfare or environmental quality.

§ 1508.25 Scope.

Scope consists of the range of actions, alternatives, and impacts to be considered in an environmental impact statement. The scope of an individual statement may depend on its relationships to other statements (§§1502.20 and 1508.28). To determine the scope of environmental impact statements, agencies shall consider 3 types of actions, 3 types of alternatives, and 3 types of impacts. They include:

(a) Actions (other than unconnected single actions) which may be:

(1) Connected actions, which means that they are closely related and therefore should be discussed in the same impact statement. Actions are connected if they:

(i) Automatically trigger other actions which may require environmental impact statements.

(ii) Cannot or will not proceed unless other actions are taken previously or simultaneously.

(iii) Are interdependent parts of a larger action and depend on the larger action for their justification.

(2) Cumulative actions, which when viewed with other proposed actions have cumulatively significant impacts and should therefore be discussed in the same impact statement.

(3) Similar actions, which when viewed with other reasonably foreseeable or proposed agency actions, have similarities that provide a basis for evaluating their environmental consequences together, such as common timing or geography. An agency may wish to analyze these actions in the same impact statement. It should do so when the best way to assess adequately the combined impacts of similar actions or reasonable alternatives to such actions is to treat them in a single impact statement.

(b) Alternatives, which include: (1) No action alternative.

(2) Other reasonable courses of actions.

(3) Mitigation measures (not in the proposed action).

(c) Impacts, which may be: (1) Direct; (2) indirect; (3) cumulative.

§ 1508.26 Special expertise.

"Special expertise" means statutory responsibility, agency mission, or related program experience.

§ 1508.27 Significantly.

"Significantly" as used in NEPA requires considerations of both context and intensity:

(a) *Context.* This means that the significance of an action must be analyzed in several contexts such as society as a whole (human, national), the affected region, the affected interests, and the locality. Significance varies with the setting of the proposed action. For instance, in the case of a site-specific action, significance would usually depend upon the effects in the locale rather than in the world as a whole. Both short- and long-term effects are relevant.

(b) *Intensity.* This refers to the severity of impact. Responsible officials must bear in mind that more than one agency may make decisions about partial aspects of a major action. The following should be considered in evaluating intensity:

(1) Impacts that may be both beneficial and adverse. A significant effect may exist even if the Federal agency

believes that on balance the effect will be beneficial.

(2) The degree to which the proposed action affects public health or safety.

(3) Unique characteristics of the geographic area such as proximity to historic or cultural resources, park lands, prime farmlands, wetlands, wild and scenic rivers, or ecologically critical areas.

(4) The degree to which the effects on the quality of the human environment are likely to be highly controversial.

(5) The degree to which the possible effects on the human environment are highly uncertain or involve unique or unknown risks.

(6) The degree to which the action may establish a precedent for future actions with significant effects or represents a decision in principle about a future consideration.

(7) Whether the action is related to other actions with individually insignificant but cumulatively significant impacts. Significance exists if it is reasonable to anticipate a cumulatively significant impact on the environment. Significance cannot be avoided by terming an action temporary or by breaking it down into small component parts.

(8) The degree to which the action may adversely affect districts, sites, highways, structures, or objects listed in or eligible for listing in the National Register of Historic Places or may cause loss or destruction of significant scientific, cultural, or historical resources.

(9) The degree to which the action may adversely affect an endangered or threatened species or its habitat that has been determined to be critical under the Endangered Species Act of 1973.

(10) Whether the action threatens a violation of Federal, State, or local law or requirements imposed for the protection of the environment.

[43 FR 56003, Nov. 29, 1978; 44 FR 874, Jan. 3, 1979]

§ 1508.28 Tiering.

"Tiering" refers to the coverage of general matters in broader environmental impact statements (such as national program or policy statements) with subsequent narrower statements or environmental analyses (such as regional or basinwide program statements or ultimately site-specific statements) incorporating by reference the general discussions and concentrating solely on the issues specific to the statement subsequently prepared. Tiering is appropriate when the sequence of statements or analyses is:

(a) From a program, plan, or policy environmental impact statement to a program, plan, or policy statement or analysis of lesser scope or to a site-specific statement or analysis.

(b) From an environmental impact statement on a specific action at an early stage (such as need and site selection) to a supplement (which is preferred) or a subsequent statement or analysis at a later stage (such as environmental mitigation). Tiering in such cases is appropriate when it helps the lead agency to focus on the issues which are ripe for decision and exclude from consideration issues already decided or not yet ripe.

INDEX

INDEX—Continued

APPENDIX C

CITATIONS OF FEDERAL AGENCY NEPA REGULATIONS

Source: Council on Environmental Quality, Twenty-second Annual Report, 1991.

NEPA Regulations Citations

As of December 15, 1991, the following federal agencies had issued departmental orders or had published NEPA regulations in the *Code of Federal Regulations* (CFR) or the *Federal Register* (FR):

Agency	Citation
Department of Agriculture	7 CFR Part 3100; 7 CFR Part 1b
Agricultural Research Service	7 CFR Part 520
Agriculture Stabilization and Conservation Program Service	7 CFR Part 799
Animal and Plant Health Inspection Service	44 FR 50381 (1979)
Cooperative State Research Service	56 FR 8156 (1991) (to be published in final form)
Farmers Home Administration	7 CFR Parts 1901, 1940
Forest Service	36 CFR Part 219, FS Manual Chapter 1950
Soil Conservation Service	7 CFR Part 650
Rural Electrification Administration	7 CFR 1794; Dir. 17.02-2
Department of Commerce	45 FR 47898 (1980)
Department of Administration	Order 216-6
Economic Development Administration	45 FR 63310 (1980), amended by 45 FR 4902 (1980), Dir. 17.02-2
National Oceanic and Atmospheric Administration (NOAA)	48 FR 14734 (1983); NOAA Directives Manual 2-10, Environmental Review Procedures, July 23, 1984
Department of Defense	32 CFR Part 214, 32 CFR Part 197

Agency	Citation
Department of the Air Force	32 CFR Part 989
Army Corps of Engineers	33 CFR Part 230, 32 CFR Chapter 11
Department of the Army	32 CFR Parts 650, 651
Department of the Navy	32 CFR Part 775
Defense Logistics Agency	DLA 1000.22
Department of Energy	45 FR 20694 (1980), amended by 47 FR 7976 (1982), 48 FR 685 (1983), 52 FR 659 (1987), 52 FR 47662 (1987)
Coal Utilization Program	10 CFR Part 305
Department of Health and Human Services (HHS)	HHS General Administration Manual; 45 FR 76519 (1980), amended by 47 FR 2414 (1982)
Food and Drug Administration	21 CFR Part 25
Department of Housing and Urban Development	24 CFR Part 50, 51, 58
Housing and Community Development Act	24 CFR Part 50, 58
Community Development Block Grant Program	24 CFR 58
Department of the Interior	516 Departmental Manual 1-7; 45 FR 27541 (1980), amended by 49 FR 21437 (1984)
Bureau of Indian Affairs	30 BIAM Supplement 1, NEPA Handbook; 46 FR 7490 (1981)
Bureau of Mines	516 WBM 1-5, Environmental Quality—National Environmental Policy Act of 1969 (March 29, 1990); 45 FR 85528 (1980)

Agency	Citation
Bureau of Land Management	H-1790-1; 46 FR 7492 (1981), amended by 48 FR 43731 (1983)
Fish and Wildlife Service	Service Administration Manual, 30 AM 2-4; 45 FR 47941 (1980), amended by 47 FR 28841 (1982), 49 FR 7881 (1984)
Geological Survey	46 FR 7485 (1981)
Minerals Management Service	51 FR 1855 (1986)
National Park Service	NPS-12, NEPA Compliance Guideline; 46 FR 1042 (1981)
Office of Surface Mining, Reclamation and Control	OMRE NEPA Handbook on Procedures for Implementing the National Environmental Policy Act
Bureau of Reclamation	National Environmental Policy Act Handbook, October 1990; 45 FR 47944 (1980), amended by 48 FR 17151 (1983)
Department of Justice	28 CFR Part 61
Drug Enforcement Agency	28 CFR Part 61, Appendix B
Immigration and Naturalization Service	28 CFR Part 61, Appendix C
Bureau of Prisons	28 CFR Part 61, Appendix A
Office of Justice, Assistance, Research and Statistics	28 CFR Part 61, Appendix D
Department of Labor	29 CFR Part 11
Department of State	22 CFR Part 161
Department of the Treasury	45 FR 1828 (1980)
Department of Transportation	Order DOT 5610.1c; 44 FR 56420 (1979)

Agency	Citation
Coast Guard	Commandant Instruction M 16475.1B, July 12, 1985; 45 FR 32816 (1980), amended by 50 FR 32944 (1985)
Federal Aviation Administration	Order FAA 1050.1D (agency-wide order); FAA 5050.41, Airport Environmental Handbook, October 8, 1985 (applies to airport grant program only); 45 FR 2244 (1980), as amended by 49 FR 28501 (1984)
Federal Highway Administration	23 CFR Parts 650, 771
Federal Railroad Administration	Procedures for Considering Environmental Impacts; 45 FR 40854 (1980)
Maritime Administration	Maritime Administrative Order 600. 1, July 23, 1985
National Highway Traffic Safety Administration	49 CFR Part 520
Saint Lawrence Seaway Development Corporation	Order SLS 10 5610.1C, May 26, 1981, 56 FR 28795 (1981)
Urban Mass Transportation Administration	23 CFR Part 771
	Copies of unpublished DOT procedures may be requested from: Environmental Division Room 9217 U.S. Department of Transportation, Washington, DC 20590
Department of Veteran's Affairs	38 CFR Part 26
Advisory Council on Historic Preservation	36 CFR Part 805

Agency	Citation
Agency for International Development	22 CFR Part 216
Arms Control Disarmament Agency	45 FR 69510 (1980)
Central Intelligence Agency	44 FR 45431 (1979)
Committee for Purchase from Blind and Other Severely Handicapped	41 CFR Part 51-6
Consumer Product Safety Commission	16 CFR Part 1021
Delaware River Basin Commission	18 CFR Part 401, subpart D
Environmental Protection Agency	40 CFR Part 6, Part 227
Export–Import Bank	12 CFR Part 408
Farm Credit Administration	12 CFR Chapter VI
Federal Communications Commission	47 CFR Part I, subpart I
Federal Emergency Management Agency	44 CFR Part 10
Federal Energy Regulatory Commission	18 CFR Part 2.80, Part 380
Federal Maritime Commission	46 CFR Part 504
Federal Trade Commission	16 CFR Part 1, Subpart I
General Services Administration	PBS P 1095.4B, Preparation of Environmental Assessments and Environmental Impact Statements; 50 FR 7648 (1985)

Agency	Citation
International Boundary and Water Commission	46 FR 44083 (1981)
International Communication Agency	44 FR 45489 (1979)
Interstate Commerce Commission	49 CFR Part 1105, amended by 54 FR 9822 (1989)
Marine Mammal Commission	50 CFR Part 530
National Aeronautics and Space Administration (NASA)	14 CFR Part 1204, 1216
National Capital Planning Commission	44 FR 64923 (1979); Manual of Environmental Policies and Procedures
National Institutes of Health	General Administration Manual, 30-10
National Science Foundation	45 CFR Part 640
Nuclear Regulatory Commission	10 CFR Part 51
Overseas Private Investment Corporation	44 FR 51385 (1979); Foreign Assistance Act, §§ 231, 237m, 239g
Pennsylvania Avenue Development Corporation	36 CFR Part 907
Securities and Exchange Commission	17 CFR Part 200, subpart K
Small Business Administration	45 FR 7358 (1980)

Agency	Citation
Tennessee Valley Authority	Procedures For Compliance With The National Environmental Policy Act, April 12, 1983; 45 FR 54511 (1980), amended by 47 FR 54586 (1982), 48 FR 19264 (1983)
United States Postal Service	39 CFR Part 775
Water Resources Council	18 CFR Part 707

FEDERAL DATA SOURCES

Source: Council on Environmental Quality, Twenty-second Annual Report, 1991.

Federal Interagency Initiatives to Coordinate Environmental Data and Analysis

Program	Lead Agency	Coverage
Interagency Committee on Environmental Trends	CEQ	In 1991 CEQ established an interagency committee with participation of all agencies involved in environmental data to undertake development of environmental information framework.
National Acid Precipitation Assessment Program (NAPAP)	CEQ	NAPAP, established by the 1980 Acid Precipitation Act, coordinates federally funded research and assessment to develop a firm scientific basis for acid rain policies. The Clean Air Act Amendments of 1990 extended NAPAP, which publishes reports on the State of Science and Technology and on Integrated Assessment; NAPAP monitors the influence of the Clean Air Act on environmental impacts.
Environmental Monitoring and Assessment Program (EMAP)	EPA	A long-term monitoring program, initiated in 1988, to provide regional and national assessments of status and trends of U.S. ecological resources. During the next 5 years, EMAP monitoring networks will estimate trends in national ecological indicators on a regional basis; monitor indicators of pollutant exposure and habitat condition; seek associations between human-induced stresses and ecological condition; and provide periodic statistical summaries and interpretive reports on status and trends for resource managers and the public.
EPA Center for Environmental Statistics	EPA	EPA Development Staff established the Center for Environmental Statistics to gather environmental data from variety of sources, address data quality and statistical issues, and use data to provide information on environmental conditions.

Federal Interagency Initiatives to Coordinate Environmental Data and Analysis (Continued)

Program	Lead Agency	Coverage
Intergovernmental Monitoring Task Force	EPA and USGS	Established in 1991 by the EPA Office of Water and USGS, coordinating with NOAA, FWS, COE, USDA, OMB, and DOE, the task force identifies and recommends solutions to water quality monitoring problems, establishes a national framework for water quality monitoring, agrees on environmental indicators, and shares water quality data through system linkages and increased quality assurance and control.
Federal Geographic Data Committee (FGDC)	DOI	FGDC, with 14 departments and agencies, promotes coordinated development, use, and dissemination of surveying, mapping, and related spatial data; established by OMB in October 1990 by OMB Circular A-16, "Coordination of Surveying, Mapping, and Related Spatial Data Activities," stressing coordination and standards.
Ocean Pollution Data and Information Network (OPDIN)	NOAA	A 10-year interagency effort by 11 federal agencies to improve accessibility and usefulness of data and to increase communication and coordination on ocean and Great Lakes pollution information, the Network publishes information guides and responds to requests for information.
FCCSET/CEES/ IWGDMGC	OSTP	Within the President's Office of Science and Technology Policy (OSTP), the Federal Coordinating Council for Science, Engineering and Technology (FCCSET) established the Committee on Earth and Environmental Sciences (CEES) as a forum to coordinate interagency programs, including the

Federal Interagency Initiatives to Coordinate Environmental Data and Analysis (Continued)

Program	Lead Agency	Coverage
FCCSET/CEES/ IWGDMGC (Continued)		U.S. Global Change Research Program (USGCRP). Working with CEES, the Interagency Working Group on Data Management for Global Change (IWGDMGC) is developing recommendations on interagency management of research data. In 1991 the National Research Council drafted a "U.S. Strategy for Global Change Data and Information Management" currently under review.
President's Water Quality Initiative	USDA	Addresses nonpoint source contamination of surface water and groundwater by agricultural chemicals; seven USDA agencies, USGS, NOAA, and EPA compose the Working Group on Water Quality to coordinate efforts, such as National Water Quality Assessment (NAWQA) program created by USGS to assess water quality impacted by pesticides, nutrients, and sediment.
Interagency Advisory Committee on Water Data	USGS	Established by the Secretary of the Interior in 1964, the committee coordinates federal water data programs; chaired by Chief Hydrologist of USGS, represents 30 federal agencies that collect or use water data; six subcommittees are Ground Water, Hydrology (surface water), Sedimentation, Water Data and Information Exchange, Water Quality, and Water-Use Information.

Note: CEQ = President's Council on Environmental Quality; EPA = U.S. Environmental Protection Agency; NOAA = U.S. Department of Commerce, National Oceanic and Atmospheric Administration; OMB = President's Office of Management and Budget; OSTP = President's Office of Science and Technology Policy; USDA = U.S. Department of Agriculture; and USGS = Department of the Interior, U.S. Geological Survey.
Source: CEQ (1991).

Environmental Statistics Programs Managed by Agencies of the U.S. Government

Program	Sponsor	Coverage	Frequency/ Most Recent Report
Major Uses of Land in the United States	Department of Agriculture (USDA) Economic Research Service	Estimated acreage and inventories of major U.S. Land uses.	Intervals coincide with Census of Agriculture; 1991 report with 1989 data.
National Resources Inventory	USDA Soil Conservation Service	Data on status, condition, trends of soil, water, related natural resources of rural nonfederal U.S. land.	5-year cycle; 1989 report with 1987 data.
Forest Insect and Disease Conditions in the United States	USDA Forest Service	Data for federal, state, private forestlands in U.S.; data analyses by region, ownership, type of insect/disease, area affected, dollars lost; trend data available.	Annual collection; 1991 report with 1990 data.
Forest Inventory and Analysis	USDA Forest Service	Inventory with trend information on extent, condition, ownership, and composition of U.S. forests; wildlife habitat, forage production, and other resource characteristics.	Complete inventory on 12-year cycle; 1991 report with 1991 data.

Environmental Statistics Programs Managed by Agencies of the U.S. Government (Continued)

Program	Sponsor	Coverage	Frequency/ Most Recent Report
Land Areas of the National Forest System	USDA Forest Service	Data on extent and character-istics of forest, range, and re-lated lands in the National Forest System.	Annual reports.
Tree Planting in the United States	USDA Forest Service	Summary of tree planting in U.S.	Annual reports; 1991 report with 1990 data.
Survey of Pollution Abatement Costs and Expenditures	Department of Commerce (DOC) Bureau of the Census	Annual operating costs and capital expenditures for pollution abate-ment activities in manufactur-ing industries.	Annual collection.
Classified Shellfishing Waters	DOC National Oceanic and Atmospheric Administration (NOAA)	Monitors classi-fied shellfish-ing waters as indicator of bacterial water quality nation-wide; classifies waters for com-mercial harvest of oysters, clams, and mussels based on actual or poten-tial pollution sources and coli-form bacteria levels in surface waters.	Surveys each 5 years; most recent 1990.

Environmental Statistics Programs Managed by Agencies of the U.S. Government (*Continued*)

Program	Sponsor	Coverage	Frequency/ Most Recent Report
Fisheries Statistics Program	DOC/NOAA	National statistics (biological, economic, and sociological) on domestic commercial and recreational fisheries: commercial landings, number of vessels and fishermen, annual processed products; annual world catch of shrimp, shrimp imports, and recreational finfish; saltwater catch.	Daily/monthly/ yearly data collection.
National Climatic Data Center (NCDC)	DOC/NOAA	NCDC collects, processes, archives, and disseminates worldwide meteorological and climatological data from a global network of stations; coverage is global, land and sea, primarily of U.S. dependencies, especially for summarized data.	Responds to over 33,000 user requests per year; records data from mid-nineteenth century to present.
National Data Center (NODC)	DOC/NOAA	NODC collects, processes, archives, and	Responds to over 10,000 user requests per

Environmental Statistics Programs Managed by Agencies of the U.S. Government (*Continued*)

Program	Sponsor	Coverage	Frequency/ Most Recent Report
National Data Center (NODC) (*Continued*)		disseminates such worldwide oceanographic data as marine biology, marine pollution, wind and waves, surface and subsurface currents, and temperature.	year; records data from late-nineteenth century to present.
National Geophysical Data Center (NGDC)	DOC/NOAA	NGDC collects, processes, archives, and disseminates such worldwide geophysical data as solid earth geophysics, earthquake seismology, geomagnetic surveys, marine geology and geophysics, solar–terrestrial physics, and glaciology.	Responds to 11,000 user requests per year; records data from mid-nineteenth century to to present.
National Coastal Pollutant Discharge Inventory Program	DOC/NOAA	Compiles pollutant-loading estimates for point, nonpoint, and riverine sources in coastal counties, or 200-mile Exclusive Economic Zone that discharges to the	East, West, and Gulf Coast estimates are for 1982, 1984, and 1987, respectively. In 1992, East Coast will be updated to 1989.

Environmental Statistics Programs Managed by Agencies of the U.S. Government (*Continued*)

Program	Sponsor	Coverage	Frequency/ Most Recent Report
National Coastal Pollutant Discharge Inventory Program (*Continued*)		estuarine, coastal, and oceanic waters of the contiguous U.S. excluding the Great Lakes.	
Month and State Current Emissions Trends	Department of Energy (DOE) Argonne National Laboratory	Emissions estimates for NO_x, SO_2, and VOCs by month and state from 1975 to the present for 68 emission source groups.	Monthly; 1975 to present.
National Energy Information Center	DOE Information Administration	Collects and publishes data; prepares analyses on energy production, consumption, prices, and resources; makes projections of energy supply and demand.	Annual reports.
Carbon Dioxide Information Analysis Center	DOE Oak Ridge National Laboratory	Compiles, evaluates, and distributes information related to carbon dioxide (CO_2).	Data collection ranges from hourly to decadal.
Integrated Data Base Program	DOE	Maintains data on all spent radioactive fuel and waste in the U.S.	Annual data collection and reporting.

Environmental Statistics Programs Managed by Agencies of the U.S. Government (Continued)

Program	Sponsor	Coverage	Frequency/ Most Recent Report
National Wetlands Inventory	Department of the Interior (DOI) U.S. Fish and Wildlife Service (FWS)	Develops comprehensive information on the characteristics and extent of wetlands resources in U.S.; wetlands map coverage for 70% of lower 48 states, 22% of Alaska, and all of Hawaii, Puerto Rico, and Guam.	Continuous wetlands trend for lower 48 states; statistical estimates of U.S. wetlands acreage beginning in 1990.
Gap Analysis	DOI/FWS	Develops standardized distribution maps of surface vegetation, terrestrial vertebrates, and endangered species in the lower 48 states and Hawaii.	First analysis, 1988; national assessment to be completed in 1998, to be updated at 5-year intervals.
National Contaminant Biomonitoring Program	DOI/FWS	Documents temporal and geographic trends in concentrations of persistent environmental contaminants that may threaten fish and wildlife; covers major U.S. rivers and Great Lakes.	2- to 4-year intervals.

Environmental Statistics Programs Managed by Agencies of the U.S. Government (*Continued*)

Program	Sponsor	Coverage	Frequency/ Most Recent Report
North American Breeding Bird Survey	DOI/FWS	Provides uniform basis for assessing long-term trends in avian populations throughout North America; total number of individuals by species, survey route, and state.	2-year intervals.
Waterfowl Breeding Population and Habitat Survey	DOI/FWS	Provides annual breeding population estimates; measures breeding habitat changes over major portion of duck breeding range in North America.	Annual surveys and reports.
Statistics and Information Services	DOI/Bureau of Mines	Collects, analyzes, and publishes data on metal production and consumption, including scrap and waste.	Monthly, quarterly, and annually.
Public Land Statistics	DOI/Bureau of Land Management (BLM)	Collects summary statistics of land ownership in the United States and BLM natural resource manage-	Annual reports.

Environmental Statistics Programs Managed by Agencies of the U.S. Government (*Continued*)

Program	Sponsor	Coverage	Frequency/ Most Recent Report
Public Land Statistics (*Continued*)		ment programs at the geographic state level.	
National Stream Quality Accounting Network (NASQAN) and National Hydrologic Benchmark Network	DOI/U.S. Geological Survey (USGS)	NASQAN provides nationally uniform basis for assessing large-scale, long-term trends in physical, chemical, and biological characteristics of waters; monitors for pH, alkalinity, sulfate, nitrate, phosphorus, calcium, magnesium, sodium, potassium, chloride, suspended sediment, fecal coliform bacteria, fecal streptococcal bacteria, dissolved oxygen, dissolved oxygen deficit, and trace elements; Benchmark Network monitors water quality in surface waters largely unaffected by human activities.	Yearly data summaries for each state.
National Trends Network	DOI/USGS	Monitors deposition under National Acid Precipitation Assessment Pro-	

Environmental Statistics Programs Managed by Agencies of the U.S. Government (*Continued*)

Program	Sponsor	Coverage	Frequency/ Most Recent Report
National Trends Network (*Continued*)		gram (NAPAP); 150 stations predominantly in rural areas, in combination with USDA National Atmospheric Deposition Program (NADP); more than 200 sites nationwide.	
National Water Conditions Reporting System	DOI/USGS	Collects and analyzes streamflow data, groundwater levels, reservoir contents, and limited water quality data from 5 sites on major rivers.	Published in monthly newsletter, *National Water Conditions.*
National Water Use Information Program	DOI/USGS	Determines U.S. fresh and saline surface water and groundwater withdrawn, for what purposes, water consumed during use, and water returned to source after use.	National compilations every 5 years; most recent 1990; next report 1992.
Water Resources Assessment Program	DOI/USGS	Summarizes statistics on state and national water resources	Biennial; 1990 report covered water supply and use.

Environmental Statistics Programs Managed by Agencies of the U.S. Government (*Continued*)

Program	Sponsor	Coverage	Frequency/ Most Recent Report
Water Resources Assessment Program (*Continued*)		for USGS biennial report, National Water Summary; each report oriented to water resource theme, such as groundwater quality.	
Earth Observing System	National Aeronautics and Space Administration (NASA)	Measures key environmental variables using series of unmanned satellites; part of NASA Mission to Planet Earth program; EOSDIS, a data and information system, will coordinate with the Global Change Data and Information System, a major component of which is the NOAA data and information system.	
National Air Pollution Control Program	Environmental Protection Agency(EPA) Office of Air Quality Planning and Standards	Collects and analyzes data on ambient air quality and air pollution levels and compares them to National Ambient Air Quality Standards (NAAQS).	Annual reports on air quality and emission estimates.

Environmental Statistics Programs Managed by Agencies of the U.S. Government (Continued)

Program	Sponsor	Coverage	Frequency/ Most Recent Report
Comprehensive Environmental Response, Compensation and Liability Information System	EPA Office of Emergency and Remedial Response	Contains information on over 30,000 abandoned or uncontrolled hazardous waste sites.	Updated on-line.
Environmental Radiation Ambient Monitoring System (ERAMS)	EPA Office of Radiation Programs	Monitors radiation in air, drinking water, surface water, and milk.	Sampling intervals from twice weekly to bi-annual, based on analyses, at 332 stations; since 1973.
Hazardous and Non-hazardous Surveys	EPA Office of Solid Waste	Collect data through survey and regulated entity reports on hazardous and nonhazardous ("solid") wastes generation and management; information on regulated entities and waste volumes they generate and manage.	Current data are from 1985 biennial report and two 1986 surveys.
Toxics Release Inventory	EPA Office of Toxic Substances	Mandatory annual inventory of releases of 328 toxic chemicals to air, water, land, and off-site	Annual reports.

Environmental Statistics Programs Managed by Agencies of the U.S. Government (Continued)

Program	Sponsor	Coverage	Frequency/ Most Recent Report
Toxics Release Inventory (Continued)		disposal from more than 17,000 manufacturing facilities across the country.	
Water Pollution Control Act [Section 305(b)] Assessments	EPA Office of Water	Compiles state reports on water quality status of surface and groundwater, as required by Section 305(b), Federal Water Pollution Control Act; states prepare assessments using various monitoring data.	1990 biennial report as of 1988.
Ambient Water Monitoring Program/ STORET	EPA Office of Water	Largest database for water quality information; data contained in STORET with over 170 million data points from states and federal agencies, such as USGS, on surface water and ground water quality, sediments, streamflow, and fish tissue contamination.	Updated on-line.
Public Water System Supervision Program	EPA Office of Water	Contains information about public water supplies (PWSs)	Quarterly state and EPA regional reports.

Environmental Statistics Programs Managed by Agencies of the U.S. Government (*Continued*)

Program	Sponsor	Coverage	Frequency/ Most Recent Report
Public Water System Supervision Program (*Continued*)		and their compliance with monitoring requirements, maximum contaminant level (MCL) regulations, and other requirements of the Safe Drinking Water Act; data are stored in the Federal Reporting Data System.	
National Pollutant Discharge Elimination System Program	EPA Office of Water	Tracks permit compliance and enforcement status of facilities covered by water pollution permits; information is contained in the Permit Compliance System (PCS).	Monthly facility reports entered on an ongoing basis.
Municipal Construction Program	EPA Office of Water	Inventory of existing or proposed publicly owned treatment works (POTWS) that need construction or renovation to meet Clean Water Act requirements; information is maintained in the Needs Survey System.	2-year update from each state; biennial report submitted to Congress.

Environmental Statistics Programs Managed by Agencies of the U.S. Government (*Continued*)

Program	Sponsor	Coverage	Frequency/ Most Recent Report
Coastal and Ocean Protection Programs	EPA Office of Water	Covers environmental data (water quality, biological, permitting, environmental impact data) for discharges and pollutant loadings to coastal waters as well as ocean dumping; information contained in Ocean Data Evaluation System (ODES).	Biennial reports to Congress for National Estuary Program; annual reports to the Ocean Dumping Program; special reports on coastal programs.

Note: Neither this table nor the source document, which describes 72 federal programs, are exhaustive. For instance, USDA also maintains mission-oriented statistics in such areas as crops, snowpack, soil erosion, national forests management, and wildfires. DOC Bureau of Census maintains social, demographic, and economic statistics relevant to the environment. NOAA maintains statistics on marine resources and coastal wetlands. BLM maintains statistics for BLM lands, including condition, wildlife, minerals, and use; and NPS collects comparable statistics on the status of national parks. The Bureau of Mines collects, interprets, and publishes data on production, consumption, and trade of over 100 minerals. FWS maintains data on FWS lands and conducts surveys of fishing, hunting, and wildlife-associated recreation every 5 years, with the most recent report in 1991. USGS maps national land use and land cover, and EPA conducts regional and other pollution surveys. The Department of Transportation compiles highway and other transportation statistics, and the U.S. Coast Guard maintains data on marine pollution spills.

Source: U.S. Environmental Protection Agency, Center for Environmental Statistics, *Guide to Selected National Environmental Statistics in the U.S. Government (draft)* (Washington, DC: EPA, 1992).

State of the Environment Reports and Statistical Compendia Published by International Organizations

Commission of the European Communities (CEC)

CEC, *The State of the Environment in the European Community 1986* (Luxembourg: CEC, 1987). Earlier editions were published in 1977 and 1979.

Eurostat, *Basic Statistics of the Community,* annual (Luxembourg: CEC, 1990).

Eurostat, *Environment Statistics 1989,* annual (Luxembourg: CEC, 1990).

Organization for Economic Cooperation and Development (OECD)

OECD, *Environmental Data Compendium 1991* (Paris: OECD, 1991). Earlier versions were published in 1984 (pilot), 1985, and 1987.

OECD, *The State of the Environment* (Paris: OECD, 1991). Previous editions were published in 1979 and 1985.

United Nations

A. L. Dahl and L. L. Baumgart, *The State of the Environment in the South Pacific* (Geneva, Switzerland: UNEP, 1983).

United Nations Economic and Social Commission for Asia and the Pacific, *State of the Environment in Asia and the Pacific* (Bangkok: UNESCAP, 1985).

United Nations Environment Programme (UNEP), *Environmental Data Report 1991–92,* biennial (Oxford, England: Basil Blackwell, 1991). This report is coproduced with Kings College, London; the World Resources Institute in Washington, D.C.; and the U.K. Department of the Environment. It was intended to complement the more easily understood *World Resources Report.*

UNEP, *State of the World Environments,* annual (Nairobi: UNEP, 1991). The 1982 and 1987 reports were also published as, respectively, M. W. Holdgate, M. Kassas, and G. F. White, eds., *The World Environment 1972–1982* (Dublin: Tycooly International, 1982); and E. El-Hinnawi and M.H. Hashmi, *The State of the Environment* (Guildford, England: Butterworth-Heinemann, 1987).

UNEP, *The State of the Marine Environment: UNEP Regional Seas Programme* (Nairobi: UNEP, 1990). The following regions were covered for the years indicated: West and Central Africa (1984), East Africa (1980, 1982, 1984, and 1985), Kuwait (1985), South Asia (1985), East Asia (1985), Red Sea and Gulf of Aden (1985), Indian Ocean (1985), and Oceania and the Pacific (1985).

United Nations Statistical Office and United Nations Economic Commission for Europe, *Environmental Statistics in Europe and North America: An Experimental Compendium* (New York: UNSO and UNECE, 1987).

State of the Environment Reports and Statistical Compendia Published by International Organizations (*Continued*)

World Bank

World Bank, *The World Bank and the Environment* (Washington, DC: World Bank, 1990).

World Bank, *World Development Report 1991,* annual (Oxford, England: Oxford University Press, 1991).

World Resources Institute

A. L. Hammond, ed., *World Resources 1990–1991,* biennial (Oxford, England: Oxford University Press, 1990). This publication is coproduced with UNEP and the International Institute for Environment and Development.

World Resources Institute, *The 1992 Information Please Environmental Almanac* (Boston: Houghton Mifflin, 1991).

Source: J. Parker and C. Hope, "The state of the environment: a survey of reports from around the world." *Environment* 34(1):20 (Heldref Publications).

ORGANIZATION CHARTS OF SELECTED FEDERAL AGENCIES WITH NEPA RESPONSIBILITIES

Source: The United States Government Manual 1994/1995.

ENVIRONMENTAL PROTECTION AGENCY

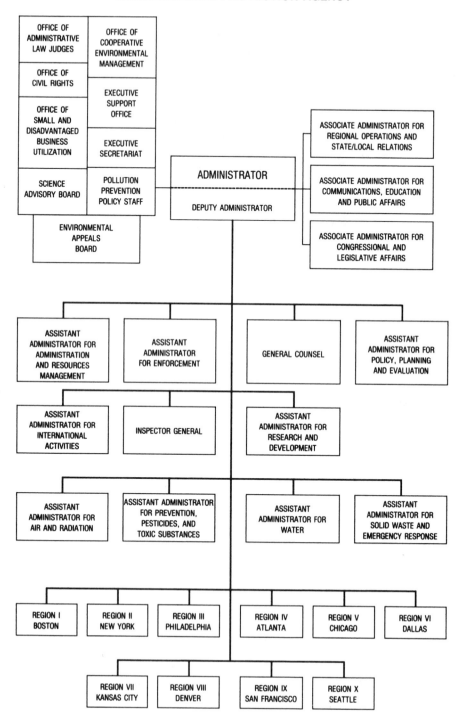

OFFICE OF MANAGEMENT AND BUDGET

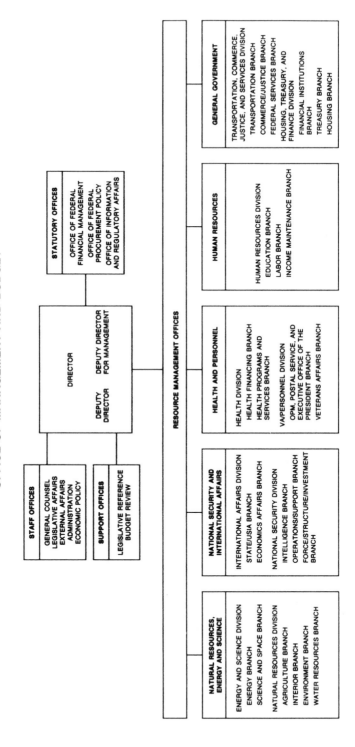

STAFF OFFICES

GENERAL COUNSEL
LEGISLATIVE AFFAIRS
EXTERNAL AFFAIRS
ADMINISTRATION
ECONOMIC POLICY

SUPPORT OFFICES

LEGISLATIVE REFERENCE
BUDGET REVIEW

DIRECTOR

DEPUTY DIRECTOR   DEPUTY DIRECTOR
             FOR MANAGEMENT

STATUTORY OFFICES

OFFICE OF FEDERAL
FINANCIAL MANAGEMENT
OFFICE OF FEDERAL
PROCUREMENT POLICY
OFFICE OF INFORMATION
AND REGULATORY AFFAIRS

RESOURCE MANAGEMENT OFFICES

**NATURAL RESOURCES,
ENERGY AND SCIENCE**

ENERGY AND SCIENCE DIVISION
ENERGY BRANCH
SCIENCE AND SPACE BRANCH
NATURAL RESOURCES DIVISION
AGRICULTURE BRANCH
INTERIOR BRANCH
ENVIRONMENT BRANCH
WATER RESOURCES BRANCH

**NATIONAL SECURITY AND
INTERNATIONAL AFFAIRS**

INTERNATIONAL AFFAIRS DIVISION
STATE/USIA BRANCH
ECONOMICS AFFAIRS BRANCH
NATIONAL SECURITY DIVISION
INTELLIGENCE BRANCH
OPERATIONS/SUPPORT BRANCH
FORCE/STRUCTURE/INVESTMENT
BRANCH

HEALTH AND PERSONNEL

HEALTH DIVISION
HEALTH FINANCING BRANCH
HEALTH PROGRAMS AND
SERVICES BRANCH
VA/PERSONNEL DIVISION
OPM, POSTAL SERVICE, AND
EXECUTIVE OFFICE OF THE
PRESIDENT BRANCH
VETERANS AFFAIRS BRANCH

HUMAN RESOURCES

HUMAN RESOURCES DIVISION
EDUCATION BRANCH
LABOR BRANCH
INCOME MAINTENANCE BRANCH

GENERAL GOVERNMENT

TRANSPORTATION, COMMERCE,
JUSTICE, AND SERVICES DIVISION
TRANSPORTATION BRANCH
COMMERCE/JUSTICE BRANCH
FEDERAL SERVICES BRANCH
HOUSING, TREASURY, AND
FINANCE DIVISION
FINANCIAL INSTITUTIONS
BRANCH
TREASURY BRANCH
HOUSING BRANCH

DEPARTMENT OF AGRICULTURE

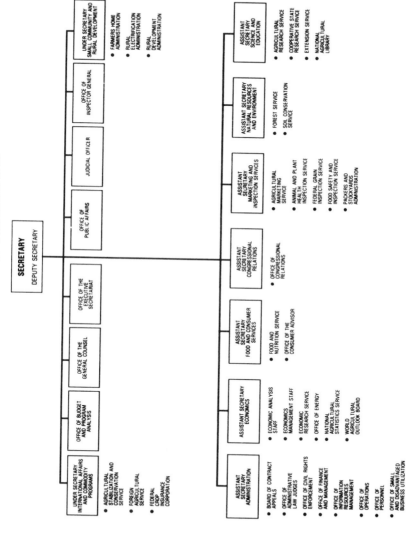

SECRETARY

DEPUTY SECRETARY

UNDER SECRETARY INTERNATIONAL AFFAIRS AND COMMODITY PROGRAMS
- AGRICULTURAL STABILIZATION AND CONSERVATION SERVICE
- FOREIGN AGRICULTURAL SERVICE
- FEDERAL CROP INSURANCE CORPORATION

OFFICE OF BUDGET AND PROGRAM ANALYSIS

OFFICE OF THE GENERAL COUNSEL

OFFICE OF THE EXECUTIVE SECRETARIAT

OFFICE OF PUBLIC AFFAIRS

JUDICIAL OFFICER

OFFICE OF INSPECTOR GENERAL

UNDER SECRETARY SMALL COMMUNITY AND RURAL DEVELOPMENT
- FARMERS HOME ADMINISTRATION
- RURAL ELECTRIFICATION ADMINISTRATION
- RURAL DEVELOPMENT ADMINISTRATION

ASSISTANT SECRETARY ADMINISTRATION
- BOARD OF CONTRACT APPEALS
- OFFICE OF ADMINISTRATIVE LAW JUDGES
- OFFICE OF CIVIL RIGHTS ENFORCEMENT
- OFFICE OF FINANCE AND MANAGEMENT
- OFFICE OF INFORMATION RESOURCES MANAGEMENT
- OFFICE OF OPERATIONS
- OFFICE OF PERSONNEL
- OFFICE OF SMALL AND DISADVANTAGED BUSINESS UTILIZATION

ASSISTANT SECRETARY ECONOMICS
- ECONOMIC ANALYSIS STAFF
- ECONOMICS MANAGEMENT STAFF
- ECONOMIC RESEARCH SERVICE
- OFFICE OF ENERGY
- NATIONAL AGRICULTURAL STATISTICS SERVICE
- WORLD AGRICULTURAL OUTLOOK BOARD

ASSISTANT SECRETARY FOOD AND CONSUMER SERVICES
- FOOD AND NUTRITION SERVICE
- OFFICE OF THE CONSUMER ADVISOR

ASSISTANT SECRETARY CONGRESSIONAL RELATIONS
- OFFICE OF CONGRESSIONAL RELATIONS

ASSISTANT SECRETARY MARKETING AND INSPECTION SERVICES
- AGRICULTURAL MARKETING SERVICE
- ANIMAL AND PLANT HEALTH INSPECTION SERVICE
- FEDERAL GRAIN INSPECTION SERVICE
- FOOD SAFETY AND INSPECTION SERVICE
- PACKERS AND STOCKYARDS ADMINISTRATION

ASSISTANT SECRETARY NATURAL RESOURCES AND ENVIRONMENT
- FOREST SERVICE
- SOIL CONSERVATION SERVICE

ASSISTANT SECRETARY SCIENCE AND EDUCATION
- AGRICULTURAL RESEARCH SERVICE
- COOPERATIVE STATE RESEARCH SERVICE
- EXTENSION SERVICE
- NATIONAL AGRICULTURAL LIBRARY

DEPARTMENT OF COMMERCE

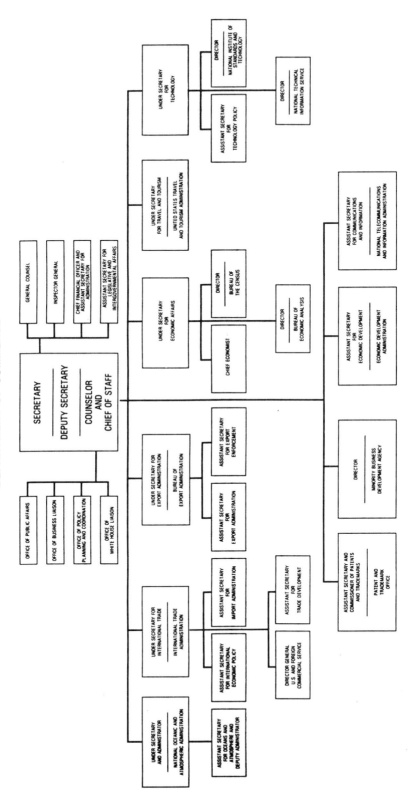

SECRETARY

DEPUTY SECRETARY

COUNSELOR AND CHIEF OF STAFF

GENERAL COUNSEL

INSPECTOR GENERAL

CHIEF FINANCIAL OFFICER AND ASSISTANT SECRETARY FOR ADMINISTRATION

ASSISTANT SECRETARY FOR LEGISLATIVE AND INTERGOVERNMENTAL AFFAIRS

OFFICE OF PUBLIC AFFAIRS

OFFICE OF BUSINESS LIAISON

OFFICE OF POLICY PLANNING AND COORDINATION

OFFICE OF WHITE HOUSE LIAISON

UNDER SECRETARY FOR TECHNOLOGY

DIRECTOR
NATIONAL INSTITUTE OF STANDARDS AND TECHNOLOGY

ASSISTANT SECRETARY FOR TECHNOLOGY POLICY

DIRECTOR
NATIONAL TECHNICAL INFORMATION SERVICE

UNDER SECRETARY FOR TRAVEL AND TOURISM
UNITED STATES TRAVEL AND TOURISM ADMINISTRATION

UNDER SECRETARY FOR ECONOMIC AFFAIRS

CHIEF ECONOMIST

DIRECTOR
BUREAU OF THE CENSUS

DIRECTOR
BUREAU OF ECONOMIC ANALYSIS

ASSISTANT SECRETARY FOR COMMUNICATIONS AND INFORMATION
NATIONAL TELECOMMUNICATIONS AND INFORMATION ADMINISTRATION

ASSISTANT SECRETARY FOR ECONOMIC DEVELOPMENT
ECONOMIC DEVELOPMENT ADMINISTRATION

UNDER SECRETARY FOR EXPORT ADMINISTRATION
BUREAU OF EXPORT ADMINISTRATION

ASSISTANT SECRETARY FOR EXPORT ENFORCEMENT

ASSISTANT SECRETARY FOR EXPORT ADMINISTRATION

DIRECTOR
MINORITY BUSINESS DEVELOPMENT AGENCY

UNDER SECRETARY FOR INTERNATIONAL TRADE
INTERNATIONAL TRADE ADMINISTRATION

ASSISTANT SECRETARY FOR INTERNATIONAL ECONOMIC POLICY

ASSISTANT SECRETARY FOR IMPORT ADMINISTRATION

DIRECTOR GENERAL
U.S. AND FOREIGN COMMERCIAL SERVICE

ASSISTANT SECRETARY FOR TRADE DEVELOPMENT

ASSISTANT SECRETARY AND COMMISSIONER OF PATENTS AND TRADEMARKS
PATENT AND TRADEMARK OFFICE

UNDER SECRETARY FOR OCEANS AND ATMOSPHERE AND ADMINISTRATION
NATIONAL OCEANIC AND ATMOSPHERIC ADMINISTRATION

ASSISTANT SECRETARY FOR OCEANS AND ATMOSPHERE AND DEPUTY ADMINISTRATOR

DEPARTMENT OF DEFENSE

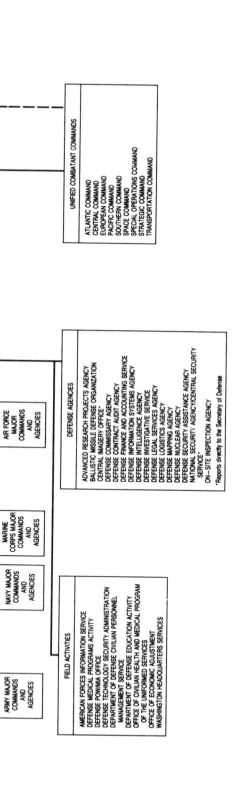

SECRETARY OF DEFENSE
DEPUTY SECRETARY OF DEFENSE

DEPARTMENT OF THE ARMY
SECRETARY OF THE ARMY

UNDER SECRETARY AND ASSISTANT SECRETARIES OF THE ARMY

CHIEF OF STAFF ARMY

ARMY MAJOR COMMANDS AND AGENCIES

DEPARTMENT OF THE NAVY
SECRETARY OF THE NAVY

UNDER SECRETARY AND ASSISTANT SECRETARIES OF THE NAVY

CHIEF OF NAVAL OPERATIONS

COMMANDANT OF MARINE CORPS

NAVY MAJOR COMMANDS AND AGENCIES

MARINE CORPS MAJOR COMMANDS AND AGENCIES

DEPARTMENT OF THE AIR FORCE
SECRETARY OF THE AIR FORCE

UNDER SECRETARY AND ASSISTANT SECRETARIES OF THE AIR FORCE

CHIEF OF STAFF AIR FORCE

AIR FORCE MAJOR COMMANDS AND AGENCIES

OFFICE OF THE SECRETARY OF DEFENSE

UNDER SECRETARIES ASSISTANT SECRETARIES OF DEFENSE AND EQUIVALENTS

INSPECTOR GENERAL

JOINT CHIEFS OF STAFF
CHAIRMAN, JOINT CHIEFS OF STAFF

THE JOINT STAFF

VICE CHAIRMAN, JOINT CHIEFS OF STAFF
CHIEF OF STAFF, ARMY
CHIEF OF NAVAL OPERATIONS
CHIEF OF STAFF, AIR FORCE
COMMANDANT, MARINE CORPS

UNIFIED COMBATANT COMMANDS

ATLANTIC COMMAND
CENTRAL COMMAND
EUROPEAN COMMAND
PACIFIC COMMAND
SOUTHERN COMMAND
SPACE COMMAND
SPECIAL OPERATIONS COMMAND
STRATEGIC COMMAND
TRANSPORTATION COMMAND

FIELD ACTIVITIES

AMERICAN FORCES INFORMATION SERVICE
DEFENSE MEDICAL PROGRAMS ACTIVITY
DEFENSE POW/MIA OFFICE
DEFENSE TECHNOLOGY SECURITY ADMINISTRATION
DEPARTMENT OF DEFENSE CIVILIAN PERSONNEL MANAGEMENT SERVICE
DEPARTMENT OF DEFENSE EDUCATION ACTIVITY
OFFICE OF CIVILIAN HEALTH AND MEDICAL PROGRAM OF THE UNIFORMED SERVICES
OFFICE OF ECONOMIC ADJUSTMENT
WASHINGTON HEADQUARTERS SERVICES

DEFENSE AGENCIES

ADVANCED RESEARCH PROJECTS AGENCY
BALLISTIC MISSILE DEFENSE ORGANIZATION
CENTRAL IMAGERY OFFICE*
DEFENSE COMMISSARY AGENCY
DEFENSE CONTRACT AUDIT AGENCY
DEFENSE FINANCE AND ACCOUNTING SERVICE
DEFENSE INFORMATION SYSTEMS AGENCY
DEFENSE INTELLIGENCE AGENCY
DEFENSE INVESTIGATIVE SERVICE
DEFENSE LEGAL SERVICES AGENCY
DEFENSE LOGISTICS AGENCY
DEFENSE MAPPING AGENCY
DEFENSE NUCLEAR AGENCY
DEFENSE SECURITY ASSISTANCE AGENCY
NATIONAL SECURITY AGENCY/CENTRAL SECURITY SERVICE*
ON—SITE INSPECTION AGENCY

*Reports directly to the Secretary of Defense

DEPARTMENT OF THE ARMY

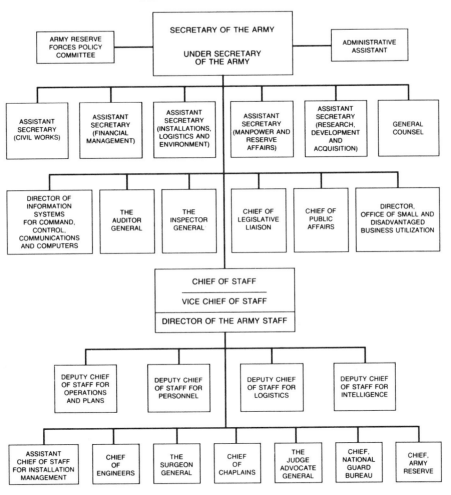

DEPARTMENT OF THE NAVY

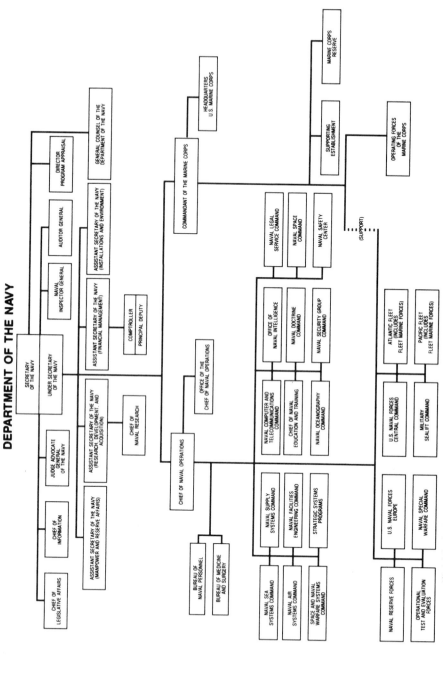

*Also includes other designated activities not shown on the chart which are under the command or supervision of the organizations depicted.

SECRETARY OF THE NAVY

UNDER SECRETARY OF THE NAVY

CHIEF OF LEGISLATIVE AFFAIRS

CHIEF OF INFORMATION

JUDGE ADVOCATE GENERAL OF THE NAVY

ASSISTANT SECRETARY OF THE NAVY (MANPOWER AND RESERVE AFFAIRS)

ASSISTANT SECRETARY OF THE NAVY (RESEARCH, DEVELOPMENT, AND ACQUISITION)

NAVAL INSPECTOR GENERAL

AUDITOR GENERAL

ASSISTANT SECRETARY OF THE NAVY (FINANCIAL MANAGEMENT)

ASSISTANT SECRETARY OF THE NAVY (INSTALLATIONS AND ENVIRONMENT)

DIRECTOR PROGRAM APPRAISAL

GENERAL COUNSEL OF THE DEPARTMENT OF THE NAVY

COMPTROLLER PRINCIPAL DEPUTY

CHIEF OF NAVAL RESEARCH

OFFICE OF THE CHIEF OF NAVAL OPERATIONS

COMMANDANT OF THE MARINE CORPS

HEADQUARTERS U.S. MARINE CORPS

CHIEF OF NAVAL OPERATIONS

BUREAU OF NAVAL PERSONNEL

BUREAU OF MEDICINE AND SURGERY

NAVAL SEA SYSTEMS COMMAND

NAVAL AIR SYSTEMS COMMAND

SPACE AND NAVAL WARFARE SYSTEMS COMMAND

NAVAL SUPPLY SYSTEMS COMMAND

NAVAL FACILITIES ENGINEERING COMMAND

STRATEGIC SYSTEMS PROGRAMS

NAVAL COMPUTER AND TELECOMMUNICATIONS COMMAND

CHIEF OF NAVAL EDUCATION AND TRAINING

NAVAL OCEANOGRAPHY COMMAND

OFFICE OF NAVAL INTELLIGENCE

NAVAL DOCTRINE COMMAND

NAVAL SECURITY GROUP COMMAND

NAVAL LEGAL SERVICE COMMAND

NAVAL SPACE COMMAND

NAVAL SAFETY CENTER

(SUPPORT)

NAVAL RESERVE FORCES

OPERATIONAL TEST AND EVALUATION FORCES

U.S. NAVAL FORCES EUROPE

NAVAL SPECIAL WARFARE COMMAND

U.S. NAVAL FORCES CENTRAL COMMAND

MILITARY SEALIFT COMMAND

ATLANTIC FLEET (INCLUDES FLEET MARINE FORCES)

PACIFIC FLEET (INCLUDES FLEET MARINE FORCES)

SUPPORTING ESTABLISHMENT

OPERATING FORCES OF THE MARINE CORPS

MARINE CORPS RESERVE

EXECUTIVE

SHORE ESTABLISHMENT*

OPERATING FORCES*

DEPARTMENT OF ENERGY

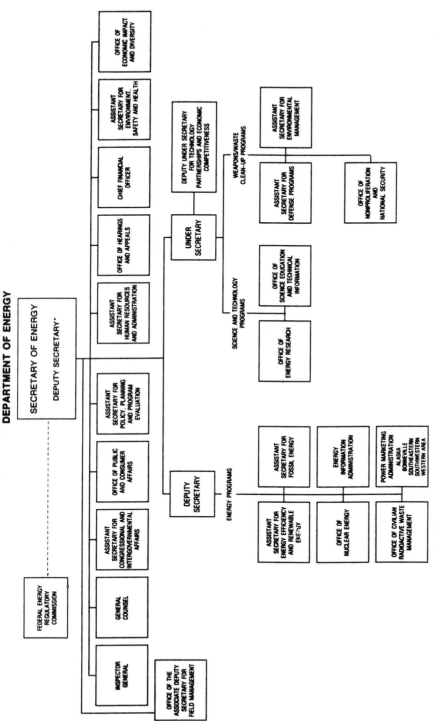

SECRETARY OF ENERGY

DEPUTY SECRETARY*

FEDERAL ENERGY REGULATORY COMMISSION

INSPECTOR GENERAL

GENERAL COUNSEL

OFFICE OF THE ASSOCIATE DEPUTY SECRETARY FOR FIELD MANAGEMENT

ASSISTANT SECRETARY FOR CONGRESSIONAL AND INTERGOVERNMENTAL AFFAIRS

OFFICE OF PUBLIC AND CONSUMER AFFAIRS

ASSISTANT SECRETARY FOR POLICY, PLANNING AND PROGRAM EVALUATION

ASSISTANT SECRETARY HUMAN RESOURCES AND ADMINISTRATION

OFFICE OF HEARINGS AND APPEALS

CHIEF FINANCIAL OFFICER

ASSISTANT SECRETARY FOR ENVIRONMENT, SAFETY AND HEALTH

OFFICE OF ECONOMIC IMPACT AND DIVERSITY

DEPUTY UNDER SECRETARY FOR TECHNOLOGY PARTNERSHIPS AND ECONOMIC COMPETITIVENESS

UNDER SECRETARY

DEPUTY SECRETARY

ENERGY PROGRAMS

ASSISTANT SECRETARY FOR ENERGY EFFICIENCY AND RENEWABLE ENERGY

ASSISTANT SECRETARY FOR FOSSIL ENERGY

OFFICE OF NUCLEAR ENERGY

ENERGY INFORMATION ADMINISTRATION

OFFICE OF CIVILIAN RADIOACTIVE WASTE MANAGEMENT

POWER MARKETING ADMINISTRATION
ALASKA
BONNEVILLE
SOUTHEASTERN
SOUTHWESTERN
WESTERN AREA

SCIENCE AND TECHNOLOGY PROGRAMS

OFFICE OF ENERGY RESEARCH

OFFICE OF SCIENCE EDUCATION AND TECHNICAL INFORMATION

WEAPONS/WASTE CLEAN-UP PROGRAMS

ASSISTANT SECRETARY FOR DEFENSE PROGRAMS

ASSISTANT SECRETARY FOR ENVIRONMENTAL MANAGEMENT

OFFICE OF NONPROLIFERATION AND NATIONAL SECURITY

* Deputy Secretary oversees energy programs and serves as Chief Operating Officer of the Department within the Office of the Secretary.

465

DEPARTMENT OF HOUSING AND URBAN DEVELOPMENT

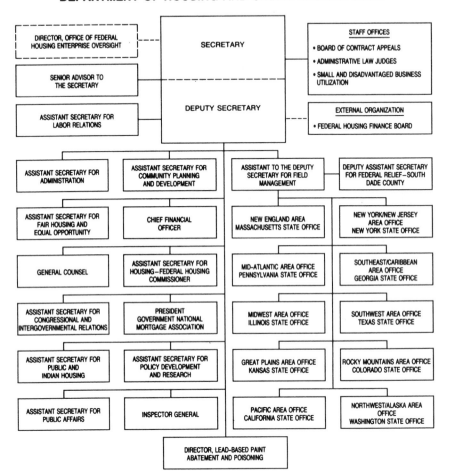

DEPARTMENT OF THE INTERIOR

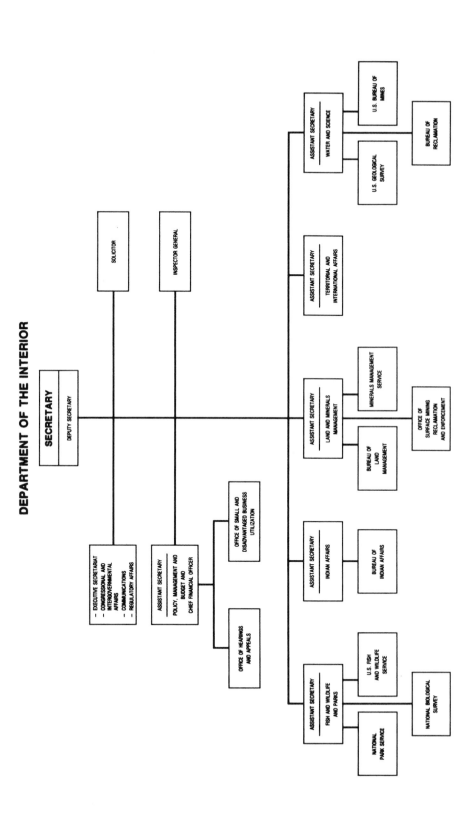

DEPARTMENT OF TRANSPORTATION

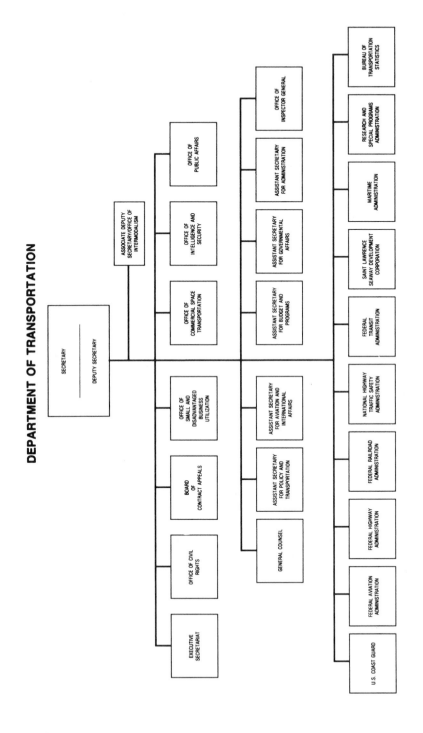

SECRETARY

DEPUTY SECRETARY

ASSOCIATE DEPUTY SECRETARY/OFFICE OF INTERMODALISM

EXECUTIVE SECRETARIAT

OFFICE OF CIVIL RIGHTS

BOARD OF CONTRACT APPEALS

OFFICE OF SMALL AND DISADVANTAGED BUSINESS UTILIZATION

OFFICE OF COMMERCIAL SPACE TRANSPORTATION

OFFICE OF INTELLIGENCE AND SECURITY

OFFICE OF PUBLIC AFFAIRS

GENERAL COUNSEL

ASSISTANT SECRETARY FOR POLICY AND TRANSPORTATION

ASSISTANT SECRETARY FOR AVIATION AND INTERNATIONAL AFFAIRS

ASSISTANT SECRETARY FOR BUDGET AND PROGRAMS

ASSISTANT SECRETARY FOR GOVERNMENTAL AFFAIRS

ASSISTANT SECRETARY FOR ADMINISTRATION

OFFICE OF INSPECTOR GENERAL

U.S. COAST GUARD

FEDERAL AVIATION ADMINISTRATION

FEDERAL HIGHWAY ADMINISTRATION

FEDERAL RAILROAD ADMINISTRATION

NATIONAL HIGHWAY TRAFFIC SAFETY ADMINISTRATION

FEDERAL TRANSIT ADMINISTRATION

SAINT LAWRENCE SEAWAY DEVELOPMENT CORPORATION

MARITIME ADMINISTRATION

RESEARCH AND SPECIAL PROGRAMS ADMINISTRATION

BUREAU OF TRANSPORTATION STATISTICS

INDEX